Principles of
Synchronous
Digital
Hierarchy

Principles of Synchronous Digital Hierarchy

RAJESH KUMAR JAIN

CRC Press
Taylor & Francis Group
Boca Raton London New York

CRC Press is an imprint of the
Taylor & Francis Group, an **informa** business

CRC Press
Taylor & Francis Group
6000 Broken Sound Parkway NW, Suite 300
Boca Raton, FL 33487-2742

First issued in paperback 2017

ISBN-13: 978-1-4665-1726-4 (hbk)
ISBN-13: 978-1-138-07749-2 (pbk)

Library of Congress Cataloging-in-Publication Data

Jain, Rajesh Kumar.
  Principles of synchronous digital hierarchy / Rajesh Kumar Jain.
    p. cm.
  Includes bibliographical references and index.
  ISBN 978-1-4665-1726-4 (hardback)
  1. Synchronous digital hierarchy (Data transmission) I. Title.

TK5105.42.J35 2012
004.6--dc23   2012021349

Visit the Taylor & Francis Web site at
http://www.taylorandfrancis.com

and the CRC Press Web site at
http://www.crcpress.com

Contents

Preface

Dear friends,

I am an engineer by qualification and a signaling and telecommunication professional by occupation. I work in the Signaling and Telecommunication Department of the Indian Railways. Our job is to install and maintain various signaling and telecommunication equipment required for railway operations. We have a phenomenal variety of signaling equipment with us, from the oldest semaphore signaling systems to the latest electronic interlocking systems and a variety of traffic control, management, protection, and data logging systems. On the telecommunication front, we are engaged in the installation, operation, and maintenance of many types of equipment in most of the segments of popular telecommunication technologies. We have telephone exchanges, omnibus control circuits, microwave radio links, optical fiber cables carrying long-distance synchronous digital hierarchy (SDH) links, data communication networks running critical applications, ISM band wireless links including short-distance DSSS and long-distance OFDM links, GSM/R (GSM/R is a variation of GSM technology especially suited to railway applications) for mobile communication between the running train and the ground administration, and so forth.

There is no doubt that it is beyond the capabilities of a normal human being to keep track of the fine details of all the signaling and telecommunication technologies, current and upcoming, in the current scenario of superfast technological developments.

A few years back in my career, destiny threw me to a posting in Railtel Corporation of India Ltd. (RCIL), a public-sector company under the Ministry of Railways of the Government of India. Although RCIL was engaged in many types of telecommunication businesses, its main revenue stream was the telecommunication bandwidth across the country. Their network was built on railways' optical fiber cable by deploying SDH/WDM systems. It was a sudden jolt to my ego of technical competence when I could not tackle the nuts and bolts of the SDH technology. Until then, I had been engaged in the installation, operations, and maintenance of many types of signaling and telecommunication systems, and the telecommunication systems included mainly electronic exchanges, microwave radio links, and data communication networks, aside from other minor variations. I had some exposure to the OFC systems and carrier technologies, but what was expected of me was a thorough knowledge of the SDH technology and associated systems and the ability to manage the systems and services effectively. Despite striving hard to understand the subject, I found myself nowhere near the competency level that I would generally like to have had. I continued to discharge my duties and did it quite well, somehow managing with a makeshift knowledge of the SDH systems, which I gathered from various training programs and through whatever literature I could get hold of from the bookshops or the Internet. However, despite my best efforts, I could not get good literature on the subject that could explain the concepts aside from providing functional details. The many training programs proved inadequate as well, as the trainers themselves were not very competent.

It was a boon to my desire to learn when I was sent on forced leave for a period of seven months because of my promotion without a post being available in the higher grade. After passing the initial few months in reading fiction, which I always wanted to read but could never do as I was always hard-pressed for time throughout my career, the idea of strengthening my understanding of SDH struck me. I again started searching for good literature on SDH, but I could not find any book on the subject that could impart the necessary understanding. I needed some support on the fundamentals of telecommunication technologies as well—after years of being away from fine details of many technologies, a kind of refresher course was required, besides the knowledge of fundamentals of new technologies. I started gathering all the training materials on the subject, visited bookshops on all available opportunities, bought many books, and gathered as much of related information as I could. Referring to the books on fundamentals side by side with the training materials, white papers, and tutorials available on the Internet, the concepts started taking shape in my mind. Encouraged, I started going deeper into the subject until I finally found myself confident in not only discussions on the subject but also in handling the equipments.

Thus, I decided to properly organize my learning and put it in the form of notes, which could be used by our training institutes to impart knowledge to people who desired to learn. While preparing the notes, I was reminded of the difficulties that I had had to face in understanding the particular subject. This led me to include the fundamentals of the underlying technology to enable the reader to understand the subject without having to worry about knowing these fundamentals. The process went on with every subject, and I found that the volume of notes was growing with every topic. This was the time when the idea of converting the whole text into a book struck me. I started working toward it and found that I had to include more details on the fundamentals of underlying telecommunication technologies to make it a self-contained text on the subject of SDH for telecommunication professionals.

The first lesson I learned was that it was impossible to understand SDH if your PDH fundamentals are weak. Thus, I included a detailed chapter on PDH. To understand PDH, one has to know such things as HDB-3, clock, time division multiplexing, pulse code modulation, etc. Hence, I included chapters on line coding, clock, PCM, and TDM to provide ready reference to the fundamentals. Bit errors are a measure of the quality of digital signal. To appreciate what bit errors are and their causes, one has to know the properties of transmission media and their effect on the quality of digital signal. One chapter has therefore been dedicated to signal impairments caused by transmission media. Although it is expected that telecommunication professionals would never really forget the concepts of analog-to-digital conversion and vice versa, a chapter has been included to provide a quick refresher. The subject of digital multiplexing cannot be appreciated well in the absence of the concepts of the first multiplexing technique, which was analog frequency division multiplexing (FDM); hence, the introductory chapter is dedicated to multiplexing and analog multiplexing basics. How and why digital technology established itself, completely wiping out analog technology, have also been addressed in brief to help maintain the flow of concepts. Operations and maintenance features are among the most important features of digital communication systems. These features of PDH and SDH have been dealt with in separate chapters. In fact, the OM features of the PDH systems are the foundation of SDH. SDH is, in fact, far ahead of its predecessors in terms of capacity, flexibility, and management capabilities. Many types of topologies are possible in SDH, and these have been explained in brief aside from the protection switching features. (The protection switching feature of SDH facilitates automatic and instantaneous traffic restoration in case of a breakdown.)

The phenomenal growth of data traffic in the recent years has forced SDH to develop features for efficient transport of data communication traffic. One chapter has been devoted to the data-carrying capabilities of SDH and these new improvements. The transmission media deserve a little discussion as well to refresh the fundamentals and provide a glimpse of new technologies, and accordingly, a chapter has been devoted to it. Although the subject of the work was primarily SDH, to cover all the variations of digital multiplexing, I also included a chapter on optical transport networks (OTN), the latest transport platform.

I have thus attempted to present a comprehensive learning tool to professionals in the field of telecommunication transport technologies. A telecommunication engineer who left college years back should not need any reference other than this to develop a strong understanding of digital multiplexing and transport technologies.

The book has been written in a very simple language and with easy to understand illustrations. In fact, in many instances, the illustrations appear to be too simple and repetitive, which may not appeal to a learned reader. However, my earnest request to such readers is to bear with me in the interest of the persons who do not have an equally strong foundation. In any case, the repetitive illustrations will only make your concepts more deeply ingrained.

I have taken the utmost care that all the material and facts presented are correct and authentic and that editing and typing mistakes have been eliminated, and I highly regret any lapse. I will be grateful if you can give me feedback on this.

I wish you all happy reading.

Author

Rajesh Kumar Jain is chief signal engineer in the Signaling and Telecommunication Department of Indian Railways. He did his engineering in electronics and telecommunication from Government Engineering College Jabalpur (India) in 1979. The institute is a pioneer in engineering education in the field of electronics and telecommunication in India, being the first college to offer this course since 1948. He worked in DRDO (Defense Research and Development Organization) as a scientist, where he was involved in the development of RADAR systems. He joined Indian Railways Services of Signal Engineers (IRSSE) in 1984. Since then, he has worked in various capacities as a maintenance and project engineer, leading a team of a few hundred technical staff to ensure efficient and reliable operations and maintenance of the signaling and telecommunication systems of Indian Railways. Throughout his career he was posted on the most challenging positions. His approach toward work has always been a problem-solving approach that led to innovation and implementation of many improvements in the design and maintenance practices of signal and telecommunication systems of Indian Railways. Jain handled the toughest and heaviest traffic section of Indian Railways, i.e., the Mumbai Suburban section for a period of eight years at a stretch. He innovated, developed and implemented a number of modifications of the signaling systems that ensured safe and efficient running of more than 1500 trains in the Mumbai–Kalyan section every day. He has to his credit a number of honors bestowed upon him by Indian Railways. Jain won the most coveted Minister's award in the year 2001 for his contributions to the organization. He has presented many papers in various technical forums on the subjects relating to signaling, telecommunication and lightning protection systems, and all of them were highly applauded.

The author can be contacted through e-mail at rajeshjain2001@indiatimes.com.

1

Introduction

As far as the landline telephone is concerned, the world of a telephone subscriber has not changed much from the day the first telephone service was introduced. The technology of the service, however, has undergone phenomenal changes. From the simple battery and telephone set, we have moved to complex switching, multiplexing, and transport technologies. This chapter will demonstrate where we began and where we are today.

1.1 You and Me

When it comes to telecommunication, I am always reminded of the game we used to play as little children. In those days, there were telephones, but in very limited numbers, and we were not allowed to touch them, for fear that we would drop and break them. So we used to develop our own telephone by connecting two empty cans with a long cotton thread, by drilling tiny holes in them. We would then move apart, each holding a can, until the thread tightened. Then one of us would speak into the can while the other held theirs close to his/her ear (Figure 1.1).

We could never be sure whether what we heard was the result of transmission of sound waves via the thread or of direct shouting. However, the tin boxes did vibrate a little and certainly made some sounds.

The fascination with the telephone has increased exponentially throughout the world. Let us briefly review how the telephone was developed and what was its fundamental technology.

1.2 "Mr. Watson, Come Here. I Want to See You."

These were the words of Alexander Graham Bell to his assistant Mr. Watson that made history on March 10, 1876 on his invention of the telephone. Mr. Watson was in the next room.

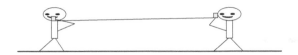

FIGURE 1.1
Children's communication through cotton thread.

Although the invention of the telephone by Alexander Graham Bell and Elisha Gray was almost simultaneous, Bell won the patent rights after a long legal battle and is now accepted as its inventor.

1.3 Technology of the Telephone

Although the materials of the basic components have changed, the technology used by Bell in his first telephone is still in use today.

The three basic components of a telephone are

(i)   Transmitter or microphone

(ii) Receiver or earpiece

(iii) Metallic wire

Figure 1.2 shows the principle.
Let us have a brief look at how these components function.

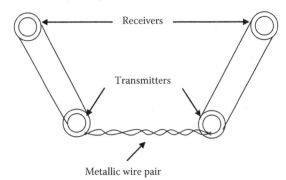

FIGURE 1.2
The basic telephone technology. (If you are wondering why there is only one pair of wires, and why only transmitters are connected to each other and not the receivers, here is the answer. Although there is only one pair of wires connecting the two telephones, the transmitted voice is separated from the received voice by means of a circuit called "Hybrid." When the voice is received by the telephone, Hybrid directs it to the receiver (or earpiece), and when the person speaks, the voice is guided to the transmitter by the Hybrid circuit.)

1.3.1 Transmitter

The principle of operation of the transmitter is illustrated in Figure 1.3.

Fine carbon granules are filled between two carbon electrodes. One of the carbon electrodes is fixed, while the other is movable. An aluminum diaphragm is attached with the help of a soft rubber to the movable electrode. The movable electrode is suspended on thin mica flanges with the transmitter body. DC voltage is applied between the electrodes, which results in the flow of a DC current.

When a person speaks in front of the diaphragm, the sound waves generated by the pressure of the voice move the diaphragm to and fro. The displacement of the diaphragm is proportionate to the pressure of the sound waves. The diaphragm in turn moves the movable electrode. The movement of the electrode causes the pressure on the carbon granules to increase or decrease in accordance with the sound pressure. This causes the resistance between the two electrodes to vary, which in turn causes the current flowing through the circuit to vary in accordance with the incident sound pressure. Thus, the speech of the person is converted into electrical current variations. When this current is applied to a receiver (depicted in the figure as resistance R), which can convert the electrical variations back into voice, the conversation is achieved. Since these electrical variations could be carried on metallic wires to long distances (a few miles), the telephone became feasible.

The present-day transmitter is very different from that described above in the construction details and the materials used, but the basic principles of operation remain the same.

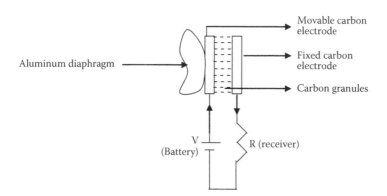

FIGURE 1.3
The principles of functioning of a transmitter (or microphone).

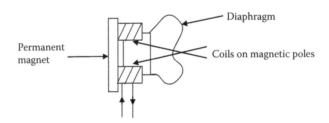

FIGURE 1.4
The principle of operation of a telephone receiver.

1.3.2 Receiver

The principle of operation of the receiver is just the opposite of the transmitter. Figure 1.4 illustrates this.

A permanent magnet is fixed with two poles that carry copper wire coils with a high number of turns. A diaphragm is attached to the assembly by means of soft rubber. The current received from the transmitter is fed to the coil.

The magnet normally exerts a pull on the diaphragm. When current received from the transmitter is fed to the coils, this current produces electromagnetic flux. This flux also causes a pull on the diaphragm. Since the flux created by electromagnetic coils varies in accordance with the variations in the incoming current, the pull on the diaphragm also varies in accordance with the current received from the transmitter. This variation in the pull on the diaphragm produces sound waves, which reproduce the speech of the person speaking into the transmitter.

Again, the modern receiver has undergone substantial changes in the construction details and the material used; but the basic principles of operation remain the same.

1.3.3 Metallic Wires

To connect the transmitter to the receiver requires a pair of metallic wires. Although any metal can be used for this purpose, copper is the most appropriate one, due to its low resistance. In the early days, overhead wires of galvanized iron and aluminum were also used, but underground copper cable became the standard medium. Low resistance allows long distances to be covered with a given amount of transmitted power or current.

Figure 1.5 shows the arrangement where a person "A" is speaking (transmitting) and a person "B" is listening (receiving) from a distance. However, in a practical situation, these roles need to be reversible. The arrangement with involvement of both the persons "A" and "B" is as shown in Figure 1.6. Thus, the use of a single pair of copper wires was standardized for a telephone conversation between two persons.

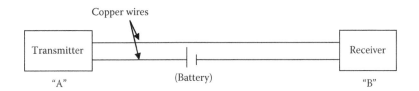

FIGURE 1.5
Person "A" is transmitting (speaking) while person "B" is receiving (listening).

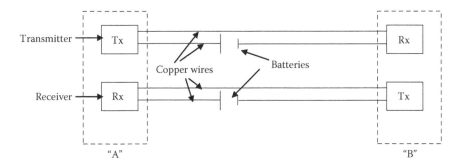

FIGURE 1.6
Person "A" and "B" both are transmitting (speaking) and receiving (listening).

Copper wire is the most expensive of the three components. Long distances were involved, and it was expensive to bury the wires: they had to be bunched in a strong cable, trenches dug, and roads cut. Thus, it was felt necessary to reduce this cost as much as possible. One way to achieve this was to use only a single pair of copper wires instead of two. By means of a circuit called "Hybrid," a single pair of copper wires carries the voice signals from both directions. The Hybrid directs the received signal to the receiver and sends the transmitted signals on the single pair. Figure 1.7 illustrates

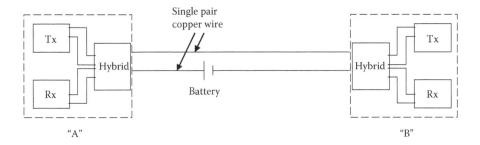

FIGURE 1.7
Use of a single pair for telephonic communication between two persons, with the help of the "Hybrid" circuit.

the concept. (The description of the device called Hybrid is quite interesting but is out of the scope of this text. The readers who are interested in reading further about it may refer to any basic text on telephony.)

1.4 Telephone Exchanges

After the telephone was invented, with its establishment came the need for a telephone exchange. As the name suggests, "telephone exchange" was a facility that could connect a person's telephone to different persons at different times. Of course, all those who could be connected had to have a pair of wires leading from their home or office to the telephone exchange.

Initially, the exchanges were manual, where an operator received the call from the calling person (the "calling party" or subscriber) and connected him to the desired person (the "called party"). As the number of subscribers grew, it became impossible to manage manual interconnection, and the "automatic exchange" was born.

1.4.1 Manual Exchange

To facilitate the interconnectivity between all the subscribers of the telephone service, a pair of copper wires was provided between the exchange and the home/office of each subscriber. Each subscriber was provided with a telephone set consisting of a transmitter and a receiver. The telephone set was also provided with a switch that operated when the handset was picked up from the "base unit," sending a signal to the operator in the exchange, calling for his attention. The operator would ask the calling party for the name of the desired called party and then connect both parties to each other with the help of "cords" fitted with "plugs," which were inserted in the respective "jacks." Before connecting to the called party, the operator would signal him with a ring, talk to the person, inform him about the calling party, and then connect them. Figure 1.8 shows the general layout of a manual exchange in a locality.

Inside the exchange facility, all the cables/wires (pairs) of the subscribers were terminated on boards fitted with jacks. Each subscriber was connected to a separate jack. The jacks were fitted with brass connectors, which provided a connection to the subscriber. A plug would be inserted into one of the jacks, which also carried brass contacts. The subscriber's wire connection was thus extended from jacks to the plugs. The plugs on the other side of the board were connected with a cord. The cord, a few meters long, consisted of a number of wires, with a second plug connected at the other end.

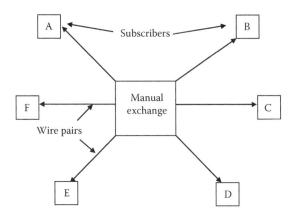

FIGURE 1.8
General layout of a manual exchange.

Thus, to establish a connection between two subscribers, the two plugs of a cord were inserted in the jacks pertaining to the interested (calling and called) subscribers.

Additional connections on the jacks, plugs, and cords enabled calling signals from the calling party, ringing signals to the called party, call monitoring, disconnection of the parties when the conversation was over, and billing.

This, in brief, demonstrates the fundamentals of telephones and telephone systems.

1.4.2 Automatic Exchanges

As the number of telephone subscribers grew, it became impossible to manage the calls with manual exchanges. The size of the jack panel became unmanageable, with dozens of operators moving from one end to the other with long cords in their hands, criss-crossing each other. The only solution was an automatic exchange. The need gave rise to the invention, and the automatic exchange was born.

The general layout of an automatic exchange is the same as that of a manual exchange (Figure 1.8). The difference is the machinery inside.

The automatic connection between the calling and the called party was established by using electromagnetic devices. The first such equipment that remained in use until recently was called "Strowger exchange," after the name of the inventor, Almon B. Strowger.

Strowger exchanges were able to handle between 4000 and 5000 subscribers. As the demand grew, the "crossbar" type of exchange was developed, which could handle greater capacity, and required more complicated devices.

The real breakthrough came after the invention of semiconductor switching devices, which gave birth to the solid-state exchanges (as the name suggests, there were no moving parts in these types of exchanges, in contrast to the electromechanical Strowger and crossbar exchanges). These solid-state exchanges provided a much larger density of population of subscribers and greater ease of operation.

The present technology is, however, many steps ahead of the early solid-state exchanges. Today's exchanges are also "solid state," as they too do not have any moving parts, but they differ in the method of processing and use of devices. Microprocessors are used to process and establish the connections. No physical connection is established, but the subscribers talk through the processor. Other items such as signaling, billing, etc. are also microprocessor-controlled. These exchanges are called "stored program-controlled" digital electronic exchanges.

The latest technology is again a completely new concept. With VOIP (voice over Internet protocol), the voice is digitized and processed as data. This process allows the advantages of "data communication" technology to be passed on to voice telephony.

Description of how exchanges function is beyond the scope of this text, as this work is intended to explain digital multiplexing technologies. The brief discussion above serves as an introduction to this closely related subject of telephony.

1.5 Long-Distance Communication

As subscribers to the local telephone service increased, it was a logical development to expand the service further.

In long-distance telecommunication, the basic principles remain the same. The only difference is in the amount of signal loss and signal distortion in the communication link (or so-called channel). The shorter the link, the fewer the losses, the longer the link, the higher the losses. One way to make up for the losses is to increase the transmitted power, but there are technological constraints to arbitrarily increasing the transmitted power. Hence, a second option of using a low-loss "medium of transmission" (known as "media") is also explored. In practice, a combination of both, i.e., a higher transmitted power and a low loss media, is adopted.

The best low-loss media available in the early days of telephone development was copper wire, but thick wires of copper were very costly, and their high value made them prone to theft. Initially, another low loss media, aluminum, was tried, but it was not as effective as copper as far as losses were concerned. It was a better conductor than metals such as iron and was less costly than copper. The aluminum wires were suspended on poles between cities, but as aluminum lacked sufficient tensile strength, a steel conductor was used in the center, surrounded by aluminum wires. This was called ACSR wire (aluminum conductor steel reinforced).

Subsequently however, the ACSR wire with its limited capacity gave way to copper coaxial cables and microwave radios, etc. (This will be amplified in subsequent sections.)

1.6 Need for Multiplexing

We have seen in the previous sections that various methods were in use for long-distance communication. They remained very costly, however, because of the great distances involved. Creation of this infrastructure involved huge costs. It became apparent that it was not economically viable to use the entire "intercity" infrastructure of ACSR wires, or any other media, for only a single telephone call at a time.

Means were therefore developed to carry a large number of simultaneous telephone conversations on a single pair of wires. The technique evolved for achieving multiple simultaneous communications on a single physical media is called multiplexing.

1.7 Techniques of Multiplexing

The need for multiplexing required analysis of the frequency spectrum of speech. It was found that most components of speech are within a frequency band of 0 Hz to 4 kHz. Although the range of frequencies audible to humans is from 20 Hz to 23 kHz, frequencies higher than 4 kHz consist mostly of musical notes and other high-pitched sounds. On further analysis, it was established that if the frequencies in the range of 0.3 to 3.4 kHz are picked up from the human voice, speech of a perfectly acceptable quality can be faithfully reconstructed.

A means was developed of stacking many such frequency bands, one above the other, and transmitting them as a single block on a single media. Speech channels were stacked one above the other with the support of carrier frequencies, which were nearly 4 kHz apart from each other. At the receiving end they were separated. Multiple, simultaneous communication—multiplexing—was achieved. Figure 1.9 gives a schematic depiction of the concept.

As Figure 1.9 shows, there are four channels multiplexed together to form a single combined channel. Each channel of bandwidth 0.3 to 3.4 kHz rides on a specific frequency, which is called a "carrier frequency," because it actually "carries" that channel. The voice frequency band of 0.3 to 3.4 kHz is called the "base-band."

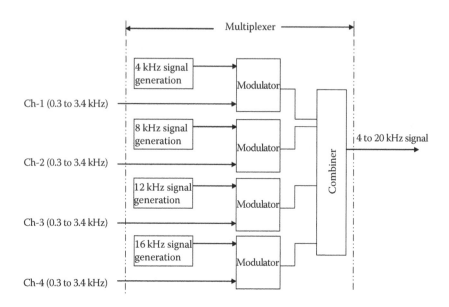

FIGURE 1.9
Multiplexing of four channels with the support of carrier frequencies.

A signal of the desired frequency, say 4 kHz, is generated by an oscillator. It is then fed to an amplifier whose gain is controlled by the base-band signal of 0.3 to 3.4 kHz. The result is a signal whose frequency is 4 kHz and its amplitude varies according to the base-band signal. The "carrier frequency" or simply "carrier" is then said to have been "modulated," and the process is called "amplitude modulation."

Thus the four carrier frequencies carry four individual channels at frequencies of 4, 8, 12, and 16 kHz. All these signals are fed to a combiner circuit. The combined output is a multiplexed signal of frequency range 4 to 20 kHz (each carrier frequency with 4-kHz bandwidth).

This process of multiplexing is called "frequency division multiplexing," or FDM, because channels are multiplexed by dividing the frequency band into small segments.

When this multiplexed signal is received at the other end in the receiver, it is passed through band pass filters to separate the 4-kHz wide bands, and a detector circuit removes the carrier frequency from the signal to retrieve the original base-band signal separately for each channel. Figure 1.10 illustrates the concept.

Although the frequency band of 0.3 to 3.4 kHz is actually only 3.1 kHz, not 4 kHz, and 3.1-kHz bandwidth is enough to carry the desired voice signals, some margin is allowed to accommodate the spreading of the bandwidth of

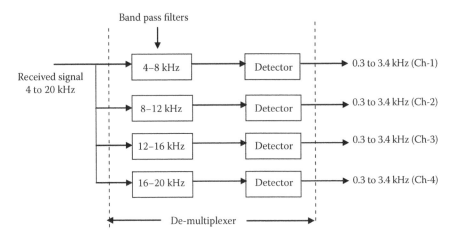

FIGURE 1.10
De-multiplexing of channels at the receiver.

individual channels in the transmission media. Thus, a flat 4-kHz bandwidth is used for each channel.

1.8 Multiplexing Structure of FDM Schemes

In the previous section, we have seen the conceptual design of a four-channel multiplexer. In practice the minimum number of channels defined for a multiplexer is 12. This group of 12 channels is called a "basic group" and is of $12 \times 4 = 48$-kHz bandwidth. The basic group's frequency range is defined as 60 to 108 kHz (and not 4 to 52 kHz), and the first channel is on the top frequency, as per the standards defining body in telecommunication, the International Telecommunication Union, Telecom Standards group (ITU-T). This basic group is called first basic group. The second, third, fourth, and fifth basic groups combine with it to form a "super group." The grouping structure is shown in Figure 1.11.

The basic group supplies 12 channels, the super group combines 5 basic groups to give 60 channels, and if more channels were required to be multiplexed on the same pairs of wires, a "basic master group" was formed by combining 5 super groups to provide 300 channels. The multiplexing continued to form higher-order groups. Systems were developed and deployed for multiplexing up to 10,800 channels.

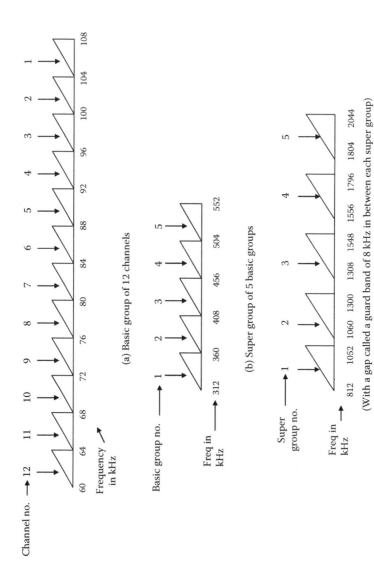

(a) Basic group of 12 channels

(b) Super group of 5 basic groups

(c) Basic master group made of 5 super groups

(With a gap called a guard band of 8 kHz in between each super group)

FIGURE 1.11
Multiplexing structure of FDM schemes.

1.9 Capacities of Various Analog (FDM) Multiplexing Systems

The analog or the FDM systems were developed in a number of categories, primarily depending upon the type of media used. The systems and their capacities differ according to the types of media, copper wires, coaxial cables, and wireless. (The wireless systems were mostly in the microwave's range, and were known as Microwave or MW systems.)

1.9.1 Copper Wire Pairs

Underground copper wires in two pairs (one pair for transmission and one pair for reception) were used to deploy a 12-channel system. The diameter of the copper conductor was normally kept at 0.9 mm. As many systems as the number of quads (a quad is a set of two pairs of wires) could be deployed in a cable. The systems needed repeaters at every 6 km or so. Higher distances were always a problem due to increased noise levels with each added repeater. The overall reliability was very poor as a large number of repeaters was required to cover the long distances involved.

1.9.2 Coaxial Cables

The advent of coaxial cables provided a considerable relief in accommodating a higher number of channels because of their ability to carry much higher bandwidths. The coaxial cable remained the predominant communication media until the advent and deployment of digital systems. Coaxial cables could carry up to 10,800 channels, but they too needed a large number of repeaters, and the larger the number of channels the system carried, the greater was the number of repeaters required. A system carrying 2700 channels required repeaters at every 4.5 km, whereas a system carrying 10,800 channels required repeaters at every 1.5 km. The problems of noise addition at every repeater and poor reliability due to the large number of repeaters continued to be the nightmare of telecom engineers. A 1000-km-long system carrying 10,800 channels would have 630 repeaters. Some respite, however, was offered by the microwave systems.

1.9.3 Microwave Systems

The wireless or radio systems (generally known as microwave systems, as they operated mostly in the microwave range of frequencies) reduced the number of repeaters required and thus the amount of noise in the long-distance links systems were developed to carry up to 6000 channels. With 100-m towers, many located on hills, it was possible to reduce repeater spacing to between 50 and 100 km. Other problems of the terrestrial links (coaxial cables, etc.) were avoided, such as damage by civil works agencies and extreme weather. This led to a great improvement in the communication quality.

With the advent of digital technology, all these problems of analog systems were thrown into sharp focus. It became apparent that the analog systems had no future, and by now they have been completely replaced by digital systems.

1.10 Digital Multiplexing

The digital technology has revolutionized the telecommunication scenario. The two main advantages of the digital systems are

(i) Repetition of signal without adding any noise, and
(ii) Time division multiplexing has been made possible

Although there are several other advantages of the digital systems, these two advantages far outweigh all others. They have led to the possibilities of adding large numbers of repeaters through noiseless regeneration, and accommodating a very large number of channels in a single transmission link through the method of time division multiplexing. These subjects are dealt with in detail in subsequent chapters, and only a brief description of the prevailing technologies follows here.

The digital multiplexing technologies are plesiochronous digital hierarchy (PDH), synchronous digital hierarchy (SDH), coarse wavelength division multiplexing (CWDM), dense wavelength division multiplexing (DWDM), and optical transport network (OTN).

1.10.1 Plesiochronous Digital Hierarchy

This technology was the first to come into practice with the advent of digitization of the voice. The initial systems were called "pulse code modulation" (PCM, so named because of the modulation technique used). PCM was a 24-channel system and required repeaters every 2 km or so. The systems were further developed up to PDH level 4, aggregating capacities of up to 1920 channels.

Soon, these capacities started to fall short as demand grew and the network management and synchronization issues added to the need for development of new systems. Finally, a new system SDH was developed with much better management features, automatic route protection, and much larger capacities in terms of channels.

PDH systems are now confined to the customer access points. For backbone carrier networks, they are no more being deployed or manufactured, and the systems that are still in operation are being phased out.

Sr. No.	SDH Module	No. of E1s	No. of speech channels
1	STM-1	63	1890
2	STM-4	252	7560
3	STM-16	1008	30,240
4	STM-64	4024	120,960

FIGURE 1.12
Channel capacities of various levels of SDH.

1.10.2 Synchronous Digital Hierarchy

As digital technology developed further and the transmission medium changed to optical fiber, more efficient multiplexing technology became possible. The new systems were envisaged to have a much larger channel capacity, automatic protection of circuits against interruptions in any path, and a host of improved management features. (The system was first developed in North America with the name synchronous optical networks, or SONET. Subsequently, ITU-T defined its standards in the name of SDH. Currently, North America and Japan are using SONET standards whereas the rest of the world uses SDH standards.)

The basic SDH system is STM-1 (synchronous transport module 1), which carries 63 E1s (One E1 is equal to 30 speech channels). The next SDH level is STM-4, which is four times the STM-1 in capacity. Similarly, STM-16 and STM-64 have 16 and 64 times the capacity of STM-1. The numbers of channels they carry are given in Figure 1.12.

The standards for STM-256 have also been defined, but so far, systems up to STM-64 only are more popular. To date, the need for higher versions of STMs is not great, as the availability of CWDM/DWDM multiplies capacity.

1.10.3 Coarse Wavelength Division Multiplexing

This multiplexing technology is applicable only to optical fiber based systems. Normally, in optical fiber cable (OFC) transmission, only one wavelength is used to carry signals from STM-1 to STM-256, but CWDM systems can use more than one wavelength.

A few wavelengths, say 2, 4, or 10, are transmitted together, carrying different STM streams. These wavelengths are multiplexed together by the CWDM multiplexer. This is nothing but the analog FDM with the difference that here it is called wavelength instead of frequency. The frequencies are very high, being in the "near-infrared" spectrum. It is called "coarse" because of the use of a limited number of wavelengths, as against "dense" (DWDM), which uses a much higher number of channels.

1.10.4 Dense Wavelength Division Multiplexing

The philosophy, principles, and advantages of DWDM are the same as those mentioned above for CWDM, with the difference that DWDM packs many more wavelengths, typically 80 and higher, with the present systems being available up to 160 wavelengths.

The DWDM technology has once again revolutionized the telecom scenario, and it would not be an exaggeration to say that it has revolutionized the SDH revolution. Apart from improving their quality and reliability, the OFC, SDH, and DWDM are the technologies responsible for the very inexpensive rates of telephony and data services.

1.10.5 Optical Transport Network

A new telecommunication platform recommended by ITU-T, which has recently begun manufacture and deployment, is OTN. OTN has been designed to optimize the transport of SDH-based services along with data communication and DWDM services. The advantages of all the transport systems such as SDH, data communication, and DWDM have been exploited to develop this new unified transport platform.

Review Questions

1. Who invented the telephone and when?
2. Describe the basic concept of voice transmission and reception in a telephone.
3. What is the role of various components in a transmitter and a receiver?
4. How many wires are required for a telephone communication to take place? Explain why.
5. What is a "Hybrid" and what is its application?
6. List the main differences between manual and automatic exchanges. What is meant by "Strowger?"
7. How does long-distance communication differ from local area communication through telephone exchanges and what is meant by multiplexing?
8. Describe the principles of analog multiplexing technology. What was the significance of carrier frequency?

9. Why is the frequency band of 0.3 to 3.4 kHz chosen for voice channels?

10. What are the main components of analog multiplexing and de-multiplexing systems?

11. How is grouping of channels done in FDM and why? What is the need of a guard band?

12. Up to how many channels could be multiplexed in an FDM system?

13. Make a table comparing the number of channels and repeater spacing in various analog systems.

14. Compare the advantages and disadvantages of copper wire, coaxial cable, and microwave systems.

15. What are the developments in digital multiplexing technologies and how do they score over analog multiplexing technologies?

Critical Thinking Questions

1. Is it possible to achieve telephone communication using only one wire? If yes, how?

2. What is meant by an "exchange?" Make a layout of an exchange that is different from the one given in this chapter.

3. Try to work out an automatic exchange of four lines using any electronic or electromagnetic devices that you know of.

4. Give an example of multiplexing in a field other than telecommunication.

5. Out of the two technologies, SDH and WDM, which one is better and why? Are they complementary to each other, and if so, why?

Bibliography

1. L.E. Frenzel, *Principles of Electronic Communication Systems*, Tata McGraw-Hill, India, 2008.
2. S. Ramo, J.R. Whinnery, and T. Van Duzer, *Fields and Waves in Communication Electronics*, John Wiley and Sons, 1994.
3. J.D. Ryder, *Network Lines and Fields*, Prentice-Hall, 1975.
4. J.C. Bellamy, *Digital Telephony*, John Wiley and Sons, Singapore, 2003.

5. T. Vishwanathan, *Telecommunication Switching Systems and Networks*, Prentice-Hall India, New Delhi, 2001.
6. F.E. Terman, *Electronic and Radio Engineering*, Fourth Edition, McGraw-Hill Kogakusha, Tokyo, Japan, 1955.
7. ITU-T Recommendation G.322, *General Characteristics Recommended for Systems on Symmetric Pair Cables*, International Telecommunication Union, Geneva, Switzerland.
8. ITU-T Recommendation G.334, *18 MHz Systems on Standardized 2.6/9.5 mm Coaxial Cable Pairs*, International Telecommunication Union, Geneva, Switzerland.
9. A.F. Molisch, *Wireless Communications*, John Wiley and Sons, UK, 2005.
10. B.P. Lathi, *Modern Digital and Analog Communication Systems*, Oxford University Press, 1998.
11. P.V. Shreekanth, *Digital Microwave Communication Systems with Selected Topics in Mobile Communications*, University Press, 2003.
12. P.N. Das, *An Introduction to Automatic Telephony*, Modern Book Agency, Calcutta, India, 1975.
13. B. Govindarajalu, *Computer Architecture and Organization*, Tata McGraw-Hill, India, 2006.

2

Advent of Digital Technology

In the previous chapter, we have seen how telecommunication began, the basic components of a telecommunication system, how the need for expansion of telecom services drove the world to develop manual and automatic exchanges, and how they worked. Further proliferation of services led to the development of long-distance telephony for intercity, interstate, and finally international calls. The various media used for long-distance communication, the need for multiplexing, analog multiplexing principles and applications, and modern digital multiplexing technologies were discussed.

This chapter concerns the problems and limitations associated with analog systems. What led to the development of digital technology? How were the problems of analog systems overcome by digital technology? What are the other advantages of digital technology, and what are the limitations and disadvantages of digital technology?

2.1 Analog Communication Systems

The analog communication system is the natural communication system. We perform all our actions in analog form, i.e., continuous form. As speech continuously varies in intensity (except the pauses), vision and hearing are continuous. By contrast, an electrical switch has only two possible functions or states, "on" or "off," and as such can be treated as having a discrete (or digital) function, and the device, the electrical switch, can be regarded as a digital device.

In the process of the development of telecommunication, it was obvious to use the voice signal "as it was," and establish a communication between two interested parties using suitable devices such as a microphone, telephone line, receiver, and amplifier. Early communication systems were established this way, and they worked well for decades. However, with increased demand and the need for longer distances to be covered, the poor quality of received speech became a serious issue and the subject of considerable effort on the part of the telephone engineers.

The major factors leading to speech quality deterioration were noise, distortion, attenuation (reduction in signal strength with distance), cross talk (one call partially coupled to another in progress on another pair of wires in the cable), and interference. The shortcomings of the equipment, the amplifiers, filters, transducers, etc., and the transmission media, were the main contributing factors. A number of techniques were developed to overcome the adverse effects of these factors on the quality of speech. However, apart from quality issues, the capacity of analog communication systems were starting to show signs of saturation.

2.2 Problems in Analog Communication Systems

To appreciate and understand the problems and issues related to analog systems, it is helpful to consider the very basic analog system consisting of a microphone, an amplifier, a channel consisting of a pair of copper wires, and a receiver. Although the systems developed for multi-channel capacities and long-distance capabilities were much more complex, involving many types of media such as coaxial cables, microwave, etc., and complex multiplexing technologies including sophisticated repeaters, the problems and their effects can be best understood by this basic model. (In fact the only part of the transmission network which remains without an amplifier is the one between the subscriber's telephone and the local exchange.)

A basic analog communication model using base-band, i.e., 0.3 to 3.4 kHz audio frequency transmission on a copper pair, is shown in Figure 2.1. Only one-way communication has been shown for the purpose of simplicity. The microphone gives the voice input to the amplifier, which is transmitted on the transmission line after amplification. The receiver tries to filter out the corruptions of the signal caused by the amplifier and the channel.

Now let us examine the various sources of corruption of the signal while it makes its journey from the microphone at one station to the loudspeaker at the other station, with reference to Figure 2.1.

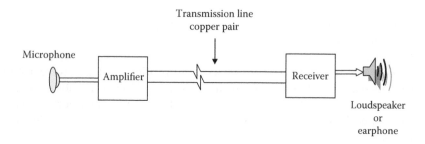

FIGURE 2.1

Basic analog communication model using base-band (audio frequency) transmission on a copper wire pair.

2.2.1 Attenuation

As the voice signal traverses the channel (transmission line), it loses strength due to ohmic loss in the channel caused by the resistance of the copper wire. This is the reason for deployment of an amplifier at the transmitting end, to boost the signal sufficiently, such that the received signal, after suffering losses due to attenuation, is still within the threshold limit of the receiver. The amplifier solves one problem but adds many more, such as distortion and noise, which is the cost to be paid for the gain. Nevertheless the use of the amplifier is a must (we will see about the problems added by the amplifier later in this chapter). Figure 2.2 illustrates the attenuation phenomenon.

The signal shown in Figure 2.2 is a sinusoidal signal. In reality the speech signal, or any other natural signal for that matter, is never a perfect sinusoid. However, the sinusoidal representation of signals facilitates an easy understanding, and modeling using it is perfectly applicable on actual natural signals of speech/video and others. The voice signal may look something close to what is shown in Figure 2.3.

The attenuation is measured in decibels per kilometer. (A decibel is a measurement unit universally adopted for attenuation. For a detailed discussion

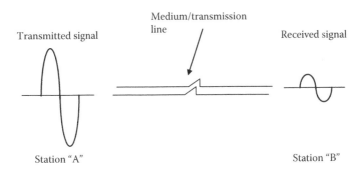

FIGURE 2.2
Signal attenuation due to losses in the channel.

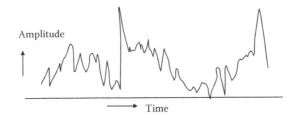

FIGURE 2.3
Actual speech signal representation.

of decibel, see Section 6.9.) With the use of an amplifier, the attenuation permitted depends upon the signal-to-noise ratio required by the receiver along with the sensitivity threshold of the receiver. Without the use of an amplifier, a loss of up to nearly 8 db is permissible between the talker and the listener in the speech signal in the medium, while retaining a good quality of audibility.

Attenuation would have been less of a problem if the amplifier did not "add" problems such as distortion and noise, which is multiplied many times by a large number of repeaters (each repeater has an amplifier) in a long communication link. We will examine this aspect in detail in subsequent sections.

2.2.2 Noise and Interference

Noise has probably always been the single most bothersome factor for the design and maintenance engineers of telecommunication systems, and it continues to be so today.

Noise is defined as "unwanted electrical interference" with the telecommunication signal.

Referring to Figure 2.1 again, in our basic communication model (the terms telecommunication and communication are used synonymously in this text), the noise is generated in each and every part of the link, be it microphone, channel, receiver (the receiver has an amplifier in addition to other processing circuitry), or the loudspeaker. Let us see the effect of noise on the signal.

Figure 2.4 above shows the input-output relationship of a device with respect to an analog signal. The input analog signal is absolutely clean (assumed to be a sine wave for simplicity of expression) but the output signal is corrupted by the noise generated by the device.

Now imagine a situation where a piece of equipment, such as an amplifier or receiver, is made up of a large number of such devices. The noise generated by all such devices accumulates at the output of each such piece. The trouble multiplies in a long communication link where many such pieces of

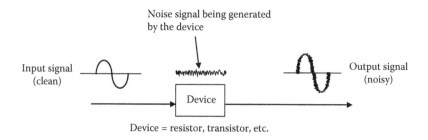

FIGURE 2.4
Corruption of an analog signal caused by the noise generated in a device.

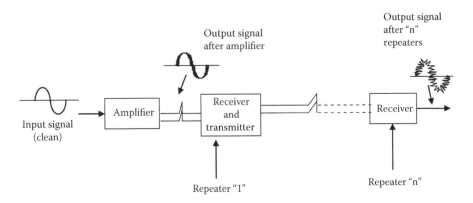

FIGURE 2.5
The noise multiplication effect in a long communication link with "n" repeaters.

equipment come into play. Moreover, all the amplifiers keep on amplifying the noise along with the signal amplification. Figure 2.5 depicts the situation.

Figure 2.5 depicts a long communication link, having a large number of repeaters. (The signal strength goes down with distance due to attenuation and is boosted by amplification at the repeater.) Finally the noise becomes so much that the link cannot be extended any further, depending upon the acceptable quality of the audio. Thus, a practical limit is reached in the length of the communication link.

All noise reduction measures extend the limit, but a "limit" as such remains.

2.2.2.1 Sources of Noise

The type of noise we have just seen in Figure 2.4 is generated by each and every device, equipment, or channel through their resistive components. It is generated by the random motion of electrons and atoms inside the components, due to heat. The higher the temperature the more is the noise, and the lower the temperature the less is the noise. This type of noise, is called "thermal noise." Since ohmic resistance is present in all devices and in metallic wire channels, this noise is generated by every component of the link. It is completely random in nature, and no correlation can be established in the noise values at any two instants of time. Thermal noise is evenly distributed throughout the frequency band beginning with a few hertz to several Gigahertz. Due to this more or less constant amplitude throughout the frequency spectrum, it is also called "white noise." The white light is composed of all the frequencies of the visible light spectrum, which gives white noise its name.

Other sources of noise include almost everything that can generate electromagnetic waves. Noise is created by atmospheric activities such as lightning discharges and thunderstorms. Solar and cosmic radiation also create noise. However, the second biggest source of noise is industrial activity, which may include the ignition of automobile engines, aircraft, switching surges of motors and power transmission grids, and leakages from high voltage lines.

2.2.2.2 Relevance of Type of Noise

As mentioned above, white noise and industrial noise are the major contributors to the noise activities as far as communication systems are concerned. While the noise generated by electronic devices is generally a white noise which is of low power but high band width, and is always present, the industrial noise is generally of impulse type, inducing impulses of various magnitudes and durations, occasionally interferes with the communication signal.

The white noise is of prime importance in analog communication, as impulses occur only for short duration during speech, which can be tolerated without any significant loss of information and without much discomfort to the conversing parties. However, in the case of data communication, whether it is purely a digital signal or data over analog lines through modems, the impulse noise is of prime importance as it can change the very meaning of the data by simply altering a few bits.

Noise implications are generally quantified in terms of signal-to-noise ratio. The signal-to-noise ratio, S/N, is defined as the ratio of signal power to noise power.

$$\text{Signal-to-noise ratio SNR} = \text{signal power/noise power} = S/N$$

A signal-to-noise ratio of 40 to 45 db is generally acceptable for speech circuits.

However, in pure speech circuits, the signal-to-noise ratio may not represent the actual annoyance level of the listener because the same level of noise, which may be tolerable during conversation when the speech levels are high, may become very annoying during low speech levels and during pauses when the speech level is very low or zero. Thus, actual noise assessment in speech circuits is checked by "test tone" to noise ratio at a particular frequency with constant amplitude of test tone.

2.2.3 Distortion

As the meaning of the word goes, the shape of the signal changes with respect to the original signal during transmission, and since this change is not a welcome change, it is called distortion. It is caused by the amplifier as well as by the channel and by the receiver equipment. If equipment or a channel introduces a change in amplitude and a constant delay, there is no distortion because the shape of the transmitted signal is preserved. Distortion occurs when, with or without the amplitude change, the shape of the signal changes, and moreover, the delay is not constant for all frequencies. The distortion falls into two categories, "amplitude distortion" and "phase or delay distortion." Let us look at them in a little more detail.

2.2.3.1 Amplitude Distortion

Amplitude distortion is encountered by a signal when the frequency response of the system components (e.g., transmitter equipment or channel) is not constant for all the frequencies.

In the case of an amplifier, the inductance and capacitance associated with the active components, or those otherwise used discretely in the circuit, make the response (amplification or gain) sensitive to frequency. This is because the impedance of inductors and capacitors is highly dependent on frequency. The impedance of an inductor is proportional to the frequency, and that of the capacitor is inversely proportional to the frequency. Different frequencies are amplified to different magnitudes. Figure 2.6 shows the typical frequency response of an audio amplifier and the amplitude distortion of the signal.

As can be seen in Figure 2.6a, the gain of the amplifier is more at central frequencies and less at low and high frequencies. Figure 2.6b shows the input signal having medium and high frequency components. The high frequency in this case is falling at the knee (A) of the frequency response curve.

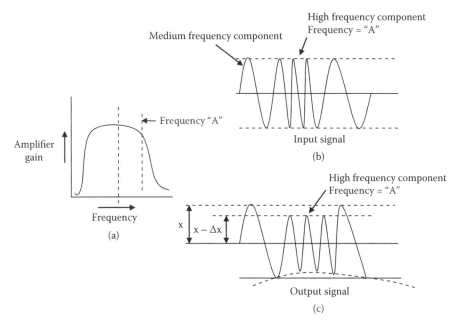

FIGURE 2.6
Amplitude distortion in an amplifier. (a) Amplifier's frequency response characteristic; (b) typical audio signal with varying frequency contents; and (c) distorted output signal when the audio signal of (a) is applied at the input of the amplifier.

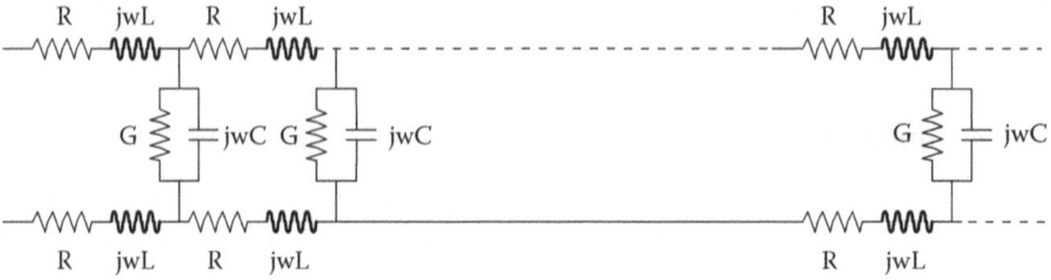

R = Series resistance
jwL = Impedance offered by the inductor of value "L"
jwC = Conductance offered by the capacitor of value "C"
G = Leakage conductance

FIGURE 2.7
A channel or transmission line can be modeled to constitute very small elements of resistance, inductance, leakage conductance, and capacitance.

The gain starts falling on these frequencies, with the result that the output signal is distorted in shape as shown in Figure 2.6c, having lower amplitude at higher frequencies.

In the case of a channel the phenomenon is almost the same, except that instead of gain, the channel introduces losses or attenuation. The attenuation varies with frequency to give rise to the amplitude distortion, similar to the one shown in Figure 2.6. The inductive and capacitive reactance are in fact much more predominant in case of a channel (particularly the metallic media like copper wire pair or coaxial cable) as compared to those in the amplifier. Figure 2.7 shows that a channel of copper wire pair has resistance and inductance all along its length due to the presence of wire. It has capacitance and leakage conductance because of parallelism with the second conductor. These components are uniformly distributed all along the line.

Since the resistance, inductance, and capacitance are distributed continuously on the channel, the channel can be modelled as being made up of very small segments consisting of resistance, inductance, capacitance, and leakage conductance elements all through its length.

Such a channel full of reactive elements gives rise to amplitude distortion similar to the one shown in Figure 2.6 by causing different magnitudes of attenuation at different frequencies of the signal.

2.2.3.2 Phase Distortion or Delay Distortion

The transmission line or channel consists of various components. Out of these components the inductance and capacitive impedance are sensitive to frequency. The inductive impedance value is "jwL" and the capacitive impedance is $1/jwC$, both of which have "w," i.e., $2\pi f$ (f is the frequency) in their impedance values. Owing to these frequency-sensitive components, the velocity of propagation of the signal in the transmission line becomes a complex issue. Frequencies do not all travel at the same velocity through the medium of the transmission line, giving rise to a change in the shape of the received waveform, as compared to the transmitted waveform at the sending end. This change of waveform is called "phase distortion" or "delay distortion."

In the case of a twisted pair of telephone cables, the velocity of propagation can be given approximately by the following equation:

$$V = \sqrt{2w/CR}$$

where the terms w, C and R have similar meaning as brought out in Figure 2.7 and "V" is the velocity of propagation of the input signal wave in the transmission line.

It can be clearly seen that the velocity is directly proportional to the square root of the frequency. Thus, higher frequencies travel faster than the lower frequencies, giving rise to distortion in the received waveform, called "phase distortion" or "delay distortion."

It is important to note that some amount of propagation delay will always be there for all frequencies. As long as this delay is constant, there is no delay distortion, and only when the delay is nonuniform over the range of transmitted frequencies does the delay distortion occur.

At times one more cause which adds to the delay distortion is the reflection of the propagating wave inside the channel due to impedance inaccuracies. This leads to multiple waves arriving at the receiver causing distortion of the phase.

The effect of phase or delay distortion is negligible on speech circuits because the human ear cannot diagnose such a minute difference. However, in the case of a video transmission, the effect becomes very significant as it causes different time delays to each picture element, causing the picture to appear smeared.

The phase or delay distortion is illustrated in Figure 2.8. The delay distortion shown is only the one caused by high frequency components travelling faster. The delay common to all frequencies has not been shown, to facilitate easy understanding of the concept.

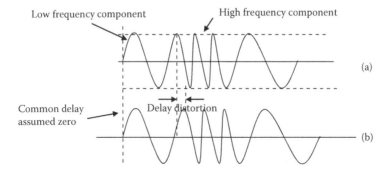

FIGURE 2.8
"Phase" or "delay" distortion in the received signal.

All these distortions that we have seen so far are also called linear distortions, as none of them is caused by the nonlinear characteristics of the system. In the next section we will see about the nonlinear distortion.

The above discussion about phase or delay distortion applies equally to the amplifier and other equipment, except that its value is less than that in the channel.

Fortunately, distortion has a good feature: it can be predicted, unlike noise, which is random. This leads to deployment of equalizers for overcoming the adverse effect of distortion in the received signal. The equalizers have a characteristic opposite to that of the transmitter equipment and channel, as far as the distortion is concerned, and are thus able to minimize distortion to a great extent.

2.2.3.3 Harmonic Distortion

When the range of the input signal is such that it occupies the nonlinear portion of the transfer characteristic of the system (i.e., transmitter or the channel), harmonics are generated. The situation is illustrated in Figure 2.9.

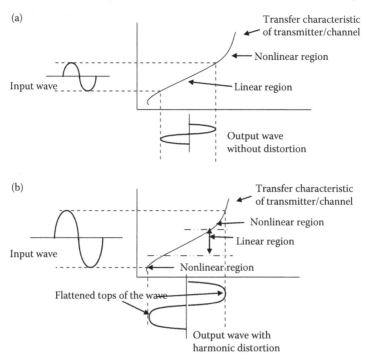

FIGURE 2.9
Harmonic distortion caused by the nonlinear characteristic of the system (transmitter or channel). In (a), output is nondistorted because the input signal falls within the linear range of transfer characteristic, whereas the output in (b) is distorted because input signal is large enough to occupy both linear and nonlinear portions of transfer characteristic.

Figure 2.9a shows the input and output waveforms when the signal operates within the linear region of the transfer characteristic, of the transmitter or the channel, causing no distortion to the output; Figure 2.9b shows the input and output waveform when the range of the input signal goes beyond the linear region of the transfer characteristic. The resulting output waveform in the second case is distorted.

This type of distortion is called harmonic distortion. The nonlinear operation results in the generation of second, third, fourth, and so on orders of harmonic wave components. If all these components are added up, they will give rise to the shape of the distorted waveform obtained at the output as shown in Figure 2.9b. (A detailed discussion about the harmonics and others is beyond the scope of this text. The interested readers can find good details in any text containing Fourier series analysis of signals and systems.)

The second-order harmonic signal is of frequency 2f, if the original signal is of frequency "f," and similarly third-order 3f, fourth-order 4f, etc.

Since the original speech signal consists of many frequencies, the harmonics that are generated are also of many frequencies. These harmonics interact with each other and produce what is called "inter-modulation distortion."

Overall it becomes a very complex theoretical phenomenon. But, take a deep breath, the solution is simple. All these harmonic components can be brought within acceptable limits by using "low-pass" or "band-limiting" filters. The limits of harmonic distortion are a measure of the quality of the system. Many of us who have ever tried to buy a hi-fi sound system must have come across this factor, where the unwanted higher frequencies (basically harmonics) are filtered out with the use of such filters.

2.2.4 Cross-Talk, Echoes, and Singing

Cross-talk, echoes, and singing are a few more impairments that hamper the quality of communication, particularly in a speech circuit.

Cross talk takes place when the signal of one speech channel gets coupled to its adjacent channel, perhaps in a multi-pair telephone cable, or in a multiplexed carrier system. We are able to hear some other conversation while talking on our telephone. This happens due to unwanted coupling of signals between two pairs of wires of the cable due to poor insulation between pairs, inadequate carrier suppression in a multiplexed environment, or due to defective components permitting overlapping frequency bands.

Echo is produced when the transmitted signal becomes partially coupled back to the receiving pair of the channel. This coupling back of signal is called reflection. Reflection takes place due to impedance mismatch amongst various components of the system such as the transmitter, the hybrid, and the transmission line ("hybrid" is a circuit used for converting the two-wire telephone circuit to a four-wire channel having separate pairs for transmitting and receiving the signal or vice versa).

Singing takes place when the echo is continuously coupled.

2.3 What Is Done about These Problems

Engineers applied a phenomenal amount of effort to minimize these problems. Although the problems could not be completely eliminated, considerable success was achieved in reducing them to varying degrees in each case.

Attenuation is the impairment where such efforts have not met with much success. The improvements in cable parameters could not achieve much reduction in the attenuation per unit length of the cable. Further reduction needed the diameter of the copper conductor to be increased, which involved higher costs. A sizable improvement was achieved in the cables optimized for the "voice frequency" circuits through the use of lumped inductances in series with the conductors at fixed intervals. These inductors were called "loading coils," and provided compensation to the cable capacitance, reducing the attenuation in the higher frequency area of the voice frequencies.

On the noise and interference front, commendable improvements were achieved. The development of low noise amplifiers and receivers, and various techniques to reduce the electro-magnetic interference (EMI), helped in reducing this menace to a great extent. Installing telecom equipment away from the heavy machinery or electrical switching installations also helped.

Distortion, as mentioned earlier, was compensated to a great extent by the use of equalizers. Band pass filters minimized the harmonic distortion.

Cross talk, echo, and singing are almost completely manageable if the equipment is properly designed and maintained. Echo suppressors have been successfully developed and deployed.

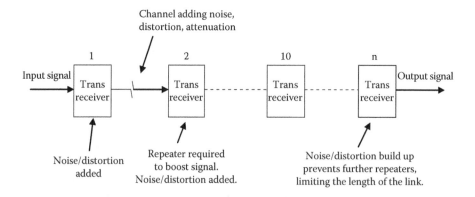

FIGURE 2.10
A long link with many repeaters is finally limited in length by the noise and distortion build-up.

In all, the problems continued, and so did the efforts to rectify them. But ultimately the length and capacity of an analog communication link were limited severely because of these problems. Every time the signal was required to be amplified, noise was added to it, and every time it was again transmitted on the channel, it gathered more noise, distortion, and other impairments. Figure 2.10 depicts an analog communication link to illustrate the point.

The capacity of the analog link was increased to a few thousand channels by the use of coaxial cable in place of twisted pair copper cable. But the repeater spacing was always a trade-off with the number of channels. More the channels, less was the repeater spacing, and the lesser the repeater spacing, more numbers of repeaters were required for a given length, so finally the length of the link. Thus, long-distance links could not have a very high number of channels. Very high-density links could be built only for shorter distances.

2.4 Digital Advantage

Digital technology overcomes almost all the above problems with ease. So, what is the magic by which all the impairments discussed in the previous sections do not affect the digital signal? There is no magic whatsoever, all the impairments affect the digital signal as much as they do an analog signal, and in fact the effect of each of the impairments on digital signal is more as compared to that on an analog signal.

The digital signal does not fight much to overcome the effect of all those impairments, but it simply "tolerates" them. Let us see how.

2.5 Digital Signal

No natural signal is digital. Our speech or movements are analog, or continuous. To make use of digital technology, we have to convert the analog signal to a digital signal, then process and transmit it as per the requirement of the application, and finally convert it back to analog, to make it suitable for human interaction. The analog to digital and digital to analog conversion is dealt with in detail in the next chapter. For the time being, if you are not familiar with digital technology at all, please assume a square pulse as a digital signal. This will be enough for our present explanation of the digital advantage. Figure 2.11 depicts a simple digital signal in the form of a square pulse.

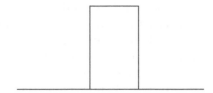

FIGURE 2.11
A digital signal represented as a square pulse.

2.6 Processing of Digital Signal

When the digital signal of Figure 2.11 is processed, i.e., filtered, amplified, transmitted, and received over the telecommunication link, it suffers from the same impairments as those for the analog signal. The worsening effects are greater in the case of a digital signal because of the square shape of the pulse. Its sharp edges are much less likely to maintain their shape than the sinusoidal shape of our assumed analog signal. (According to Fourier analysis, a square wave consists of a fundamental sine wave and an infinite number of harmonics of the fundamental sine wave. In fact if we work it back, i.e., if we take a sine wave and add to it its entire set of harmonics in proper proportions, we will get a square wave. Because of these high frequency components, the shape of the pulse suffers more impairment.)

Figure 2.12 shows a typical transmitted and received digital signal.

We can see from Figure 2.12 that the square wave pulse, when received at station "B," has completely lost its shape, besides being heavily attenuated. Then what is the advantage of digital signal?

To the rescue comes "digital processing."

The received signal is processed digitally, simply to establish whether it was a pulse or not. The receiver acknowledges the incoming signal as a pulse if its level is above a certain value. Let us say the detection threshold of the receiver is "x" volts. Now if the received signal is above "x" volts, it will be recognized as a pulse. Moreover, the square shape of the pulse does not have

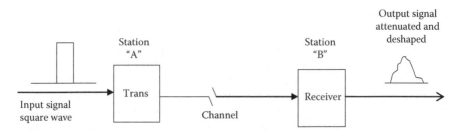

FIGURE 2.12
Transmission and reception of a digital square wave signal.

to be maintained in its original shape, as the triggering by the receiving circuitry occurs if the amplitude of the received signal is above the threshold level of the receiver, irrespective of its shape (Figure 2.13).

Thus, all that the receiver is required to do is generate a square pulse locally (exactly as per the defined and mutually agreed norms between the transmitter and receiver), and pass it on to the digital to analog converter, or to the user directly. The digital to analog converter or the user receives a clean pulse, which is an exact replica of the transmitted pulse.

Thus, even though all the impairing factors were present, it was possible to transmit and receive a square pulse without any degradation at all in its quality.

In digital technology only two types of signals are processed: a "one" and a "zero." What we have seen just now can be considered as 1, and a negative pulse or an absence of a pulse may be considered as a 0, or vice versa (there are many variations to this philosophy, the details of which can be seen in Chapter 4).

If there is a long communication link with a large number of repeaters, at every repeater the digital signal will be generated afresh or it will be "regenerated." For this reason the repeaters in digital communication links are called "regenerators." This regeneration approach permits an unlimited chain of repeaters without affecting the signal quality at all. Whatever may be the length of the digital link, there will be a crystal clear sound on the

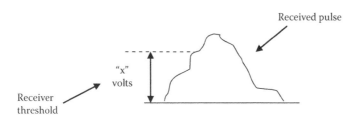

FIGURE 2.13
Recognition of a received signal as a pulse.

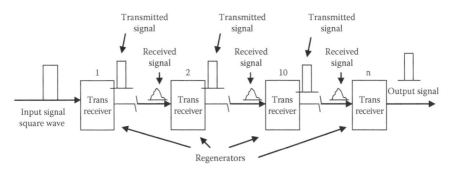

FIGURE 2.14
Regeneration of digital signal at trans-receivers (repeaters).

telephone and a very clear picture on the TV. All the problematic factors can have their effect only in between the regenerators; the multiplying effects of impairments at every repeater are simply driven away. As long as the regenerator spacing is maintained within acceptable limits for correct reception, the signal does not deteriorate at all, no matter how many times it is regenerated. The concept is illustrated in Figure 2.14.

2.7 Channel Capacity in Digital Technology

In the preceding section the most fundamental advantage of the digital signal over the analog, i.e., unlimited link length without any signal quality degradation, has been discussed.

Now let us see how the digital technology helps to overcome another most important limitation of the analog systems, i.e., the limited line capacity.

The digital technology allows the use of "time division multiplexing" (TDM). For analog to digital conversion of signal for speech, one sample of the analog signal is taken every 125 μs (we will see the whys and hows of this in detail in Chapter 3). If the "width" of the digital signal (pulse) is less than 125 μs, then another similar signal can also be accommodated in the same period of 125 μs (Figure 2.15).

Generally, the width of the pulse of each signal is much less than 125 μs. Using this principle, initially, a 24-channel system was developed by multiplexing their corresponding pulses in a common time frame of 125 μs. (In practice, eight bits, also called 1 byte, are used for each signal. These details will become clear as we read Chapter 3. However, to understand the concept, it is easier to visualize one pulse for each signal.)

Further developments have led to more and more pulses being accommodated in this time frame of 125 μs. With the help of synchronous digital hierarchy (SDH) technology and the use of optical fiber, as many as 483,840 multiplexed channels are possible in the time frame of 125 μs (in case of SONET, the American system, the number of channels presently possible

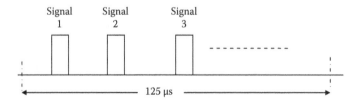

FIGURE 2.15
Accommodating more than one signal in the time period of one signal.

is 516,096), which travel on a single pair of optical fibers. Moreover, there is no distance limitation for such high capacity links. Thus, we can see that the digital technology has revolutionized the system capacities as well.

(TDM is theoretically possible in analog systems too, and in fact, it had been deployed also in a very small way before the advent of digital technology. TDM of analog signals is achieved by taking samples of the signal at frequent intervals, either the amplitudes of the pulses are proportional to the amplitudes of the samples [pulse amplitude modulation, or PAM] or the width of the pulses is made proportional to the amplitudes of the samples [pulse width modulation, or PWM]. In either case, the pulse had to be received and amplified/processed at the receiver as it is, for further transmission or conversion to the user signal. Thus, it suffered from the same problems as the continuous analog signal. The receiver needed to reconstruct the exact shape of the transmitted pulses to faithfully reproduce the signal. The repeaters did not generate clean pulses, but only reconstructed the transmitted pulse to the extent possible. In contrast, the digital system only determines whether the signal is a 1 or a 0, and then regenerates a clean signal. Thus, the analog TDM never became very popular.)

2.8 Advantages of Digital Technology

Besides the features discussed in the previous section, there are numerous other advantages of digital technology, leading to a virtual uprooting of analog technology in signal processing and transmission. The user interface remains analog, but the rest has all been changed to digital. There are some important advantages of the digital technology, in addition to the revolutionary advantages such as "regeneration" and TDM we have seen in the preceding section.

2.8.1 Universality of Components and Media

Whatever be the type of signal or application, the components of the digital processing remain almost the same, in contrast with analog technology where we needed separate electronics for dealing with speech, video, and data. Speed and capacity of the digital components is all that varies, depending upon the application.

For optimized voice transmission we needed cables to be loaded with loading coils and capacitance balancing, for video we needed RF coaxial cables with low loss characteristics. Wave-guides were required for microwave transmission from radio equipment to the antenna. In digital transmission a common set of components, say microprocessors, memories, digital integrated circuits, and embedded analog components, make up systems like synchronous digital hierarchy (SDH) equipment, which can carry all these services simultaneously and without any discrimination.

2.8.2 Cheap and Reliable Storage

It was almost impossible to store analog data for a long time and retrieve it correctly. The storage was mostly on magnetic tapes. The same storage medium provides much more reliable storage of the digital data, as the reproduction requires only retrieval of levels 1 and 0.

A variety of devices such as CD ROMs, DVDs, and Blu-ray disks are available today, which provide very high capacity storage at a fraction of the cost of storing analog data, and the reliability is very high. The storage densities are so high that electronic copies of hundreds of books can be stored on a single disk which can be easily carried in your pocket.

2.8.3 Low Cost of the Equipment

The cost of the digital signal processing equipment is much less than its analog counterparts. For example, the cost of analog frequency division multiplexing (FDM) equipment is many times higher than that of digital TDM on a per channel basis.

Because economies of scale have been achieved due to the high demand for a variety of applications, the cost of digital components is rapidly reducing, to the extent that the cost of the components of a particular capacity halves approximately every two years.

2.8.4 Ease of Signaling

In telephony applications, the signaling information is required to be carried separately in the analog systems. In digital systems the signaling information is transmitted as a part of the whole data stream, in the same format as data. At the receiving end, the signaling information is taken out through the knowledge of its position in the data stream. The signaling in analog systems had always been a problem area, giving rise to the development of different types of signaling equipment for different types of signaling techniques. In digital systems, signaling is just a protocol, which can be changed any time without requiring modifications in any of the interface hardware.

2.8.5 Use of Microprocessor

Digital technology permits use of the microprocessor for signal processing, which provides great flexibility, versatility, and universality in equipment design. Apart from the operational features, the microprocessor greatly enhances the management features of the system without adding much to the cost, as the same set of hardware is able to take up both jobs.

2.8.6 Less Precision Required

The precision required in digital components is much less than in analog components. In analog processing, preserving the exact shape of the signal is very important, as the precision of components directly affects the output quality. However, in the case of digital components, only ones and zeros are required to be detected and processed. Since these two levels have large margins for detection, a very low precision is required in digital components.

Two other advantages of digital signal processing are the feasibility of linear phase filters which eliminate the nonlinear distortion and the processing of very low frequency signals, such as seismic and oceanographic signals, which becomes very economical in the digital domain as it requires very high capacity inductors and capacitors in analog processing.

2.9 Disadvantages of Digital Technology

The biggest disadvantage of digital technology over analog technology is that digital processing needs the analog signal to be first converted to digital, and reconverted to analog signal. Thus, the process requires additional components in the form of analog to digital and digital to analog converters. But this is a very small price to pay for a much larger gain, and it is many more times compensated by the reduced cost of digital components.

Another disadvantage of the digital technology is that the digital signal is never an exact replica of an analog signal, it is only an approximation. However, the corruption of the signal caused by this approximation process (called quantization) is many orders of magnitudes less than the impairments suffered by the analog signal, particularly on long-distance links. This is more than evident to us if we know and compare the poor quality of the analog days to the present-day digital quality of our long-distance telephone calls.

The current situation is that digital technology has completely replaced analog technology in processing and transmission. The analog technology, however, continues to play an important role at the entry and exit points of the systems, where the signals are required to interact with humans. In short, analog technology today acts as an interface between people and the digital technology.

Review Questions

1. What is the need of digital technology? Why did we have analog technology if digital technology is good?

2. What is the need of multiplexing in telecommunication systems?

3. Draw a simple model of an analog communication system showing all the essential components.

4. Which are the sources of corruption of a signal in a communication system?

5. Show with the help of graphical representation the difference between a natural signal and a sinusoidal signal.

6. What do you understand by attenuation? Why does it take place and what are its units?

7. What are the problems added by the amplifier in a communication system?

8. What is noise? How is it generated and what are its effects on telecommunication?

9. Show the effect of noise on the signal through a graphic representation.

10. How does noise build up in a link?

11. What are the factors that limit the length of an analog communication link?

12. What are the sources and types of noises?

13. What do you understand by SNR and test tone to noise ratio?

14. What is distortion and what are its types?

15. Explain the amplitude distortion with a sketch.

16. Compare an amplifier with a transmission line for various factors responsible for corruption of a signal.

17. What is phase of delay distortion? Explain with a sketch.

18. Bring out the difference between linear and nonlinear distortion.

19. What are the good and bad features of noise and distortion in comparison to each other?

20. Explain harmonic distortion with a sketch. How does it differ from inter-modulation distortion?

21. How are the problems due to noise, interference, and distortion tackled?

22. What is a repeater? How does it help in increasing the link length?

23. Why is the digital signal not affected by the noise, interference, and distortion?

24. What are the frequency components of a square wave? Illustrate with a sketch.

25. What is responsible for the digital advantage of the shape of the digital signal or the digital processing? And how?

26. Explain the process of regeneration.

27. What is the effect of digital signal processing on channel capacity? Please explain.

28. What are the advantages and disadvantages of digital technology?

Critical Thinking Questions

1. Why is the frequency band of 0.3 to 3.4 kHz chosen for voice channels?

2. How does noise differ from interference?

3. Draw sketches showing the effect of various types of noises on analog and digital signals.

4. What are the factors that affect the velocity of propagation of a signal? How do they affect it and why?

5. Compare the various factors affecting the transmission quality. Which are the worst?

6. How is it possible to overcome these factors without digitization and achieve similar capacities?

7. Try to design in principle an analog TDM system for 1000 speech circuits.

8. List the advantages of digital technology in the order of their merit.

Bibliography

1. J.G. Proakis and D.G. Manolakis, *Digital Signal Processing Principles, Algorithms, and Applications*, Third Edition, Prentice-Hall India, New Delhi, 2005.
2. W. Stallings, *Data and Computer Communications*, Seventh Edition, Prentice-Hall India, New Delhi, 2003.

3. P. Moulton and J. Moulton, *The Telecommunications Survival Guide, Understanding and Applying Telecommunication Technologies to Save Money and Develop New Business*, Pearson Education, 2001.
4. G. Kennedy and B. Davis, *Electronic Communication Systems*, Fourth Edition, Tata McGraw-Hill, New Delhi, 2005.
5. H. Taub and D.L. Schilling, *Principles of Communication Systems*, Second Edition, Tata McGraw-Hill, New Delhi, 2005.
6. B.P. Lathi, *Modern Digital and Analog Communication Systems*, Oxford University Press, 1998.
7. B. Govindarajalu, *Computer Architecture and Organization*, Tata McGraw-Hill, 2006.
8. D.E. Comer, *Computer Networks and Internets*, Second Edition, Pearson Education Asia, 2000.
9. S. Salivahanan, A. Vallavaraj, and C. Gnanapriya, *Digital Signal Processing*, Tata McGraw-Hill Publishing, 2000.
10. T. Vishwanathan, *Telecommunication Switching Systems and Networks*, Prentice-Hall India, New Delhi, 2001.
11. NIIT, *Introduction to Digital Communication Systems*, Prentice-Hall India, New Delhi, 2004.
12. L. Rabiner and B.H. Juang, *Fundamentals of Speech Recognition*, Prentice Hall, 1993.
13. F.E. Terman, *Electronic and Radio Engineering*, Fourth Edition, McGraw-Hill Kogakusha, Tokyo, Japan, 1955.
14. A.S. Tanenbaum, *Computer Networks*, Fourth Edition, Pearson Education, 2003.
15. S. Ramo, J.R. Whinnery, and T. Van Duzer, *Fields and Waves in Communication Electronics*, John Wiley and Sons, 1994.
16. J.L. Flanagan, *Speech Analysis, Synthesis and Perception*, Springer-Verlag, 1972.

3

Analog-to-Digital Conversion
and TDM Principles

We have seen in the last chapter that time division multiplexing (TDM) technology is responsible for the high number of voice channels that can be multiplexed into a single transmission medium. Although the TDM techniques are not strictly restricted to the digital world, as they can very well be deployed on analog links too, with the help of sampling and applying pulse amplitude modulation, pulse width modulation, and pulse position modulation techniques (PAM, PWM, or PPM, respectively, discussed later in this chapter) are not useful for long-distance links, owing to the accumulation of the impairments described in Chapter 2. Thus, today whenever we talk of TDM, it is with reference to digital TDM.

To be able to understand these digital TDM details, we will have to understand the underlying fundamentals of sampling theory and analog-to-digital (A/D) conversion. We will discuss these subjects in this chapter to the extent required for building the understanding.

3.1 Analog and Digital Signals

The analog signal is a continuous signal. All natural signals are analog signals. For example the speech of human beings, all types of sounds and our movements etc all are analog signals.

With this definition, the analog signal can be classified into two categories, namely "continuous time analog signal" and "discrete time analog signal." Here are examples of both types.

3.1.1 Continuous Time Analog Signal

Continuous time analog signal is our normal analog signal of speech, a sinusoidal wave, etc., where the signal is continuously present with respect to time. Figure 3.1 shows one such signal.

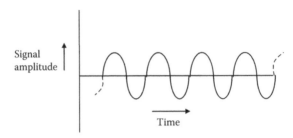

FIGURE 3.1
Continuous time analog signal.

3.1.2 Discrete Time Analog Signal

As the name suggests, the signal is not continuously present with respect to time, but it appears at a particular time and then disappears and reappears again.

The signal shown in Figure 3.2 is not continuous with respect to time, but still it is an analog signal.

Now compare this with a digital signal.

3.1.3 Digital Signal

No natural signal is digital. Digital signal is obtained by converting an analog signal to digital format. The digital format deals with digits, which are 1 and 0, or say, "yes" or "no" or simply "present" or "absent." There are no intermediate values. The question is, if it is a signal of fixed value, then how it can represent an actual signal, which is a variable value signal (else it will serve no purpose)? The answer is that the actual signal is built from the combinations of these 1s and 0s.

For example, the following sequence of digits represents a digital signal:

1 0 0 1 0 1 0 1

But these numbers cannot be used for transmission. Hence the sequence is converted into a pulse train as shown in Figure 3.3, where each 1 is represented by a pulse and a 0 by the absence of a pulse.

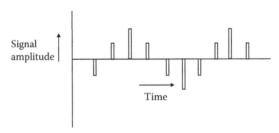

FIGURE 3.2
Discrete time analog signal.

FIGURE 3.3
A pulse train of digital signals representing a 1 by a pulse and a 0 by the absence of a pulse.

This pulse train is an eight bit sequence which can generate

$$2^8 = 256 \text{ combinations}$$

(Each such pulse/no pulse is called a "bit." Eight bits combine to be called a "byte.")

Thus, this eight-bit sequence, if arranged in different combinations of 1s or 0s can represent up to 256 signals. Each of these combinations can be used by the receiver of the digital signal to generate a "meaningful" analog signal by means of a device called a "digital-to-analog (D/A) converter." But where does this sequence originate? The answer is, from an "A/D converter" of course. We will see the A/D and D/A conversion in detail, shortly in this chapter.

The smallest value of this 8-bit signal is the one represented by the sequence

$$0\,0\,0\,0\,0\,0\,0\,0$$

and the largest one is represented by

$$1\,1\,1\,1\,1\,1\,1\,1$$

All the other combinations lie in between.

The amplitudes of the analog signals represented by some of these sequences are shown in Figure 3.3. The amplitude of the sequence "01111111" is half that of the sequence "11111111," which is the largest signal strength represented by the 8-bit sequence. Similarly, the signal strength represented by the sequence "00111111" is one-fourth of that of the maximum (Figure 3.4).

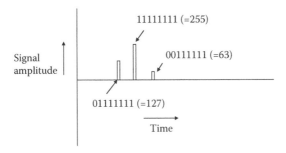

FIGURE 3.4
The levels represented by various combinations of sequences of bits.

These sequences produce one discrete time signal each at an intermediate stage of the receiver output from where they are processed by additional circuitry to form a continuous time analog signal.

3.1.4 Digital Signal Processing

As described in the previous chapter, the "regenerator" in a digital communication link produces a "clean" pulse from the "corrupt" pulse that it receives. How does it do it? This magic performed by the digital processing is the most wonderful feature of digital signal processing. Although it is a vast subject, a glimpse into this aspect will make you appreciate its beauty and simplicity.

Figure 3.5a shows the transmission of a clean pulse, which gets corrupt due to various impairments, before reaching the receiver. Figure 3.5b shows a typical arrangement which demonstrates how easily the corrupt pulse is converted to a clean pulse by the processor. A device made up of a simple transistor and RC circuits known as "flip-flop" can be used for this purpose. On reception of a "clock" signal it transfers the information available on its input to the output. As far as flip-flop is concerned, any input higher than its threshold level "X" is a 1 irrespective of the shape of the pulse. Thus, it transfers a 1 at the output, which is a "clean" locally generated pulse. The transfer takes place on the rising or falling edge of the clock, to make it happen at precise timings. (In practical systems the received pulse is invariably

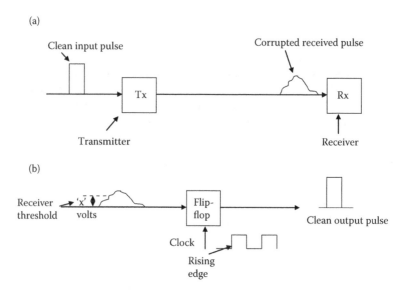

FIGURE 3.5
(a) Pulse transmission and reception. (b) Production of a clean pulse from the input corrupt pulse.

processed for improving its shape, by filtering, equalizing, etc. to remove the noise and unwanted frequency components and to minimize the effect of other transmission impairments, prior to its processing for detection.)

With this brief introduction of analog and digital signals and a glimpse of digital signal processing, we will now move on to the subject of sampling theory, which establishes a few of the most fundamental concepts of digital multiplexing technologies.

3.2 Sampling Theorem

When we wish to convert a continuous analog signal into a discrete analog signal (maybe for finally converting it into a digital signal), we need to conduct sampling of the analog signal. As the word suggests, sampling involves taking samples of the original analog signal at fixed time intervals. These samples can be obtained by multiplying the analog signal by an impulse train (see Figure 3.6).

The sampling process is shown in Figure 3.6. The process of sampling is done for processing the signal, either in its discrete analog form, or in a digital form. In any case the signal is required to be converted back to the analog continuous signal for use by human beings. Thus, it is necessary that the samples should be sufficiently numerous to be able to regenerate the original waveform with acceptable accuracy. At the same time they should be as few as possible, so that they take up minimum processing resources. In other words, while a lesser number of samples may affect the accuracy of the signal, a larger number of samples will put unnecessary burden on processing circuitry and, more importantly, on line capacity.

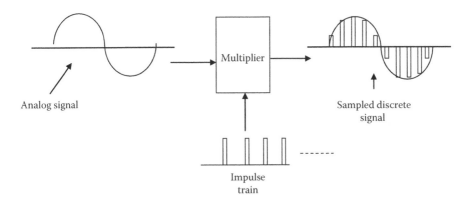

FIGURE 3.6
Sampling process.

The experiments of Harry Nyquist in 1933 showed that, to reproduce a signal exactly from its samples, the sampling frequency is required to be at least double the bandwidth of signal, i.e.,

$$Fs > 2 \times BW$$

where Fs is the sampling frequency and BW is the bandwidth of the signal.

This relationship is known as "sampling theorem" or "Nyquist sampling theorem." This sampling theorem forms the basis of all the digital communication systems in use today.

Note that the sampling frequency has to be "greater than" twice the bandwidth of the signal, and not equal to it. If the sampling frequency is equal to twice the bandwidth, the signal cannot be reproduced without distortion. "Greater than" does not mean too high but it has to be marginally higher than double the BW. For example, with speech signals the speech bandwidth is considered to be 0.3 to 3.4 kHz, i.e., 3.1 kHz. For sampling of this speech signal, a sampling rate of 8 kHz is adopted uniformly all over the world, which produces satisfactory and acceptable results in the reproduction of the signal.

The sampling theorem gave rise to the development of "pulse modulation" technologies for communication links. PAM, PWM, and PPM were deployed in the analog domain, whereas pulse code modulation (PCM) and delta modulation were developed for digital communication links.

Although the main thrust of this text is on digital technologies, we will see the introduction of the above-mentioned analog techniques in brief, to develop a clear understanding of differences amongst them.

3.3 (Analog) Pulse Modulation

The establishment of the sampling theorem promoted the application of technologies for communication links based on transmitting samples or pulse trains or discrete signals, instead of complete continuous signals. A variety of techniques were developed, namely PAM, PWM, and PPM. PAM was achieved by varying the "amplitude" of the samples in accordance with the input, in PWM, the "width" of the pulse is made proportional to the signal amplitude at the sampling instant, and in case of PPM, the "position" of pulse indicates the strength of the continuous signal at that instant.

The main advantage of the pulse modulation over continuous signal transmission was the feasibility of TDM, which permitted the use of a large number of user circuits to be carried on a common physical media. The reception quality was also better than the analog signal, as it was easier to minimize noise by proper level clipping of pulse.

The following are the types of pulse modulation techniques in greater detail.

3.3.1 Pulse Amplitude Modulation

PAM is the most obvious and simple outcome of sampling theorem. As shown in Figure 3.6 which is repeated here for convenience as Figure 3.7. The analog signal is multiplied by an impulse train to get samples which are proportional to the amplitude of the analog signal at the instant of sampling.

This type of PAM pulses are of varying magnitude and require varying transmitter power. The power required to transmit the highest amplitude pulse may be much higher than the average power. Generally it is much more convenient and cost effective, in communication systems, to operate the transmitter at fixed transmit power levels. Thus, these PAM pulses are used to modulate a carrier frequency, converting the samples to FM (frequency modulation) samples, which could be transmitted at constant power.

3.3.2 Pulse Width Modulation

The second variation of analog pulse modulation techniques is PWM.

In this system the width of the pulse is made proportional to the amplitude of the analog continuous signal. Figure 3.8 shows a PWM signal.

The width of the pulse has a finite value at zero amplitude of the analog signal. This width goes on increasing as the amplitude increases, and it decreases with reduction in the amplitude. However, the negative signals are also represented by positive pulses, because negative width of the pulse (ending before starting) cannot exist. Thus, the maximum negative amplitude is represented by the thinnest pulses.

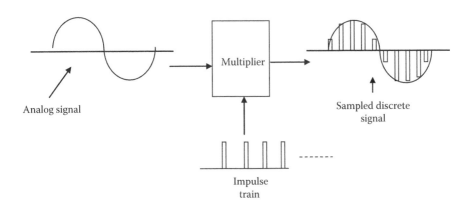

FIGURE 3.7
The sampling process to produce an analog PAM signal.

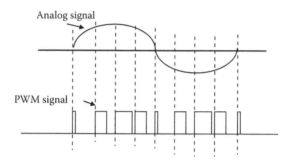

FIGURE 3.8
Pulse width modulation.

The leading edges of the PWM pulses occur at fixed intervals, but the trailing edges occur at positions whose distance from the first pulse are in proportion to the input signal strength.

The PWM also suffers from the problem of variable transmitter power. Although the amplitude of all the pulses is constant, the variable width implies variable power.

3.3.3 Pulse Position Modulation

In this type of pulse modulation, the position of the pulse with reference to a synchronized pulse indicates the amplitude of the analog signal. This is basically a variation, or an improved version, of the PWM system (Figure 3.9). The analog continuous signal has to be first converted to PWM pulses.

This system has an advantage over the PWM systems in that it requires fixed power output from the transmitter. However, its disadvantage is that it requires synchronization pulses to be sent by the transmitter and

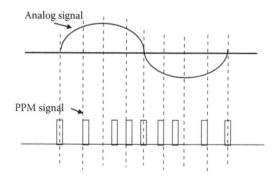

FIGURE 3.9
Pulse position modulation.

synchronization maintained at the receiver, as otherwise the differential positions of the pulses will not be known.

3.4 Digital Pulse Code Modulation

The digital pulse modulation popularly known as PCM is the most widely used type of pulse modulation technique and hence it is also the most important. In fact the whole of the long-distance communication is based upon this technology, and the present generation switches (exchanges) to deploy it for their speech coding purposes.

PCM consists of a two-stage process: sampling and quantization.

3.4.1 Sampling

This is the same process that we have seen in Section 3.2. We will not discuss the details here again; however, Figure 3.6 is redrawn here for ease of reference as Figure 3.10.

Per the Nyquist sampling theorem to reproduce the signal with minimum distortion, the frequency of sampling has to be more than double the maximum frequency encountered in the signal. The speech signal, although it may consist of frequencies up to 15 kHz, is filtered to a band of 0.3 to 3.4 kHz, before it is used by any communication network, per ITU-T standards. It has been established that this band of 0.3 to 3.4 kHz carries all the components of the speech which are necessary faithfully to convey not only the content,

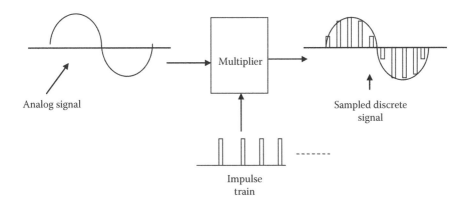

FIGURE 3.10
Sampling process.

but also the other attributes such as tone, voice identification, and emotions. Thus, the bandwidth of the speech band to be sampled is

$$BW = 3.4 - 0.3 = 3.1 \text{ kHz}$$

for which the sampling rate required per Nyquist criteria (say Fs) is

$$Fs > 6.2 \text{ kHz.}$$

Thus, keeping a small margin a sampling rate (or frequency) of 8 kHz or 8000 samples per second is considered adequate to meet all quality requirements.

These 8000 "samples" are still; only amplitude modulated analog signals, they need to be converted to digital format to obtain a PCM signal. The process of carrying out this conversion is known as quantization. Let us see what it is.

3.4.2 Quantization

Quantization is a process of converting an analog sample into a digital signal.

As stated earlier the digital signal is a pulse train of pulses of uniform amplitude. These pulses represent binary 1s or 0s, by their presence or absence, respectively (Figure 3.11).

The received signal is thus a sequence of bits (binary digits) 10110011. To convert this sequence into signal amplitude, we will have to understand a very basic fundamental concept of a binary digit system (two digit system).

In our normal decimal system we have 10 digits. The value of each of these digits is calculated as follows:

Digit position:	7	6	5	4	3	2	1
	×	×	×	×	×	×	×
Multiplication factor:	10^6	10^5	10^4	10^3	10^2	10^1	10^0

The value of any digit position "N" is obtained by multiplying the number present at that digit by $(10)^{N-1}$. For example if we have a number, say 3479, its value will be

N =	4	3	2	1	
Number	3	4	7	9	
	×	×	×	×	
Multiplier	10^3	10^2	10^1	10^0	
Digit value	3000	+400	+70	+9	= 3479

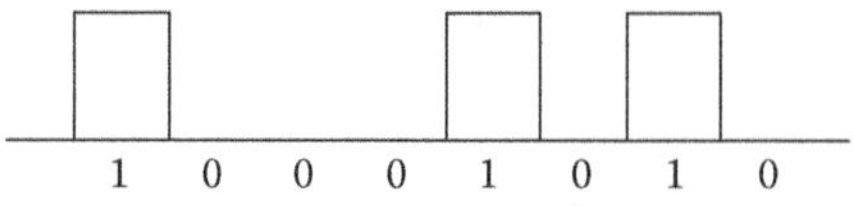

	1	0	0	0	1	0	1	0

FIGURE 3.11
A pulse train of uniform amplitude and its binary interpretation.

In binary systems the values are calculated in a similar way. But there are only two digits, thus calculations are as follows:

N =	5	4	3	2	1	
Number	1	0	1	1	0	
	×	×	×	×	×	
Multiplier	2^4	2^3	2^2	2^1	2^0	
Digit value	16	+0	+4	+2	+0	= 22 (decimal value)

The value of any digit position "N" is obtained by multiplying the number present at that digit by $(2)^{N-1}$ because it is a binary system having only 2 digits.

Although the processing is done in binary systems, for human interpretation the signal value has to be converted into a decimal system. We can see some more examples of binary sequences and their decimal values.

	2^7	2^6	2^5	2^4	2^3	2^2	2^1	2^0	
	×	×	×	×	×	×	×	×	
(i)	1	1	1	1	1	1	1	1	
	128 +	64 +	32 +	16 +	8 +	4 +	2 +	1	= 255 (decimal value)
(ii)	0	0	0	0	0	0	0	0	
	0 +	0 +	0 +	0 +	0 +	0 +	0 +	0	= 0
(iii)	1	0	1	1	1	0	0	1	
	128 +	0 +	32 +	16 +	8 +	0 +	0 +	1	= 185

The above example sequences may represent corresponding received pulse trains, and their decimal values represent the amplitudes of the analog sample that is required to be constructed from the pulse trains.

The process of quantization consists of converting the analog sample to one of these values. In the example above, we have taken pulse trains of 8 bits, which consist of 256 (0 to 255) levels, but the actual number of bits can be decided based upon the accuracy level required. The greater the number of bits the better the accuracy. In Figure 3.12, we have considered a 3-bit conversion for ease of explanation.

We can see from Figure 3.12 the process of quantization (the term quantization arises from the levels of the signal, which are in quanta of magnitude instead of being continuous). The original signal is continuously variable, but the quantized signal can assume only any one of the eight levels indicated. The maximum level can be 7 and the minimum level can be 0. The level closest to the original signal amplitude is chosen to be its representative level. We can see that the quantized signal has been distorted a

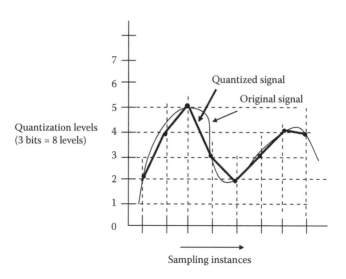

FIGURE 3.12
Quantization with 3 bits, i.e., 8 levels.

lot as compared to the original signal. However, as we increase the number of quantization levels the error reduces. With every increment in the number of bits for quantization, the error reduces by half, as the quantization level is doubled.

The process of PCM may thus be depicted as in Figure 3.13.

The normal industry standard for speech is to use 8-bit quantization (or coding, as it is normally called), which gives $2^8 = 256$ quantum levels. Thus, the whole of the dynamic range of analog signal has to be accommodated in levels starting from 0 up to 255. If we look at this coding with respect to the total information it generates, then, with 8-bit coding for each sample and 8000 samples per second, it gives rise to 64,000 bits per second. Both these values, i.e., 8-bit coding and 8000 samples per second, have become the industry standard for telephone circuits world wide and are able to deliver high quality services with complete user satisfaction.

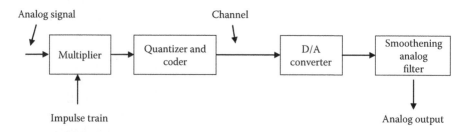

FIGURE 3.13
PCM communication link.

Some other applications however, like recorded announcements with the help PCs, make use of more precision with 16 or even 32 bit coding, and sampling rates in excess of 20 kHz or so, minimizing the quantization error to negligible proportions. The main reason for aiming for such high accuracy is the availability of very cheap storage media and the need to compensate for the distortion added by announcement amplifiers. Such systems are normally stand-alone computer systems, not requiring any long-distance transmission, so no other resources such as transmission media are required, and hence no bandwidth constraints. The bandwidth is the most precious resource in any long-distance communication system with costly multiplexers.

3.4.3 Quantization Noise

As explained in the previous section, the quantized signal is not an exact replica of the analog continuous signal, only a close approximation. The difference in the shape of a quantized signal compared with the original signal gives rise to what is known as quantization error, which in turn gives rise to some amount of distortion in the regenerated signal. This is called quantization "noise" because of the randomness of the quantization error, which is more or less like "white noise" (the amount of error will depend entirely upon the original signal amplitude, which is random). The noise increases if the number of coding bits is reduced, and vice versa. The noise is particularly bothersome at low amplitudes of signal. This will be clear from Figure 3.14.

Again a 3-bit, i.e., 8-level quantization, has been shown in Figure 3.14 for ease of understanding. Sample number 1 has amplitude of 4.3, but the nearest

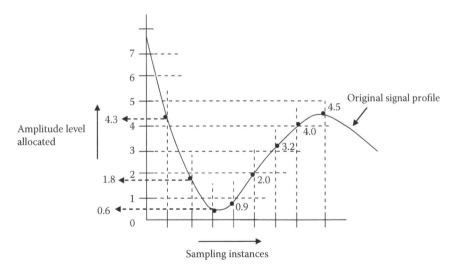

FIGURE 3.14
Quantization noise levels.

level available is 4 thus amplitude level 4 is allocated to it, accommodating a quantization error of 0.3. Sample number 2 has amplitude of 1.8, hence it is assigned the closest amplitude level 2 with a quantization error of 0.2. Let us say that amplitudes are in volts, hence all these levels will represent voltage levels. Likewise, 0.6 V of sample number 3 is assigned a voltage of 1 V and sample number 4, whose amplitude is 0.9, is also assigned a level of 1 V. The rule is that if the amplitude is less than 0.5 V midway, then it should be given the next lower level and if it is more than 0.5 V, then it should be given the next higher level. (What will happen if it is exactly 0.5? The decision may be taken at either 0 to 0.4999 or 0.5 to 0.9999, depending on the built-in accuracy of the processor.) The maximum quantization error in this case of 3-bit quantization is 0.5 V.

If we increase the number of levels to 4, keeping the maximum voltage the same, each level will get divided by 2, thus the minimum quantization level will become 0.5 V instead of 1 V, and the maximum quantization error will become 0.25 V. Thus, on increasing the quantization levels by 1 bit each time we reduce the quantization noise to half.

The point to be noted here is that the signal which has the lowest amplitude, say 0.6 V, and that with highest amplitude, say 4.5 volts, both have to suffer an accuracy loss of up to ±0.5 V. While this loss is about 11.11% for the highest amplitude signal, it is about 83.33% for the lowest amplitude signal. Thus, the effect of quantization noise is very high on the low amplitude signals. To minimize this differentiation, the low amplitude signals are boosted, compared with high amplitude signals, prior to quantization. The process is called Companding.

3.4.4 Companding

To attribute a similar percentage of quantization error to the signals irrespective of their amplitudes we can either prescribe smaller quantization steps for smaller signals and bigger steps for larger signals, or we can compress the whole of the analog signal in such a way that the smaller signals are boosted in amplitude and larger ones are reduced in amplitude. Figure 3.15 shows the first option.

As can be seen, the quantization steps close to very low amplitude signals have been made smaller and those for high amplitude signals are made larger. In fact the step size can be made proportional to the input amplitude, to give rise to a uniform percentage of quantization error to all amplitudes.

The second option of compressing the signal is shown in Figure 3.16.

The quantization levels are kept uniform, while the samples of the input signal are compressed in such a way that the signals with lower amplitude are compressed minimally and higher amplitude signals are compressed more. The compression is done before quantization. The process leads indirectly to reduction in the quantization levels to high amplitude signals.

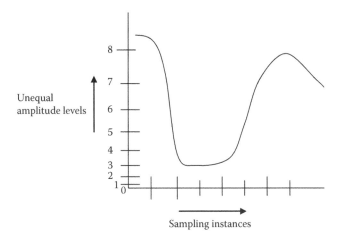

FIGURE 3.15
Employing unequal quantization steps to minimize quantization error.

In the receiver, the signal is expanded by processing it in an amplitude expander whose multiplication characteristics are exactly opposite to that of compression at the transmission end, to produce the output signal with uniform quantization error. The whole process of first compressing and then expanding the signal is called "companding." The process is shown in Figure 3.17.

The compression characteristic of Figure 3.17 can give a uniform percentage of quantization error to the samples of all sizes, if the curve is logarithmic.

For this reason the companded PCM is also known as "log PCM." In other words, a companded or log PCM attributes quantization error to the signal samples in proportion to their amplitudes.

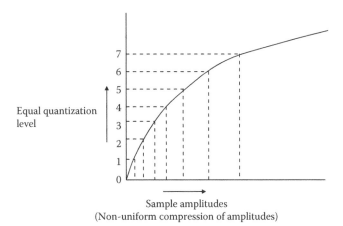

FIGURE 3.16
Characteristic of an amplitude compressor.

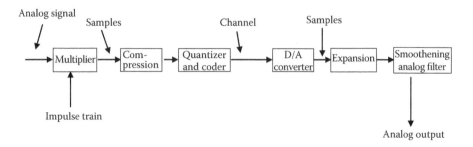

FIGURE 3.17
Digital PCM communication link with companding.

3.5 Other Digital Modulations

In addition to PCM, some other forms of digital modulations have been developed. Some of the important ones are

- Differential PCM (DPCM)
- Delta modulation (DM)
- Adaptive differential PCM (ADPCM)
- Adaptive delta modulation (ADM)

All these are variations to the standard PCM. None of them, however, is in widespread use, particularly in the mainstream communication links and systems. The primary reason for this is that although these variations offer some advantage in coding efficiency, they need different types of processing devices. Since the mainstream PCM processing devices are in universal and widespread use already, and they are cheap, designing and producing special devices for these different types of modulations is always comparatively costly. Moreover, the cost of electronics is falling day by day and the channel capacities are multiplying very fast due to advances in optical fibre and multiplexing technologies. There is therefore no pressing need for more efficient coding at the cost of increasing system complexity, which increases with the introduction of any nonstandard components and interfaces, with respect to backbone as well as access level services.

We will discuss each of these variations in brief.

3.5.1 Differential PCM

In this modulation scheme, instead of coding the full amplitude, the difference of each sample from its previous one is taken and coded. Advantage is taken of the fact that in any real speech signal the variation is generally gradual, and each amplitude is related to its previous amplitude. The difference in the sample with respect to the previous sample is normally much less than the absolute sample value in amplitude. Thus, fewer bits are required to code the information, maintaining the same level of quantization error.

For reasons described in the previous section, this modulation has not become very popular.

3.5.2 Delta Modulation

Another improvisation over the differential PCM is delta modulation. Instead of transmitting the difference in the amplitude, why not send just the information that the present sample is higher or lower than the previous one? This again takes advantage of the randomness of the speech signal, and assuming gradual variation in the signal this technique produces a highly efficient coding. Only one bit is required to be sent per sample. However, as there is a limit to the approximation, the speech signal does contain enough occasions of random abrupt variations in the amplitude. These variations are not truly reflected in the reconstructed waveform, giving rise to an unacceptably high level of quantization noise.

This variation of PCM is also not popular, for the reasons given in Section 3.5.

3.5.3 Adaptive Differential PCM

The differential PCM with an adaptive logic algorithm is called ADPCM. This can achieve better coding efficiencies than the DPCM. The ADPCM can achieve a saving of 3 to 4 bits per sample compared with the normal PCM. (The savings achieved in plain DPCM are only half of this.) Thus, the bit rate required for an ADPCM channel works out to nearly

4 bits × 8000 samples = 32 K bit per second.

ADPCM too could not succeed in finding a widespread use in the mainstream links and systems for similar reasons as other variations of PCM. Its application remains confined to cordless telephones and DECT (Digital European Cordless Telephone).

3.5.4 Adaptive Delta Modulation

The main disadvantage of delta modulation is that it is unable accurately to represent the signals with rapid amplitude changes. This disadvantage can be overcome to a certain extent if the step size is made proportional to the amplitude variation. This is called ADM, in which the step size is automatically adjusted according to the amplitude variations of the input signal.

The application of ADM too is restricted for similar reasons, as shown in Section 3.5.

3.6 A/D and D/A Converters

In the preceding discussion in this chapter, we have come across several elements of A/D conversion, yet to develop a focused understanding, it will be useful to have a comprehensive review of A/D and D/A converters.

3.6.1 A/D Converter

The A/D conversion is a three-stage process. The analog continuous signal is first converted into samples by multiplying it with an impulse train of a fixed magnitude and frequency. This process gives us samples at a fixed frequency (8 kHz in case of PCM), which are proportional to the amplitude of input continuous signal at various instants of sampling times.

The second stage is quantization of these samples. During this process the samples are each allocated the closest possible voltage level, depending upon the number of bits used for quantization (e.g., there will be 16 levels for a 4-bit quantization, 32 levels for 5 bits, and 256 levels for an 8-bit quantization). If the sample is higher than the half value of the closest level it will be assigned the next level, and if it is lower than the half value of that level, it will be assigned the same level. This stage also includes necessary compression of the signal before quantization.

The third process is called coding. Depending upon the sample's quantized amplitude the appropriate predefined code is generated for transmission to the receiving station. (This transmission is just an illustration of the principles. The practical transmission will have many more factors for consideration, for example the actual pulse amplitude and its power will depend upon the type of media and its length, receiver sensitivity, etc. Many systems will use a carrier frequency which will be modulated by the pulse code. The radio transmitters will have an RF section with entirely different parameters for RF transmission.)

These functions are shown in Figure 3.18.

Normally, the A/D converters are available as a single integrated circuit (IC). All the functional blocks shown in Figure 3.18 are integrated within the IC itself. They are available with various accuracies, for various numbers of bits and processing speeds, etc.

3.6.2 D/A Converter

At the receiving end of a digital communication link, the digital signal is required to be converted back to the analog signal to be of use to the user. Whether it is speech or video, we all listen, speak, see, and generate analog-only signals.

The process of converting a digital signal to analog signal again involves a three-stage procedure. First, the digital code has to be converted into analog

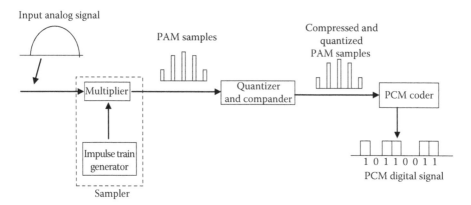

FIGURE 3.18
Functional blocks of an A/D converter.

PAM samples. These PAM samples are then passed through a "Sample and Hold" circuit to remove the glitches and interpolating the samples to form a continuous signal. Finally, a low-pass filter removes undesired frequency components from the analog signal so constructed. The process is shown in Figure 3.19.

The first stage converts the PCM code to an equivalent analog sample. This analog sample is proportional to (+ or − the quantization error) the quantized PAM sample produced by the transmitter, along with the necessary expansion factor depending upon the companding equation (the inverse of the compression at the transmission end).

The tops of all the PAM pulses are connected by interpolation in the second stage. The second stage also performs the job of removing amplitude transients called "glitches," which are generated in the PAM samples due to abrupt variations in the pulse code value (PAM amplitude).

The third stage provides a smoothing low-pass filter. The objective here is to remove from the output all the frequency components which are higher

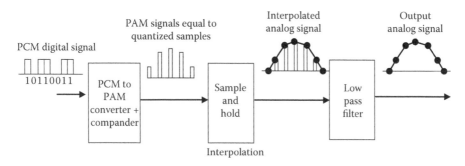

FIGURE 3.19
D/A conversion.

than double the sampling frequency. If these components are allowed to exist, they generate unwanted frequency components.

As in the case of A/D converters, the D/A converters are also available in a single IC. The accuracy again depends upon the parameters chosen and cost.

3.6.3 Accuracy of A/D and D/A Conversion

What are the factors affecting the accuracy of A/D and D/A conversion? Let us summarize them in the light of the preceding discussion.

Does higher sampling speed (or sampling frequency) improve the accuracy of A/D and D/A conversion? The answer is "no," if the Nyquist theorem/criterion is followed strictly, which states that the sampling frequency should be more than double the maximum frequency of the signal or the bandwidth. As long as this criterion is followed the analog signal can be reproduced from its samples without any distortion. Thus, the accuracy is not improved by increasing the sampling frequency in case of PCM where the BW of the speech signal is 3.1 kHz (0.3–3.4 kHz) and the sampling rate is 8 kHz, which is much more than double the BW. Reduction of sampling frequency < 8 kHz will, however, deteriorate the accuracy of conversion. (In the case of music signals, the BW of which is considered to be up to 15 kHz, the sampling is done at a frequency of 20 to 40 kHz.)

The second factor is quantization. As we have seen, quantization approximates the signal level to the nearest acceptable level and thus introduces something called quantization error or quantization noise. Thus, it causes a permanent distortion to the signal which cannot be corrected at the receiver, because what is transmitted is only a quantized signal. The original signal ceases to exist as soon as it is quantized. Therefore, there is no way it can be corrected, but the quantization error/noise can be minimized by increasing the number of levels of quantization. Increase of each bit causes the quantization levels to be doubled, halving the error. Thus, good quality A/D and consequently D/A converters can be made with a higher number of quantization levels. However, as far as PCM for speech circuits is concerned, the quantization levels are standardized at 256, which pertain to a bit rate of 8 bits per sample.

Companding, which is the process of compressing the signal at the transmitting end and expanding correspondingly at the receiver, does not compromise accuracy as long as the ITU-T standards are followed.

In case of D/A converters, the accuracy of the reproduced signal will be affected by the interpolator or sample and hold circuits. Some interpolators may produce only a staircase approximation, whereas better ones can give a more linear interpolation. Figure 3.20 shows two types of interpolators.

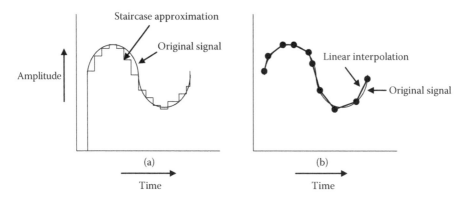

FIGURE 3.20
Two types of interpolations for D/A conversion. (a) Staircase approximation; (b) linear interpolation.

Obviously, the linear interpolator gives a better approximation of the signal, but it is more costly.

The smoothing low-pass filter at the output stage of a D/A converter is necessary to achieve acceptable accuracy, as it eliminates unwanted frequency components.

Having discussed analog and digital signals, various pulse modulation techniques, A/D and vice versa conversion, etc. in this chapter, we can now enter our core subject of interest, "TDM."

3.7 Time Division Multiplexing

TDM is a direct outcome of pulse modulation systems. When we convert a continuous signal into a discrete signal consisting of samples occurring at predefined intervals, we leave a lot of space in between the samples, which remains unused. Consider a PAM situation as shown in Figure 3.21.

The continuous signal is converted to PAM samples at sampling rate of 8000 samples per second, which means that the duration "T" between each sample is

$$T = 1/8000 \text{ s}$$
$$= 10^6/8000 \text{ μs} = 125 \text{ μs}$$

Now if the width of the sample pulse is much less than 125 μs, then we can accommodate more numbers of such samples, in between two samples of this signal, which may belong to some other signals. This is what is precisely

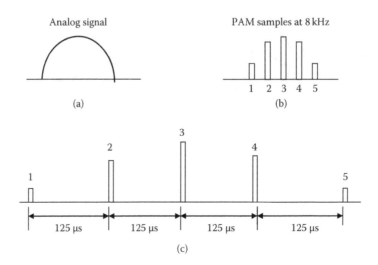

FIGURE 3.21
Time gap between PAM samples. Analog signal is shown in (a), its samples at a frequency of 8 kHz are shown in (b), and the time interval between each such sample is shown in (c).

done to achieve what is called TDM. As you can see, the time between the samples is divided to accommodate more numbers of signals. If the width of the sample pulse is 1 µs we can theoretically accommodate 125 signals or 125 channels in one transmission media through TDM, as depicted in Figure 3.22.

The samples from each signal will repeat after every 125 µs. The above example is good for illustration, but as previously stated in this chapter, the TDM of PAM or other analog modulation technologies is not in much use, basically because of their analog nature, which deprives them of the advantages of "regenerative" repeaters like in digital technology.

Multiplexing of four PAM channels is shown in Figure 3.23. The multiplexer takes the input from each channel in turn and places them in the same sequence at the output.

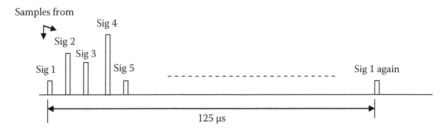

FIGURE 3.22
TDM of 125 PAM channels.

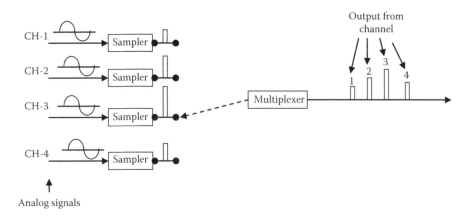

Analog signals

FIGURE 3.23
A four-channel PAM multiplexer.

3.7.1 TDM/PCM

The TDM technology deployment today is almost completely in the digital domain. The PCM signal that we have seen in the preceding discussion is multiplexed, using TDM technologies in a way similar to the one shown in Figure 3.23, but with differences that are required for digital systems.

A practical TDM/PCM system was developed in North America first, where a 24-channel multiplexer was deployed. The 24-channel PCM system was called a "DS1" or "T-1" and is still so called. Subsequently, ITU-T standardized a PCM system of 30 speech channels known as an "E1," which is universally adopted all over the world except in North America and Japan. We will discuss here this 30-channel PCM to demonstrate the principle. (This 30-channel PCM system could work up to 2 km, having a bit rate of 2048 kbps, on a copper wire pair and was designed and optimized for a voice base band signal of up to 4 kHz. The repeaters at close distance, every 2 km, were the trade-off for higher bandwidth. The system became so popular that to date it is known as a PCM instead of its official name "E1.")

We know by now that each PCM channel is coded by an 8-bit word. The samples are taken at the rate of 8000 samples per second. Thus, the channel bit rate is $8 \times 8000 = 64,000$ bits per second. If we have to accommodate 30 channels, then the bit rate becomes $30 \times 64,000 = 1920$ kbps. But what about signaling? Two channels of the same bit rate are added for signaling. Thus, the total bit rate becomes $32 \times 64,000 = 2048$ kbps or 2.048 Mbps. It is also popularly known as a 2-Mbps systems.

Sampling at the rate of 8000 samples per second gives a period of 125 µs, within which these 32 channels have to be accommodated. Thus, one time

Time slot or channel numbers

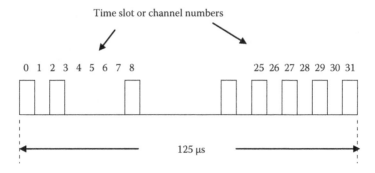

0 1 2 3 4 5 6 7 8   25 26 27 28 29 30 31

125 µs

FIGURE 3.24
Time slots allotted to each of the 32 channels.

slot out of the 125-µs period is allotted to each of these 32 channels, as shown in Figure 3.24. This period of 125 µs is called a "frame."

The multiplexing of the digital signals can be done in two ways:

(i) Bit interleaving, i.e., placing 1 bit of each channel, one after the other.

(ii) Byte interleaving, i.e., placing 1 byte of each channel one after the other.

These two methods are depicted in Figures 3.25 and 3.26 for a 4-channel multiplexer.

In case of an E1 of 32 channels, consisting of 2048 kbps bit streams, the multiplexing is as shown in Figure 3.27.

In case of E1, the multiplexing is done by byte interleaving, because the signaling and framing etc. are defined in terms of bytes and not bits. (This will become clearer in the chapter on PDH.)

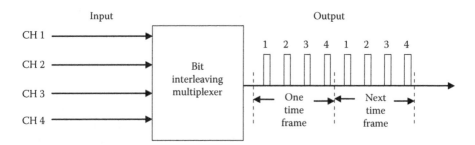

FIGURE 3.25
The bit interleaving multiplexer polls on the first bit of each channel in sequence and places a corresponding bit serially, one after the other. Then it goes for a second round of polling, and so on.

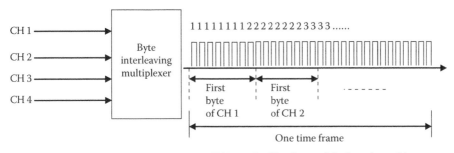

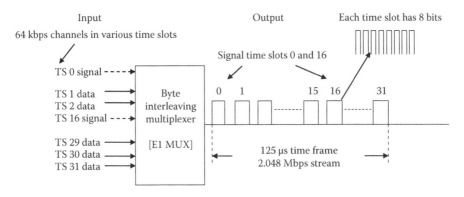

FIGURE 3.26
The byte interleaving multiplexer polls on the first byte of each channel in sequence and places a corresponding byte serially, one after the other. Then it goes for a second round of polling, and so on.

FIGURE 3.27
E1 MUX, showing the byte interleaving of 30 data channels and 2 signaling channels of 64 kbps each.

Review Questions

1. Explain the fundamental difference between an analog and a digital signal, and give examples of each type from situations in daily life.

2. How is a continuous time signal different from a discrete time signal?

3. What is the difference between a bit and a byte? In a bit sequence which bits are least significant and most significant bits?

4. The core advantage of digital technology is due to the shape of the digital signal or due to the digital signal processing. Explain.

5. What is a sample in the context of digital technology? Explain sampling process with a sketch.

6. Explain the sampling theorem. What is the significance of 8-kHz sampling rate?

7. How is PWM different from TDM?

8. Compare various features of PAM, PWM, and PPM systems.

9. Describe the stages involved in PCM. Why is this system called digital?

10. What is quantization? Work out bit values for the sequence 11101110.

11. How many quantization levels are enough for speech signal and why? Explain the quantization process with a sketch.

12. Describe the essential components of a PCM communication link. What are the major constraints?

13. What is quantization noise and how is it affected by the quantization levels?

14. What is the need of companding and how is it achieved?

15. What is "log PCM?"

16. Are there any modulation types other than the PCM deployed in digital systems? Explain them in brief. Where are they used?

17. What is the role of A/D and D/A converters in digital communication systems?

18. Show with the help of a block diagram the various functions involved in A/D conversion.

19. How is a digital signal converted to analog signal at the receiving end of a digital transmission link? Explain with the help of a block diagram.

20. What is the role of PAM and interpolation in A/D converter?

21. Summarize the factors and their effects affecting the accuracy of A/D and D/A conversion.

22. If given a free hand, how can you improve the accuracy of A/D and D/A conversion, and what will be its impact on system capacity and cost?

23. What is TDM? Why and how is it made possible?

24. Set out the salient differences between FDM and TDM.

25. What has PCM to do with TDM? Explain a TDM/PCM system with sketches.

26. What do you understand by interleaving? How is bit interleaving different from byte interleaving?

27. What is a time slot? How are the bits/bytes arranged in various time slots of an E1 multiplexer?

Critical Thinking Questions

1. Find out and list five examples of analog and digital signals from situations in daily life.
2. Can you think of any other types of variations of digital modulation techniques than the ones covered in this chapter?
3. Imagine a multiplexer with many more than 24 or 30 channels. What would be the positives and negatives?
4. If the sampling rate is increased to 16, 32, or 64 kHz, what will be the effects on the signal and process?

Bibliography

1. J.G. Proakis and D.G. Manolakis, *Digital Signal Processing Principles, Algorithms, and Applications*, Third Edition, Prentice-Hall India, New Delhi, 2005.
2. P. Moulton and J. Moulton, *The Telecommunications Survival Guide, Understanding and Applying Telecommunication Technologies to Save Money and Develop New Business*, Pearson Education, 2001.
3. W. Stallings, *Data and Computer Communications*, Seventh Edition, Prentice-Hall India, New Delhi, 2003.
4. J.L. Flanagan, *Speech Analysis, Synthesis and Perception*, Springer-Verlag, 1972.
5. H. Taub and D.L. Schilling, *Principles of Communication Systems*, Second Edition, Tata McGraw-Hill, 2005.
6. B. Govindarajalu, *Computer Architecture and Organization*, Tata McGraw-Hill, 2006.
7. S. Salivahanan, A. Vallavaraj, and C. Gnanapriya, *Digital Signal Processing*, Tata McGraw-Hill, 2000.
8. L. Rabiner and B.H. Juang, *Fundamentals of Speech Recognition*, Prentice-Hall, 1993.
9. T. Vishwanathan, *Telecommunication Switching Systems and Networks*, Prentice-Hall India, New Delhi, 2001.
10. ITU-T Recommendation G.711, *Pulse Code Modulation (PCM) of Voice Frequencies*, International Telecommunication Union, Geneva, Switzerland.

4

Line Coding and Digital Modulation

In the previous chapter, we have seen how an analog signal is converted into a digital signal. One such signal, after sampling and pulse code modulations (PCM), may look like that shown in Figure 4.1.

Each sample of the analog signal is thus converted into an eight-bit sequence of pulses (or the absence of them), called PCM. In other words, each signal is "coded" into a fixed-length bit stream. Since this coded signal is required to be sent to another place using a transmission line, the signal code has to be imposed on the transmission line. This scheme of coding the signal is called "line coding."

In the simplest form of this line coding, the bit stream shown in Figure 4.1 may itself be thought of as pulses of a certain voltage, and transmitted on the transmission line with each 1 with a particular voltage and each 0 with no voltage, or vice versa.

However, if things were that simple, the telecom engineers would have had little work to do. Transmission of the pulses of Figure 4.1, as they are on the transmission line, will create many problems. First, if there are long sequences of 0s (that means no transmission of any pulses at all), the receiver will find it very difficult to extract the timing information from the incoming signal. Second, if there are long sequences of 1s, they will create a DC imbalance called "DC Wander" in the receiver. There are many more variables such as bandwidth requirement, power requirement, ambiguity in detecting of 1s and 0s, performance monitoring ability, error probabilities in the reception of digits, and cost, which decide the choice of a particular coding scheme to be adapted for a particular situation. These factors will be discussed in this chapter.

Line coding schemes that involve transmission of pulses on the transmission line, and hence are applicable for "pulse transmission" systems, are only possible on wired media such as copper wire pairs, coaxial cable, or ACSR wires. However, when the media has to be radio waves or optical fiber, these electrical pulses cannot be transmitted as they are. They need to be modified to fit into the radiofrequency transmission media of radio waves, or the light waves of optical fiber transmission media. For this purpose, different types of codes are used to modulate the carrier frequencies of microwave radios or light waves. These modulated carrier waves carry the information to the receiver, where they are demodulated to reinstate the original signal code, and are finally converted back to the desired analog signal or digital data. Such modulation schemes are called digital modulation techniques, and again, the choice of a specific type of modulation depends upon many

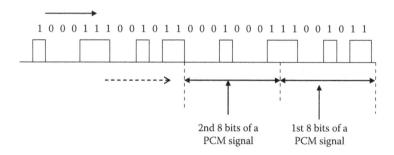

1 0 0 0 1 1 1 0 0 1 0 1 1 0 0 0 1 0 0 0 1 1 1 0 0 1 0 1 1

2nd 8 bits of a   1st 8 bits of a
PCM signal   PCM signal

FIGURE 4.1
A typical digital signal after conversion from an analog signal.

factors similar to those stated above for line coding. In this chapter, we will discuss some of the many types of modulation techniques which are relevant to developing an understanding of the subject.

4.1 Factors Affecting the Choice of Line Code

Choice of the type of line code depends upon some or all the factors mentioned below, according to the system application.

(i) Timing content

(ii) DC wander

(iii) Bandwidth requirement

(iv) Power consumption

(v) Performance monitoring features

(vi) Error probability

(vii) Cost

Let us study these factors and their implications.

4.1.1 Timing Content

A high timing content is the most desirable property in a line code. Every time the signal changes from "+ve" to 0 or "+ve" to "−ve" or vice versa, it contributes to the timing content (Figure 4.2).

Let us understand what timing content is and why it is so important.

In any digital communication system, the received signal needs to be synchronized to the transmitted signal, otherwise the received signal will be rendered meaningless (see Chapter 7 for an explanation of the concept).

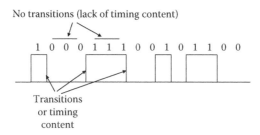

No transitions (lack of timing content)

1 0 0 0 1 1 1 0 0 1 0 1 1 0 0

Transitions
or timing
content

FIGURE 4.2
Timing content and its absence in a digital signal.

Synchronization or one-to-one correspondence of bits can be achieved by any of the following means:

(i) Transmitting a separate timing signal which can be used by the receiver to generate a clock signal, which in turn determines the bit positions, or in other words achieves synchronization. This option is costly as it requires the added cost of the separate transmission of a timing signal which reduces transmission efficiency. It also complicates the circuitry at both ends.

(ii) By using the transmitted signal itself for extracting the necessary timing information for generation of local clock. This option is generally the best and is adopted almost universally, particularly in long-distance communication systems. As it does not require separate transmission of a timing signal, it saves on bandwidth and complexity of circuits, giving rise to better economics. These advantages are not free of cost, as some penalty in terms of efficiency of coding or increased bandwidth has to be paid.

(iii) The third option is to run an independent clock at the receiver for detecting correct bit positions. This is the simplest option but it requires high precision clocks to be deployed, which are very costly. The accuracy of the received signal at high data rates continues to be a problem in spite of using highly accurate clocks. Thus this option is used only in a special application called "plesiochronous operation" (for details, see Section 7.9).

Timing recovery from the incoming signal, as is normally done, requires that the incoming signal is suitable for this purpose. The suitability of the incoming signal for timing extraction is directly proportional to the number of transitions in the signal. A transition means a change of voltage/current level from high to low or low to high, or it may be from positive to negative or vice versa (see Figure 4.2).

Thus the line code adopted should have enough transitions to allow good timing content. For example let us have a look at a PCM code shown in

1 0 1 1 0 0 0 0 0 1 1 1 1 1 1 0 1 1 1 0 0 0 0 0 0 0 0 1 1 0 - - - -

FIGURE 4.3
An arbitrary PCM code for line coding.

Figure 4.3. We can see long sequences of 1s and 0s. These long sequences do not produce any transitions, and hence create problems in timing extraction.

One of the solutions to this problem is to replace the 1s and 0s with different types of symbols. (The pulses for representing the 1s, and the "absence of pulses" for representing 0s, are called "symbols." In this case, the symbol for a 1 is a positive pulse, whereas the symbol for a 0 is a "duration of time" equal to a pulse width having "no voltage" or "no pulse.")

An example of such symbols is shown in Figure 4.4.

A 1 is represented by a full cycle of a square wave consisting of a pulse rising from negative voltage to positive and coming back to negative voltage and is followed by a pulse falling from positive polarity to negative and finishing with positive polarity. The 0 is represented by a square wave which is of opposite phase as shown in Figure 4.4. This type of coding will ensure that each 1 and 0 will have a transition by itself, thus the sequence does not matter (it may have any number of contiguous 1s or 0s, but it cannot reduce the number of transitions). The code would then look like the one shown in Figure 4.5.

In this case, the penalty paid for achieving a good number of transitions in the code is the bandwidth. A full cycle of the square wave is used in place of a half cycle for every symbol. Thus the bandwidth need is doubled. There are

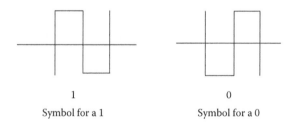

1   0

Symbol for a 1   Symbol for a 0

FIGURE 4.4
Alternative symbols for improving the timing content of the code.

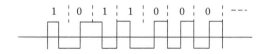

1   0   1   1   0   0   0   - - -

FIGURE 4.5
The line code of Figure 4.3, after substitutions of symbols by the code shown in Figure 4.4.

many other types of solutions in use, each paying a different type of penalty. We will discuss them later in this chapter in detail.

4.1.2 DC Wander

An inadequate number of transitions in the digital signal creates one more serious problem, known as DC Wander.

All the wire line transmission channels are AC, coupled by means of transformers and capacitors, so they do not pass DC signals. This AC coupling is necessary on account of impedance matching and elimination of DC ground loops. The copper wires of the media are also used for remote DC powering of repeater stations in some cases, which precludes DC coupling of the signal.

If the signal contains long sequences of 1s or 0s, the signal will behave more like a DC signal for the durations of these sequences. Such long sequences will not be reproduced properly by the AC coupling devices and will lead to degradation in the amplitude reference for the detection circuitry. The degradation of the amplitude reference in turn will lead to mistakes in the correct judgment of the respective levels of 1 and 0, with the result that some of the 1s may be read as 0s and vice versa. The result is a high error rate in the link.

Although there are other means available to tackle this problem, it can also be obviated by the selection of a proper line code, which is always a preferable solution.

Incidentally, the line code depicted in Figure 4.5 is also able to tackle this problem. Please note that there is an abundance of transitions, and every bit, whether it is a 1 or a 0, carries a signal transition. Thus the DC component of this code is completely balanced and hence there is no problem of DC wander if this line code is deployed.

There are some other popular line codes which also obviate the problem of DC wander, which we will examine in detail later in this chapter.

4.1.3 Bandwidth Requirement

Some line codes need more bandwidth than others. For any type of communication system, whether it is on a wire line media or radio, bandwidth is one of the most important constraints. Engineers have always therefore worked to improve the performance of the system within the constraints of the given bandwidth. With the advent of optical fiber media, this becomes less of an issue as the bandwidth available in OFC media is extremely high. However, one can not be too liberal on that, as the bandwidth is not infinity in actual practice; though generally people may like to call it infinity. In other than OFC media, from which actually all the communication systems have evolved, the available bandwidth is generally divided into many channels, each channel carrying its own signal. Thus the signal of a particular channel has to confine itself to the allotted bandwidth so as not to cause interference with the adjacent channels.

The line code of Figure 4.5 needs double the bandwidth as compared to the line code of Figure 4.3. This is because the former uses two pulses for every symbol, whereas the latter uses only one pulse per symbol. As stated in preceding sections, for higher bandwidth a penalty has to be paid to achieve the advantage of more transitions in the signal for better timing extraction. If bandwidth cannot be increased, some other type of line coding will have to be adopted which can resolve the problem of transitions in the signal.

In contrast to the normal situation, however, some applications easily permit use of higher bandwidth, such as local area networking (LAN) using Ethernet protocol, or as mentioned above, long-distance transmissions on optical fiber.

On the issue of bandwidth, it is interesting and worth knowing how much bandwidth is actually required for transmitting pulses over the transmission media.

The frequency spectrum of a square wave (pulse) signal contains an infinite number of harmonics in addition to the fundamental frequency. The signal and its frequency spectrum are shown in Figure 4.6.

A channel of infinite bandwidth would be required to transmit the pulse in its true shape. In practice systems can have only a limited bandwidth. Transmitting the pulse through a channel of limited bandwidth distorts its shape because the components belonging to the frequencies above the bandwidth of the channel are suppressed. The impulse response for a square wave pulse in a limited bandwidth channel is shown in Figure 4.7, for a pulse of width T.

As can be seen, the shape of pulse distorts greatly with ringing continuing for long, due to limited bandwidth. The narrower the bandwidth the worse is the pulse shape. This shape does not include the distortions due to

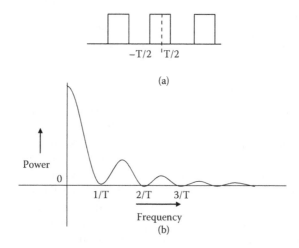

FIGURE 4.6
The square wave (pulse) signal (a) and its frequency spectrum (b).

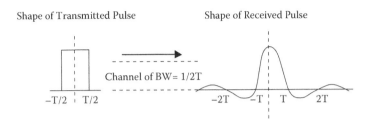

FIGURE 4.7
The shape of the received pulse distorts with limited bandwidth.

transmission media, noise, interference, and component nonlinearities. This ringing gives rise to "intersymbol interference" (ISI) because the spread of one pulse interferes with the adjacent pulses.

So if bandwidth is limited in any case, how do we overcome the problem? Fortunately, in digital communication we are not interested in the exact pulse shape. As long as we are able to truly detect whether the pulse is representing a 1 or a 0, it is enough. A second look at Figure 4.6 will show that though there are infinite numbers of harmonics in the frequency spectrum of the square wave, the amplitude of the fundamental component is much higher than other harmonics. The same is true for energy content of the frequency spectrum; more than 90% of the transmitted energy is contained in the fundamental component of the frequency (1/T). Rest of the high frequency components are confined to less than 10% transmitted energy. Hence, if only the fundamental component is transmitted and received, it will produce a true replica of the pulse, as we only need to decide whether it is a 1 or a 0. In practice it has been found that the energy available within 1/2T bandwidth is able to give us the desired sample of the pulse amplitude. Harry Nyquist had therefore established a relationship between the number of symbols and bandwidth as:

$$S \text{ max} = 2 \text{ BW}$$

where S max is the maximum signaling rate or maximum number of symbols and BW is the bandwidth of the channel. This is known as the Nyquist rate.

From the pulse shape shown in Figure 4.7, it may appear that if the pulses are repeated at the rate of 1/T in a channel of BW 1/2T, the spread of the pulses at the receiver may hamper the detection of individual pulses. However, it has been found that by replacing the square pulse at the transmitter by a specially shaped pulse called "raised cosine" and by carefully timing the sampling instances, it is easily possible to detect individual pulses without ambiguity at the Nyquist rate (refer to Section 6.4 for a detailed discussion).

4.1.4 Power Consumption

Different line codes consume different amounts of power for the same voltage levels of the pulses. Lower power consumption is always preferred if

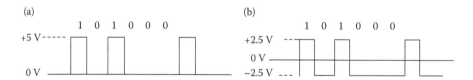

FIGURE 4.8
Two types of line codes for the same signal, having the same detection distance levels but different power levels. Signal of (a) has only positive amplitude, whereas signal of (b) has positive and negative amplitudes.

there are no other constraints. Lower power means lower ratings of the components and lower power consumption, both leading to less cost.

Let us examine two types of line codes from this perspective.

Figure 4.8a shows a line code for a particular signal, with the pulses having an amplitude of 0 or 5 V. In this case, the power required is

$$\text{For 1s} = \frac{5^2}{R} = \frac{25 \text{ W}}{R}$$

and for 0s = 0 W, thus

$$\text{Average power required} = \frac{12.5 \text{ W}}{R}$$

In case of Figure 4.8b, the same detection level of 5 V is maintained between 1s and 0s, but there are positive and negative polarities used, with 1 being at +2.5 V and 0 being at –2.5 V. The power requirement is

$$\text{For 1s} = \frac{2.5^2}{R} = \frac{6.25 \text{ W}}{R}$$

and for $\text{0s} = \frac{(-2.5^2)}{R} = \frac{6.25 \text{ W}}{R}$, thus

$$\text{Average power required} = \frac{6.25 \text{ W}}{R}$$

Thus we can see that by changing the voltage reference level we are able to achieve the same detection levels by using only half the power, which is always a welcomed step, as explained earlier.

4.1.5 Performance Monitoring Features

In any communication system, the monitoring of the system performance in the functional circuit, "in-circuit monitoring," is very important. This

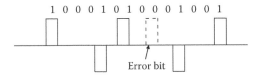

FIGURE 4.9
A line code in which all the 1s are represented by pulses of different polarities.

enables the maintenance engineers to take preventive steps for correction when the performance starts deteriorating, to avoid failures or breakdowns.

There are several techniques developed for this purpose, which include "introduction of parity bits" and "cyclic redundancy check" (CRC).

However, what could be better than the situation when the line code itself provides an inherent performance monitoring feature? There are some line codes which provide this feature through the mechanisms deployed for improving the timing content. One such code is shown in Figure 4.9. In this code, each 1 is represented by a pulse and every time a 1 occurs, the polarity of the pulse changes. Consequently, all the consecutive pulses in the bit stream are of different polarity from each other.

Thus if a bit error of 1 takes place, it will place a pulse of either positive or negative polarity next to any other pulse. This will lead to two pulses of the same polarity occurring consecutively. Since the receiver knows that in the proper signal this condition cannot arise, the error is immediately detected. Of course if two bit errors take place, causing pulses of opposite polarities to be placed in the bit stream, the pattern will be similar to the normal signal and they would not be detected. However, overall, such methods are able to give a fair account of the bit errors in the signal.

4.1.6 Error Probability

The probability of "error" in a received signal is the probability of either detecting a 1 as a 0 (or vice versa), or detecting a pulse where there is no pulse (or vice versa). The probability of causing such an error depends mainly on two factors.

(i) *Noise*: The available noise power at the receiver input, compared with the signal power (in other words signal to noise ratio, "SNR") greatly affects detection. The more the noise power (or the less the SNR) the greater is the probability of causing a detection error.

(ii) *Detection distance*: The ability of a code to distinguish the 1s from 0s for a given power. Please note that when we compare this property of two codes, the power level has to be the same, as by increasing the power level any code can improve upon this ability.

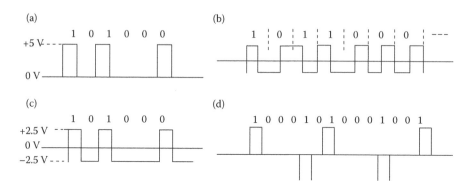

FIGURE 4.10
Different types of codes have different error probability.

Figure 4.10 depicts four types of codes, which we have already seen in the preceding section. Let us compare their error probability properties.

As explained in Section 4.1.1, the bandwidth required to transmit the line code of Figure 4.10b above is double that of the bandwidth required to transmit the line code of Figure 4.10a. The noise (the white or Gaussian noise) is proportional to the bandwidth; thus, in this case, the noise also is doubled. Accordingly, the error probability of the code of Figure 4.10b is double that of the code of Figure 4.10a. On the other hand, the code of Figure 4.10b uses positive and negative levels for pulses, hence the power requirement for the same detection distance (the overall pulse amplitude) is half of that for the line code of Figure 4.10a (see Section 4.1.4). Thus if the power level in both cases is kept the same, the signal level will be double in the case of Figure 4.10b line code, resulting in the doubling of SNR. The two factors together equalize the error probability for both cases.

The line code of Figure 4.10c will have half the error probability as compared to the line code of Figure 4.10a, on account of halving of the power requirement, as explained in Section 4.1.4. Since the bandwidth required is the same in both cases, the code of Figure 4.10c will have the clear advantage of less error probability.

The line code of Figure 4.10d will have equal probability of error as that of Figure 4.10a. This is because even though this line code has positive and negative pulses for detection of 1s, the detection of 0s is at zero level. Thus the detection distance is the same as that of the line code of Figure 4.10a. Moreover, it does not have any advantage in terms of power compared with the Figure 4.10a line code.

The line code of Figure 4.10c, therefore, is the best from the point of view of error probability or error performance, as it consumes minimum bandwidth and provides the best level distance between a 1 and a 0 for a given power.

4.1.7 Cost

Ultimately it is the cost which drives the choice of adoption of a particular code. Generally speaking, the more complex the line coding, the costlier it is. The cost of high rating components in case of higher power, the cost of higher bandwidth in case such code is chosen, and the long term cost of maintenance, also account for arriving at a decision.

Having seen the drivers of the choice of a particular type of line code, let us look at some of the popular line codes.

4.2 Types of Line Codes

We have seen a number of line codes in the preceding section during the discussion of the factors affecting the choice of a particular line code. Although they have been shown for illustration of the particular factor, most of them are actually deployed. Let us see them in more detail.

4.2.1 Unipolar (RZ) Code (On–Off)

The line code shown in Figure 4.11 is called a unipolar code, since it has the pulses only above zero, i.e., positive pulses. It is also called a RZ ("return to zero") code because after each pulse the signal returns to zero voltage level. This is the simplest type of code, and is not in much use in pulse transmission because of the disadvantages such as lack of timing content, higher power requirement, and presence of DC wander, as explained in Section 4.1. It is also known as an unbalanced code. The code is shown again in Figure 4.11. Such codes where a 1 is transmitted by a "pulse" (on) and a 0 is transmitted by "no pulse" (off), or vice versa, are also called on–off codes.

4.2.2 Polar (NRZ) Code

To reduce the power requirement while maintaining the level difference (detection distance between a 1 and a 0), the unipolar (on–off) code can be converted to a polar code, just by shifting the zero voltage line to the middle of the pulses. This way half the pulse amplitude is positive, while half of it is

FIGURE 4.11
The unipolar or on–off code.

1 0 0 0 1 1 1 0 0 1 0 1 1 0 0

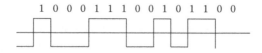

FIGURE 4.12
Polar (NRZ) code.

negative, giving rise to a 50% power saving, as explained in Section 4.1.4. The code is depicted again in Figure 4.12.

Since the signal level does not return to zero, it is called a "nonreturn to zero" (NRZ) code. It is a balanced code as the signal level is equally balanced above and below zero level. The name polar is actually confusing. As the code has positive and negative polarities, it should be actually called a bipolar code and some people do call it so; however, more popularly, it is called a polar code and the name "bipolar" is given to another code which we will discuss shortly.

As explained in Section 4.1.4, it saves 50% power as compared with the unipolar code, but it suffers as much as the unipolar code from other problems such as lack of timing content and presence of DC wander, and hence, it is not in much use in the deployed system.

4.2.3 Alternate Mark Inversion Code (Bipolar Code)

To improve the timing content of the code, all the 1s are represented by a pulse and each time the 1 occurs the polarity is reversed. The name "alternate mark inversion" (AMI) comes from this process as alternate 1s are inverted (a 1 is traditionally called a "Mark" in telegraphy language, where a 0 is called a "Space"). The code is already shown in Figure 4.9, which is reproduced here as Figure 4.13.

This process of inversion of alternate 1s helps greatly in enhancing the signal transitions for improving the timing content, as the polarity changes every time a 1 occurs. However, the problem due to long sequence of 0s continues to exist.

The problem of DC wander is somehow completely eliminated by using this code because all the 1s have positive and negative polarities alternatively and all the 0s are at zero voltage level (off). Thus a zero DC voltage is maintained on line. This representation of 1s by alternate polarities leads to naming the code as a "bipolar code."

1 0 0 0 1 0 1 0 0 0 1 0 0 1

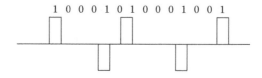

FIGURE 4.13
AMI code.

Efficiency of this code is poor because three levels of signal, i.e., "+v," "–v," and 0 ("v" is the voltage level) are used to transmit only two pieces of information, i.e., 1 or 0. The code is actually not a binary, but a "ternary code." A variation of this code, called HDB-3 ("high density bipolar"), is very popular and used in "E1" lines, which is the subject of our next section.

One of the most important features of the AMI code is its capability of error detection by means of detection of code violation, as seen in Section 4.1.5.

4.2.4 High Density Bipolar-3 Code

The problem of long sequences of 0s giving rise to lack of timing content is eliminated by a modification of the AMI code. The modified code is called HDBN ("high density bipolar N"), where N can have various values such as 3, 8, etc. We have chosen to discuss HBD-3 here because it has become an international standard, and is used in "E1" lines defined by ITU-T as a standard code.

Any sequence of N+1 0s, which, in this case, is 3 + 1 = 4, is replaced by a special code, which maintains enough transitions for timing content. This special code changes one of the 0s, of the sequence of four 0s to 1. While doing so it "violates" the AMI or Bipolar rule (which states that all the 1s will have alternating polarities. Thus when a 0 becomes 1 there will be two consecutive 1s with similar polarity, hence causing a violation of the rule). Figure 4.14 shows this violation.

As explained in Section 4.1.5, this violation could also be due to an error bit. Thus, to distinguish the deliberate violations from those due to errors, the substitution is done in such a manner that the number of "Normal" 1s between any two violations is "Odd." To achieve this, if the number of 1s after the last violation is "even" then the sequence of four 0s is replaced by a sequence of "100V" where "V" is a 1 with violation polarity, and if the number of 1s after the last violation is "odd" then the sequence of four 0s is replaced by a sequence "000V." Thus the number of 1s between two successive deliberate 1 violations is always maintained to be "odd." Such a predefined pattern is easily identified by the receiver and proper sequences are restored before decoding. Figure 4.15 depicts the pattern.

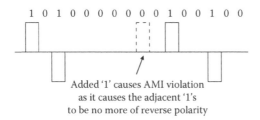

1 0 1 0 0 0 0 0 0 0 1 0 0 1 0 0

Added '1' causes AMI violation
as it causes the adjacent '1's
to be no more of reverse polarity

FIGURE 4.14
AMI violation by deliberate introduction of a 1 which replaces a 0.

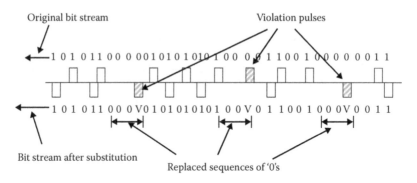

FIGURE 4.15
HDB-3 code.

Please note that the alternate violation pulses are also inverted in polarity so that DC wander is minimized and it further helps in improving transitions.

Even after introducing the deliberate violations, the error detecting capability of the AMI code is retained because violation pulses due to errors will disturb the alternating polarity of the coded violation pulses, and we will have two consecutive violation pulses of the same polarity, indicating the presence of an error bit.

In North America a similar version of HDBN code known as binary N-zero substitution (BNZS) is used, where N takes values such as 3, 6, 8, and so on. The complete process of introductions of violation pulses is otherwise the same.

4.2.5 Coded Mark Inversion Code

A different type of variant of AMI code which is able to remove the problem of long sequences of 0s is coded mark inversion (CMI) code. In this code, the 1 is represented by alternating pulses of positive and negative polarity, and a 0 is represented always by a half cycle with a rising edge situated in the middle of the cycle. This code is also one of the popular ones and is used in many of the PDH and SDH/SONET hierarchies. In PDH digital transmission hierarchy the E1 to E3 levels (i.e., up to 34 mbps) use the HDB3 code. However, at E4 level (139 mbps), it becomes difficult to detect 0 in the HDB3 code as the duration of the signal for a 0, which is detected by zero potential, becomes much less. The CMI code removes this problem by providing a pulse indicating a 0, which changes polarity from negative to positive midway between the pulses. Figure 4.16 depicts the code.

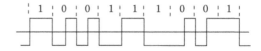

FIGURE 4.16
CMI code sequence.

Since this code uses different symbols for 1s and 0s, it is very easily distinguishable. As can be seen, the code has abundant transitions for timing extraction, and does not have any DC wander because there is a transition in every bit, be it a 1 or a 0. Since more than a single pulse is used for each symbol, the bandwidth requirement is greater, but it is easily tolerated in the optical fiber-based systems, where the bandwidth is not generally a problem.

4.2.6 Manchester Code

The code shown in Figure 4.5 is a Manchester code or a "digital biphase" or "diphase" code. It is shown again in Figure 4.17.

As explained in Section 4.1.1, this code eliminates the problem of timing content and DC wander, but consumes double the bandwidth. Hence, it can be used in systems where bandwidth is not a major issue, such as LAN or OFC systems.

There are many other types of line codes in use and each has some advantages for the specific variety of communication applications. Their description is not included here, in order to keep the focus on developing the concept, which enables the reader to understand the basic philosophy of the design of line codes.

A summary of features of the codes discussed above is shown in Figure 4.18.

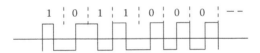

FIGURE 4.17
Manchester code.

	RZ (on-off)	NRZ	AMI	HDB 3	CMI	Manchester
Timing content	Poor	Poor	Good	Good	Very good	Very good
DC wander	Present	Present	Absent	Absent	Absent	Absent
Power requirement	High	Low	High	High	Low	Low
BW	Low	Low	Low	Low	High	High
Error probability	High	Low	High	High	High	Low
Performance monitoring	No	No	Yes	Yes	Yes	No

FIGURE 4.18
Comparative features of various line codes.

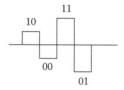

FIGURE 4.19
A multilevel code.

Cost has not been compared, as the cost for any type of code will vary from application to application. For example, deploying the Manchester code on short distance cabling of LAN is not at all costly as the available bandwidth on the used cable pairs is fully dedicated to this application. However, the cost of deploying the Manchester code on a long-distance link on radio media will be considerable, as the cost of the frequency spectrum itself will be doubled.

4.2.7 Multilevel Line Codes

To increase the bandwidth efficiency of the system, a multilevel line code is sometimes used. A multilevel line code transmits more than one bit in each symbol. A typical multilevel code is shown in Figure 4.19. The pulses may have any one of the two amplitudes. Different amplitudes represent different information, and there are 4 variations available, two in the form of +ve and –ve polarities and two in the form of different amplitudes of the pulses. Thus, each pulse has two attributes, amplitude and a polarity; hence it conveys two information or two bits. The probable bits for each type of pulse are also shown in the figure.

In this type of multilevel line code the pulse rate is called "symbol" rate or "baud" rate, which is different from the bit rate. In this case, the bit rate is double the baud rate. Thus if the voice band modem could accommodate a baud rate of 2800 bps, it can transmit bits at the rate of 5600 bps as the bit rate is 2 bits/baud. Similarly there are codes using 4, 8, or 16 levels to achieve higher bit rates. The reception becomes more difficult as the bit rate/ baud rate increases, as the receiver has to distinguish between closely spaced amplitudes. Conversely, if the same level difference has to be maintained between the adjacent symbols as that of the two-level code, the transmitted power has to be as many times higher as the number of levels.

4.3 Digital Modulation Techniques

The line coding techniques defined above are used for sending electrical pulses on the wire line systems. However, they cannot be used as they are

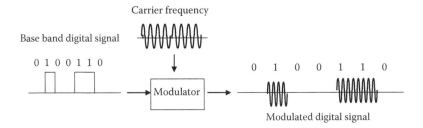

FIGURE 4.20
The modulation of a carrier frequency by a base band digital signal.

for the communication links built on radio media, or the wire line links working on carrier frequencies carrying multiple channels with frequency division multiplexing, because in these cases the transmission is through a high frequency sinusoidal signal. Thus, in such cases, instead of directly transmitting the pulses, the line code pulses are used to modulate a carrier frequency signal. The modulated signal carries the entire information of the coded sequence. At the receiving end the line code is regenerated by demodulating the carrier frequency. The code sequence in this case is called "base band" signal. The principle is illustrated in Figure 4.20.

The digital modulation techniques can be divided into the following categories:

(i) Amplitude modulation and

(ii) Angle modulation ⎱ frequency shift keying
⎰ phase shift keying
quadrature amplitude modulation.

4.3.1 Amplitude Modulation (ASK)

The typical modulation technique that we see in Figure 4.20 is an example of amplitude modulation. In this case, the amplitude of the carrier wave is varied or modulated in accordance with the base band signal. There is finite amplitude indicating the presence of the modulated carrier signal for a 1 and there is zero amplitude or simply absence of the modulated carrier wave for a 0. It is a kind of switching on and switching off operation of the carrier. Thus it is also called "on–off keying" (OOK). Another name commonly assigned to this process is ASK (amplitude shift keying).

4.3.1.1 Demodulation of Amplitude-Modulated Carrier

The detection of this OOK or ASK signal is very simple. Envelop detection just like the detection of analog amplitude modulated signals serves the

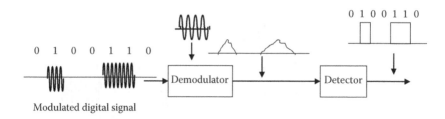

FIGURE 4.21
Detection of OOK/ASK modulated signal at the receiver.

purpose of correctly recovering the digital code. Figure 4.21 illustrates the principle. When a modulated signal is passed through a detector, the peaks of the "carrier" frequency of the received signal charge a capacitor at the output. Before the capacitor gets discharged next peak hits. Successive peaks thus recreate the original "modulating signal" back at the receiver.

The simplicity of the detector makes it a very low cost system. However, the performance of the amplitude modulation system is poor compared with other forms of modulations which fall into the category of angle modulations (to be discussed next). Power required for the OOK/ASKS system is also more than that required for frequency shift keying (FSK) or phase shift keying (PSK), for the reasons explained in Section 4.1.4 of this chapter. The power requirement is further boosted, as the amplifiers are required to be operated at lower power levels, relative to their ratings, to avoid the problem of saturation of the power amplifier stages, similar to the analog amplitude modulation systems. Hence other types of modulation techniques as stated above are more popular.

4.3.2 Frequency Shift Keying

As is evident from the name, FSK uses the base band digital code to shift the frequency of the carrier signal. The process is almost equivalent to frequency modulation of an analog system. The only difference is that in analog FM, continuous variations of carrier is in accordance with the analog signal, but in the case of FSK, a digital base band code shifts the frequency only to one step on the positive and/or one step on the negative side. A two level NRZ type of code is normally used. Figure 4.22 illustrates the principle.

FSK requires change (modulation) only in frequency, thus the transmitted power level remains constant, which allows the use of optimum power levels. In the case of ASK, it is not possible to use the transmitters at optimum power levels, as it may lead to the transmitter power amplifiers getting into the saturation region, as stated in the previous section. The only disadvantage of FSK as compared to ASK is that it requires a higher bandwidth.

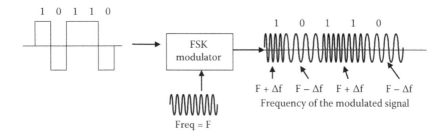

FIGURE 4.22
FSK modulation.

4.3.2.1 Demodulation of FSK Signals

Since there are two frequencies present in the FSK signal, each pertaining to a 1 or a 0, two tuned circuits can be deployed at the receiver, each one tuned for one of the two transmitted frequencies. The output of the tuned circuits is sent to an envelop detector, which produces the pulse shapes like those produced by the envelop detector of ASK signals as discussed in the previous section. Thus one envelop detector will produce all 1s, while the other one produces all 0s. Their occurrences at particular time instances are decided by the receiver clock, which converts the outputs of the detectors to the desired sequence, corresponding to the incoming signal. The principle is illustrated in Figure 4.23.

4.3.3 Phase Shift Keying

Also known as "digital phase modulation," PSK may be considered as a spe-cial case of FSK, where, instead of varying the frequency of the carrier in a pre-determined manner, only the phase of the carrier is shifted for one of the symbols, say 0, whereas the carrier is transmitted as it is for another symbol,

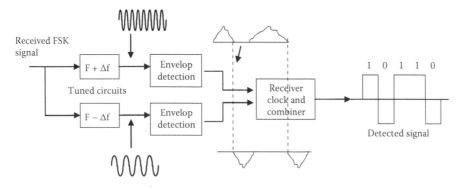

FIGURE 4.23
Demodulation of FSK signal.

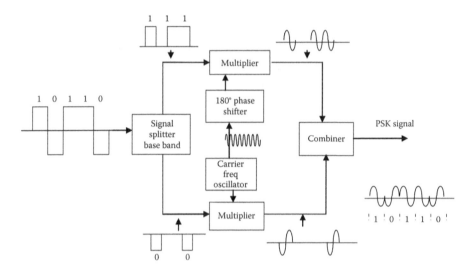

FIGURE 4.24
PSK modulator implementation.

say 1. The phase shift introduced is 180°, or simply speaking the phase is reversed. Like FSK an NRZ type of polar code is normally used in PSK. The constant envelop signal thus generated also retains the advantage of operation at higher power levels, just like FSK. The PSK modulation process is illustrated in Figure 4.24.

As can be seen in Figure 4.24, the carrier signal is shifted in phase by 180°. The original carrier is multiplied with 0s and the shifted carrier is multiplied by 1s of the line code. The amplitude-modulated type of carrier waves so generated are added together maintaining their phases as they are, to achieve the desired PSK signal, which is a constant envelop signal with phase modulations in accordance with the digital signal.

Since the information about the signal is contained in the phase of the transmitted signal, the receiver has to correctly detect the phase of the incoming signal. In other words the receiver has to be a "coherent receiver."

4.3.3.1 Demodulation of PSK

The PSK signal has to be demodulated with reference to a signal of the carrier frequency, which is coherent in phase, or phase synchronized with the transmitted signal. The received signal is multiplied with the signal generated by a coherent source on one hand, and by 180° shifted version of it on the other. Both these streams are then passed through low pass filters to get the envelops containing 1s on one limb and 0s on the other. With the help of a precise clock (synchronized with the timing contents of the received signal) the 1s and 0s are detected by sampling of appropriate instances and finally

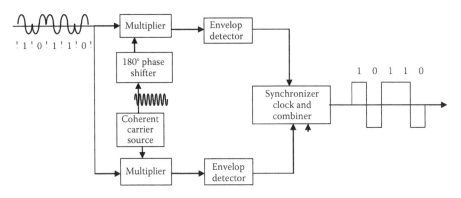

FIGURE 4.25
PSK demodulator implementation.

they are combined with their phases intact, to get the desired bit sequence of the transmitted signal.

The principle of the PSK demodulator is illustrated in Figure 4.25.

The performance of PSK is superior to ASK and FSK. The bandwidth required for PSK is also less than FSK, because FSK needs to transmit two frequencies whereas PSK needs to transmit only one frequency like ASK. Less bandwidth adds less noise to the channel and thus improves the noise immunity of the signal. Due to these advantages PSK is the most popular scheme of digital modulation. The detection of PSK has to be very precise as it has to be a coherent detection to be able to detect the phase of the signal correctly. The precision detection adds to the cost.

Although the foregoing discussion is enough to understand the concepts of digital modulation techniques, an interesting variant of PSK is discussed in brief in the next section to familiarize the reader with the terms, which they may occasionally encounter.

4.3.4 Multiple Phase Shift Keying

The phase shift keying that we just studied in the previous section is called "2 PSK," to distinguish it from the various higher level PSK schemes which were developed primarily to obtain higher data rates on low bandwidth voice band channels, through the devices called voice band "modems" (i.e., modulator-demodulator).

The other versions of PSK are 4 PSK (which is also called QPSK), 8 PSK, 16 PSK, and now even 64 PSK and 256 PSK.

In 2 PSK systems we use a phase difference of 180° or simply reverse the polarity of the carrier, to distinguish a 1 from a 0. In 4 PSK or QPSK systems the phase is shifted by 90° (i.e., quadrature phase shift, and hence the name QPSK). The phase shifted carriers are shown in Figure 4.26.

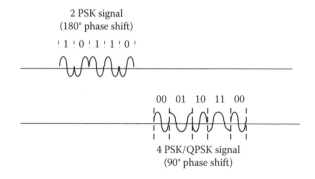

2 PSK signal
(180° phase shift)

! 1 ! 0 ! 1 ! 1 ! 0 !

00 01 10 11 00

4 PSK/QPSK signal
(90° phase shift)

FIGURE 4.26
Carrier modulation in 2 PSK and 4 PSK systems (single cycles of the waveform have been shown for clarity, but in practice, each symbol will have a large number of carrier wave cycles).

Similarly, smaller segments of phase differences are used for generating 8 PSK, 16 PSK, etc.

We can see that each phase segment, and thus each symbol represented by the particular phase, represents two bits as against one bit represented by the symbols discussed previously. Thus the bandwidth efficiency of the QPSK is double that of 2 PSK. Though the bandwidth efficiency may not be of much consequence in the backbone networks working on optical fibers, it is of prime importance in the radio networks such as short haul point-to-point links or cellular mobile telephone operations. This is due to the fact that the radio spectrum is always in short supply, because of a large number of services operating in the same local area, each demanding its own share of the limited spectrum, and of course due to the large numbers of subscribers. This scarcity of spectrum makes the techniques like QPSK, 8 PSK, or 16 PSK, etc. very attractive, as they can multiply the bit rate without increasing the bandwidth.

4.3.5 Quadrature Amplitude Modulation

The QPSK allows us to transmit two bits per symbol by separating each symbol by a phase difference of 90° (quadrature phase shift) as explained earlier, and higher PSK levels facilitate more number of bits like 8, 16, etc. per symbol by reducing the phase shift appropriately. However, there is a better approach to achieving more reliable reception, called QAM (quadrature amplitude modulation). QAM applies variation of amplitude in addition to the phase to achieve transmission of many bits per symbol. While 4, 8, 16, and 64 QAMs are already very popular, 256 QAM is also being successfully deployed.

The principle was first employed in analog communication systems. The transmission of double side band (DSB) signal required double the bandwidth compared with the single side band transmission in amplitude modulated analog signals. The problem was overcome by shifting the phase of one of the side bands by 90° (quadrature phase shift), and combining both of them for transmission on the same channel. It required half the bandwidth. The two signals could be separately detected and combined at the receiver.

A multilevel code like the one shown in Figure 4.19, which has four amplitude levels, can be used to modulate the amplitudes of the QPSK signals to obtain a 16 QAM signal. Of the four levels of amplitude, each will combine with one of the four phases in turn to give four combinations. Thus all four amplitudes will generate 4 × 4 = 16 combinations. Thus, each symbol will carry four bits of data, increasing the bit rate four times for the same bandwidth.

Due to varying amplitude (power) levels, QAM has a better detection ability as compared to a pure PSK signal with the same levels (say 16 QAM versus 16 PSK), and hence is more popular. However, the same property is also a disadvantage of QAM, as varying power levels require the use of less than optimum power levels of amplifiers to avoid saturation of the output amplifiers of the transmitters. Moreover, higher power levels are required for the same bit rate. For the bandwidth advantage these prices are, however, happily paid. Another problem with PSK and QAM modulations is that, because of closely spaced phases and amplitudes, they both need precise and thus complex circuitry for correct detection of the symbols.

For high bit rates on low bandwidth media, like that of the last mile of copper to the customer premises, or the cellular mobile telephone connectivity, QAM is currently the modulation of choice. Many systems deploy adaptive QAM modulation, which changes from 4 QAM to 16 QAM or 64 QAM dynamically, depending upon the condition of the environment and media availability.

4.3.6 Digital Modulations for Optical Fiber Transmission

Optical fiber transmission makes use of near infrared waves, generated by laser diodes or other types of laser sources. They generally operate on a single frequency (defined normally in wavelength, the most popular ones being 1310 nm and 1550 nm [nanometers]). The frequency of these waves is much higher than the frequency of our data (data rate or bit rate) so this near infrared frequency is required to be modulated in some way to carry and convey the information we intend to send.

Many types of line codes may be used as base band signal for modulating light wave carriers for transmission over fiber. However, since the bandwidth is available in abundance in optical fiber, the types of codes requiring higher bandwidth but providing better timing content, like the NRZ codes, are the preferred ones.

All the types of digital modulations discussed in this chapter are also suitable for optical fiber. The ASK, FSK, and PSK can be used to modulate the near infrared wave carrier in a similar way as they are used for modulating microwaves or other radio or carrier frequencies. While electro-optical modulators are preferred for PSK modulation, a waveguide modulator is used for ASK modulation.

4.4 Other Means of Improving the Timing Content

We have seen from the above discussion that timing content is the most important attribute of a line code, and a variety of line codes has been developed to improve upon this quality in the transmitted signal. However, it is not necessary to "build" the feature of rich timing content "into" the line code. The use of special types of coding called "block coding," or some addition or manipulation of the bits in the desired manner (using scrambler), may do the job equally well and sometimes better. Inevitably a penalty has to be paid, either in terms of lost bits or additional redundant bits, each leading to reduced efficiency.

4.4.1 Using Block Coding

To improve the timing content of the digital signal, a line code may be allowed to transmit data in blocks of a number of bits (called m) by choosing only those codes from a higher number of bits (called n) in combinations, which have the desired timing content. Let us consider a code in which m = 2 and n = 3. The two bits of m can produce 4 binary combinations and the 3 bits of n can produce 8 binary combinations. We have to choose 4 combinations from the 8 combinations to get good timing content or transitions from 1 to 0 or vice versa.

Let the data sequence be 01000010, which can be divided into four blocks of two each, such as 01 00 00 and 10.

The binary blocks possible for 2 bit and 3 bit codes are as follows:

    2-bit code 00, 01, 10, and 11

    3-bit code 000, 001, 010, 011, 100, 101, 110, and 111

We can choose the following mapping (note that the codes 000, 010, 100, and 110 will remain unused):

2-Bit Code in Data	3-Bit Code Chosen for Transmission
00	010
01	011
10	101
11	111

Thus our code sequence becomes

from 01 00 00 10 → 01000010

to 011 010 010 101 → 011010010101

We can see that the converted bit sequence contains enough numbers of transitions for achieving a much better timing content. Since the sequences are predefined, the receiver converts the sequence back to its original one.

The penalty to be paid is the transmission of 50% more pulses, which brings down the coding efficiency. However, in many cases, where the bandwidth is not an issue as in optical fiber transmission, these techniques provide excellent means of improving the signal transitions.

The above example was an illustration of the principle. In practice a block code with m = 4 and n = 5 is used for fiber-distributed data-interface (FDDI) standard, which is a protocol for Local Area Networking (LAN) on optical fiber. Higher combinations of m and n, say (m, n) (5, 6) (8, 10) and (64, 65) are in use for many applications.

4.4.2 Using Scrambler

This is one of the most important and most popular methods of improving the timing content. Let us see what it is and how it serves the purpose of improving the timing content.

As the name suggests, a scrambler is a circuit which scrambles the data. The structure of the data or the bit sequence completely changes after passing through the scrambler. The scrambling is done in a complex but well defined way such that a reverse process called descrambling can retrieve the data correctly.

Scramblers were originally invented for maintaining the secrecy of data through encryption for military applications, but the process became very useful for improving the timing content of the digital signal, and today it is adopted in most of the high-speed digital transmission systems, including those on optical fiber media.

The scrambler usually makes use of two simple logic devices known as "Shift Registers" and "Exclusive OR Gates" (exclusive OR gates are also called modulo two adders).

The function of the shift register is to delay the input signal by a defined number of bit intervals, as shown in Figure 4.27. In the two-stage shift register, the output.

B of Flip-Flop 1 is the value of A delayed by one bit interval, and the output "C" of Flip-Flop 2 is the value of B delayed by one bit interval or the value of "A" delayed by two bit intervals.

The function of exclusive OR gate and its "truth table" (the input output statement) is shown in Figure 4.28. As can be seen, the output of the Ex-OR Gate is 1 only when the inputs are different, otherwise it is zero. This property is very useful in constructing scrambler, when the Ex-OR operation (or module 2 addition) is performed along with shifted/delayed signals.

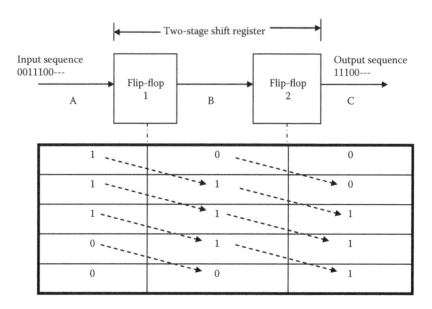

FIGURE 4.27
Functioning of a shift register.

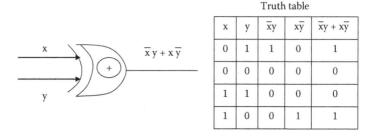

FIGURE 4.28
Function of Ex-OR gate (module 2 addition) and its truth table.

Let us construct a simple scrambler with the help of the above two elements, to understand the principle (Figure 4.29).

The signal to be transmitted is fed to an Ex-OR gate and the output is transmitted as the scrambler output. However, the output is also fed back and added to the input in a modulo two logic addition. The resulting output sequence is a much different sequence, as compared to the input sequence.

Let us take an arbitrary bit sequence at the input, say "1 1 0 0 0 1 0 1," and see what the scrambler output is. Please note that the signal at "C" is equal to the signal at "B" delayed by two bits; thus its initial value is 0 for two bits. The delay is denoted by x^2. The bit sequences "A," "B," and "C" are tabulated in Figure 4.30 for this scrambler.

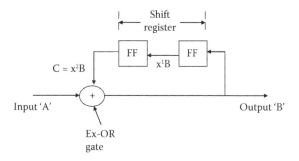

FIGURE 4.29
A simple scrambler.

As can be seen, the sequence at the scrambler output "B," is very different as compared to the input sequence "A." Please note that the long sequence of 0s present in the input bit stream is absent in the output sequence, facilitating much better timing content (better timing content also helps in reducing jitter, which is dealt with in detail in Section 6.5). This example is only an illustration of functioning principles of the scrambler. Actual scramblers use more units of delay and modulo two additions.

We have completely changed the bit stream through the use of scrambler, which will not render any meaningful signal at the receiver, unless the original sequence is recovered with the use of "descrambler," a reverse operation. Let us see if it really works out. The descrambler for the scrambler of Figure 4.30 is shown in Figure 4.31.

The input and output bit sequences of this descrambler are given in Figure 4.32. The sequence "C" is nothing but sequence "B" delayed by two bits or, as it is usually denoted, "x^2B." We can see that the recovered sequence is the same as the original sequence at the scrambler input.

A	C	B
1	0	1
1	0	1
0	1	1
0	1	1
0	1	1
1	1	0
0	1	1
1	0	1

FIGURE 4.30
The bit sequences A, B, and C for the scrambler of Figure 4.29.

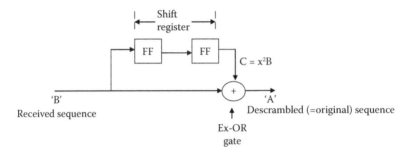

FIGURE 4.31
Descrambler for scrambler of Figure 4.30.

The output of a scrambler is generally represented by a polynomial. To calculate the polynomial for the example scrambler, let us again refer to Figure 4.29. We can see that the output "B" is the modulo two sum of the input "A," and the delayed signal "C" which is equal to "x^2B." Since initially A = B, we have

$$C = x^2 A$$

However, this signal is again fed back to the shift registers for another delay of x^2 and so on. Thus the sequence becomes

$$B = A \; (+) \; x^2 A \; (+) \; x^4 A \; (+) \; x^6 A \; (+) \; x^8 A$$

Let us check whether this polynomial gives us the same results as we have seen above in Figure 4.32.

B	C	A
1	0	1
1	0	1
1	1	0
1	1	0
1	1	0
0	1	1
1	1	0
1	0	1

FIGURE 4.32
The bit sequences of the input/output of the descrambler.

$$A = 1\ 1\ 0\ 0\ 0\ 1\ 0\ 1$$
$$+x^2A = 0\ 0\ 1\ 1\ 0\ 0\ 0\ 1\ 0\ 1$$
$$+x^4A = 0\ 0\ 0\ 0\ 1\ 1\ 0\ 0\ 0\ 1\ 0\ 1$$
$$+x^6A = 0\ 0\ 0\ 0\ 0\ 0\ 1\ 1\ 0\ 0\ 0\ 1\ 0\ 1$$
$$+x^8A = 0\ 0\ 0\ 0\ 0\ 0\ 0\ 0\ 1\ 1\ 0\ 0\ 0\ 1\ 0\ 1$$

or

$$B = 1\ 1\ 1\ 1\ 1\ 0\ 1\ 1$$

this is the same as the sequence "B" shown in Figure 4.32.

An actual scrambler used in practical SDH systems is shown in Figure 4.33.

The polynomial for this scrambler will be

$$B = A\ [1\ (+)\ x^6\ (+)\ x^7\ (+)\ \ldots]$$

The scrambler completely randomizes the data. A properly chosen scrambler removes the long sequences of 0s, to provide better timing content in the received signal. Depending upon the line code used, long sequences of 1s may also be as troublesome as those of 0s, and moreover may lead to DC imbalance. Scrambler removes long sequences of 1s either. The probability, however, remains that once in a blue moon the scrambler might produce long sequences of 0s or 1s by itself. Such a probability may be excised by predicting and blocking such sequences, as the input sequences which will produce them can be easily predicted by feeding such output sequences to the descrambler input and recording the output sequences of the descrambler. With the use of PLL (PLL stands for "Phased Locked Loop" and is explained in Section 6.5.1.1 in detail) for timing recovery, such long sequences are tolerated in practical systems without any noticeable degradation if they occur occasionally. Thus the use of the scrambler fully serves the purpose of improving the timing content. The improvement in the timing content also helps in reducing the jitter in the signal.

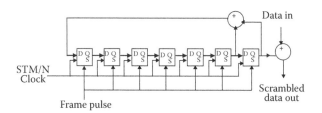

FIGURE 4.33
Scrambler used in SDH systems.

Besides the possibility of occasional long strings of 0s or 1s, the scrambler has one more disadvantage that the errors, if there are any in the data, are multiplied. Thus a single error in the transmitted data may result in multiple errors in the received data. This has to be taken as a minor penalty for the great advantages the scrambler offers of improving the timing content and minimizing jitter.

4.4.3 Adding Redundant Bits

Besides using the scrambler or adopting a particular line code, the timing content of the digital signal may also be improved by adding redundant bits to the original bit stream at the transmitting end, and removing them at the receiving end. Redundant bits can be added in many ways, some of which are as follows.

One way is to simply insert a 1 whenever the sequence of 0s gets longer. Let us say we decide to allow sequences of 0s up to 4. Thus, after every sequence of four 0s, a 1 will be inserted which can be removed by the receiver by the same logic. This method will lead to increasing the bit rate a little, but is perfectly acceptable for many data applications.

A variation of the above method could be to dedicate a few bits of the bit stream for timing. For example, every fifth bit of the bit stream could be a timing bit carrying a 1. This method will, however, need more extra bits compared with the above one which needed the timing bits occasionally.

Another option which is suitable only for speech circuits is to omit the long sequences of 0s completely from the bit stream. Though it appears strange, the analysis shows that the probability of occurrence of such long sequences of 0s is so low that dropping them will have virtually no effect on the voice quality. Naturally this cannot be applied in data communication.

Review Questions

1. What is coding in context of a digital signal, and what is line coding?
2. How is line coding different from digital modulation? How do they differ in application?
3. What are the factors that affect the choice of a line code for a particular application?
4. What do you understand by Timing Content of a digital signal? What effect does it have on the system? How can it be improved?
5. What is DC wander and what effect does it have on reception of a digital signal?
6. Explain the relationship between a line code and bandwidth requirement.

7. How can the received signal quality be improved in a limited bandwidth system?

8. What is "ringing" and "ISI," and what is their effect on the error rate of a received signal?

9. Is power consumption of the communication system a concern? How does power consumption vary with type of line code?

10. Illustrate an example of "in-circuit monitoring" feature in a line code.

11. What are the probable causes of error in the received digital signal? What is detection distance?

12. How does the cost matter in a communication system? What factors lead to increased cost?

13. Which are the line codes usually deployed in practical systems?

14. How are RZ and NRZ codes different from each other?

15. What is the difference between Unipolar, Polar, and Bipolar codes?

16. Why is AMI code so called? Please explain, with illustration.

17. Explain HDB-3 coding. What is meant by violation and how it helps improve code?

18. What are the plus and minus points of CMI and Manchester codes?

19. Compare different features of various line codes.

20. What is meant by levels in a line code? What is a multilevel line code and what is its advantage?

21. What is the difference between bit rate and baud rate?

22. Where do we need digital modulation instead of a line code?

23. Which are the popular types of digital modulation techniques?

24. What are ASK and OOK, how is the modulation and demodulation achieved?

25. How is FSK superior to ASK? How is FSK modulation and demodulation achieved? Please illustrate.

26. Is PSK superior to FSK? If so, why?

27. Explain QPSK. What are its main advantages?

28. How is QAM different from QPSK? How does QAM score over QPSK?

29. Is the digital modulation used for optical fiber different from that used for copper cable? If so what are the differences and why?

30. What are the means of improving the timing content of a code?

31. Does block coding improve the timing content of a code? If yes, how?

32. What is a scrambler? What was the primary purpose of its development? How can it be used to improve timing content of a code?

33. How is the data integrity maintained through the process of scrambling and descrambling?

34. What are the disadvantages of using redundant bits for improving the timing content?

Critical Thinking Questions

1. Can you think of some other means of improving the detection distance? What will be its implications?

2. Based on the features of various line codes, work out some new type of line codes which may have most of the plus points.

3. Will there be any problems in extending QAM to very high levels, say 256 QAM or 1024 QAM? If so, what and why?

4. Try to develop a small scrambler using different logical components than what are described in this chapter.

Bibliography

1. L.E. Frenzel, *Principles of Electronic Communication Systems*, Tata McGraw-Hill, 2008.

2. S.J. Campanella, *Digital Speech Interpolation Systems in Advanced Digital Communications, Systems and Signal Processing Techniques*, Prentice-Hall, 1985.

3. NIIT, *Introduction to Digital Communication Systems*, Prentice-Hall, India, New Delhi, 2004.

4. B.P. Lathi, *Modern Digital and Analog Communication Systems*, Oxford University Press, 1998.

5. J.H. Franz and V.K. Jain, *Optical Communications Components and Systems*, Narosa Publishing House, New Delhi, 2000.

6. D. Roddy and J. Coolen, *Electronic Communication*, Fourth Edition, Prentice-Hall, India, New Delhi, 2001.

7. S. Haykin, *An Introduction to Analog and Digital Communications*, John Wiley and Sons Asia, 1989.

8. NIIT *Advanced Digital Communication Systems*, Prentice-Hall, India, New Delhi, 2005.

9. J.C. Bellamy, *Digital Telephony*, John Wiley and Sons, Singapore, 2003.

10. H. Taub and D.L. Schilling, *Principles of Communication Systems*, Second Edition, Tata McGraw-Hill, 2005.

11. B. Sklar, *Digital Communications Fundamentals and Applications*, Pearson Education, 2002.

5

Clock

The most significant difference in the digital communication systems as compared with their analog counterparts is the presence of "clock" in the digital systems. In digital systems, the events happen with the advancement of clock step by step. If the clock stops, the digital communication world would stop. The clock in the digital systems has the same function as the normal clock (or watch) in our daily life. As in our daily life, all the events such as getting up, going to the office, eating, sleeping, etc. are governed by the time of the day, which is maintained by the clock, and so it is in the world of digital communication. All the events take place, keeping steps with the equipment clock. The clock of the digital communication equipment has the same function as that of the clock of our daily life, and that is "keeping track of the time." The only difference perhaps is that it has much smaller steps (microseconds or nanoseconds) as compared with that of the normal clock (seconds).

In the course of our discussion on various subjects in the subsequent chapters, the clock will figure repeatedly. Thus, it is worthwhile knowing a little about it.

5.1 What Is a Clock?

The clock of a digital communication system is a periodic signal (a periodic signal is the one in which the pattern repeats itself), mostly a continuous square wave signal (Figure 5.1). This signal is generated by means of an oscillator.

The type of circuitry for the oscillator depends upon the required frequency and accuracy of the clock. It may consist of simple resistance and capacitance (RC) elements, or it may include a piezoelectric crystal or an atomic source; each of them provides higher accuracy than the other respectively. We will discuss each of these types briefly in this chapter.

To understand the significance of clock in the digital world clearly, let us consider a few examples of its deployment.

Square wave clock

FIGURE 5.1
Clock (or clock signal) as a periodic continuous square wave.

5.2 Significance of the Clock

The significance of the clock in the digital communication cannot be overemphasized, as it has to be an integral part of the equipment inevitably. A few examples here will be able to explain the concept.

5.2.1 Triggering the Events

A simple set-reset (SR) flip-flop is shown in Figure 5.2. The function of this device is to set output A to 1 if the value of input S is 1 and set output A to 0 if the value of input R is 1. (Flip-flops are logic devices used extensively in digital equipment such as memory elements, shift registers, counters, etc., mostly as parts of the integrated circuits. It is presumed in the context of this book that the reader is familiar with simple digital logic devices. In case you are not, you may like to refer to one of the many text books available on the subject. However, the discussion here does not really call for this knowledge. The subject of discussion can be understood by simply presuming the function described.)

The output $\bar{A}$ is a called compliment of A (when A = 1, $\bar{A}$ = 0, and vice versa). The DC voltages corresponding to 1 or 0 may be present at input S and R for long times, but output A will change in accordance with the above rule only when a clock pulse is present. However, the clock pulse is present as well for a substantial period, as it has a finite width (Figure 5.1); thus, when exactly during the whole of the clock pulse period should the "action" of change of output take place? To avoid confusion and maintain precision, the action generally takes place at the rising edge or falling edge of the clock

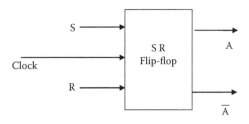

FIGURE 5.2
An SR flip-flop with a clock.

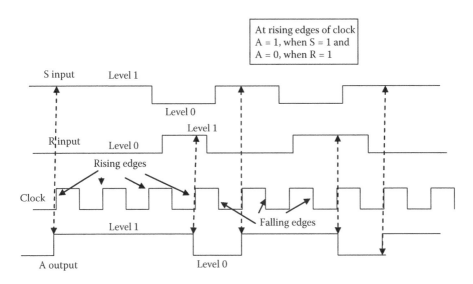

FIGURE 5.3
Output of the SR flip-flop changes in accordance with the inputs, only at the instants of rising edges of the clock signal.

pulse. Figure 5.3 illustrates the variation of output A with respect to the variation in inputs S and R, with the change taking place at the rising edges of the clock during each cycle.

As can be seen, output A changes from 0 to 1 only at the rising edge of the clock pulse, even though the S input might have been a 1 for a long time. The output again changes to 0 from 1 due to the R input being a 1, but again, the change takes place only at the event of a rising edge (also called a positive going edge) of the clock signal.

5.2.2 Reception of Digital Bit Stream

A meaningful reception of a digital bit stream (digital signal) cannot be done without a proper clock. Let us take a hypothetical transmitted bit stream to examine the concept (Figure 5.4). The bit sequence is coded with a unipolar (return to zero) line code (Figure 5.4a), which has positive voltage pulses for indicating 1s and 0 voltage levels for indicating 0s.

When this sequence is received at the receiver, in the absence of a clock, the pulses may be used to generate the received voltages through their positive voltages, but what will happen to 0s? Since there is no voltage for 0s, the only way to detect them is to "sample" the incoming signal at regular intervals and find out whether there is a voltage. This sampling has to be done at the same frequency as that of the transmitted pulses. This process is nothing but "using a clock." Even the 1s, when they fall continuously one after the other, cannot be detected without sampling in a similar fashion. Thus, the use of

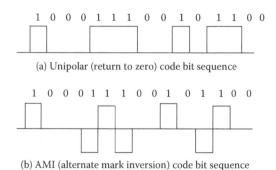

(a) Unipolar (return to zero) code bit sequence

(b) AMI (alternate mark inversion) code bit sequence

FIGURE 5.4
A hypothetical bit sequence that is transmitted as a digital signal.

clock in the receiver is inevitable. The transmitter as well has to have a clock for sending regularly spaced pulses at stable frequencies.

Even if we use an AMI line code as shown in Figure 5.4b (for line code details, see Chapter 6), a meaningful detection cannot be achieved without using a clock. In this case, all the 1s will get easily detected as they all carry a distinctly changing voltage on each and every occurrence (alternate 1s have different polarities or are "inverted"), but all the 0s have zero voltage; thus, they cannot be detected without sampling at regular intervals as described above, necessitating the use of a clock.

5.3 Clock Waveform

The job of the clock is to trigger the events at precisely regular time intervals as discussed in the previous section. Thus, any waveform that can provide this feature will be suitable for clock application. Figure 5.5 shows three types of waveforms, a square waveform, square pulses, and trigger pulses. As we have discussed earlier in this chapter, the job of triggering the events has to be performed by either the rising or the falling edges of the clock signal pulses.

Hence, any one of the type of clock signals shown in Figure 5.5a, 5.5b, or 5.5c will be suitable. The square wave waveform can perform the job of triggering on rising or falling edges of the pulses. The square pulses of Figure 5.5b are similar to the square wave of Figure 5.5a except that the duty cycle of the pulses is much less. (The duty cycle is the ratio of the pulse's "on" period (having amplitude) to the cycle duration. In the case of Figure 5.5a, the duty cycle is approximately 50%, whereas that of Figure 5.5b is only about 20%.)

Such pulses are generally used as triggering pulses as a whole and not the rising or falling edges of them.

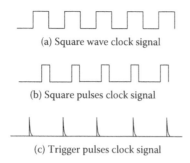

(a) Square wave clock signal

(b) Square pulses clock signal

(c) Trigger pulses clock signal

FIGURE 5.5
Various types of possible clock waveforms.

The triggering pulses of Figure 5.5c are also used as whole pulses for triggering the events, as only rising edges are clearly defined in the signal.

In the modern transmission systems, the square waveform of Figure 5.5a is normally used with a duty cycle of 50%. While the transmitter normally uses the falling edges, the receiver uses the rising edges for their operations.

5.4 Types of Clocks

The clocks can be divided into three main categories:

(i) Multivibrator clock
(ii) Crystal oscillator clock
(iii) Atomic clock

Let us see about each of them in brief.

5.4.1 Multivibrator Clock

The square waveform required for the clock signal can be generated by using a multivibrator circuit. This is the most basic, low-cost, and low-accuracy clock that can be had.

The multivibrator used is of "astable" type. (Multivibrator is a semiconductor circuit consisting of transistors, resistors, and capacitors. It is of three types: monostable, bistable, and astable. The astable multivibrator is able to generate a continuous square wave with properly designed values of components in the presence of the power supply voltage. A detailed description of the subject is available in any text on electronic circuits. The details

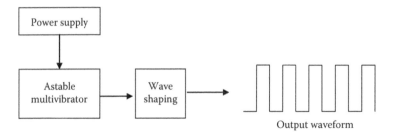

FIGURE 5.6
Square wave generation through astable multivibrator.

however are out of the scope of this text and not really important for the present discussion.)

Figure 5.6 depicts schematic diagram of an astable multivibrator and its waveform. The output waveform of multivibrator is not a true square wave and therefore requires wave shaping to get a true square wave clock.

This clock is able to generate any frequency ranging from a few cycles to many megacycles. The frequency depends upon the values of the resistor and capacitor (R and C, respectively) components. The accuracy of this clock depends upon the accuracy of the components, i.e., R and C. The frequency varies greatly with temperature and other environmental conditions that cause a variation in the values of R and C components. This is the cheapest clock with the poorest accuracy and stability, and hence, it is never used for any practical communication systems. It finds use only for experimental purposes.

5.4.2 Crystal Oscillator Clock

Any type of clock circuit is basically an oscillator. The multivibrator described above is also an oscillator. The conventional oscillators, however, have a "tuned circuit" invariably. This tuned circuit may consist of either capacitor and inductors or capacitors and resistors. Such oscillators have poor stability of frequency, as explained in the previous section, due to variations in the component values with environmental conditions.

However, if the tuned circuit of the oscillator is replaced by a "piezoelectric crystal," the frequency stability is improved tremendously. Thus, high-accuracy clocks use the piezoelectric crystal oscillators, rather than using LC or RC oscillators.

The piezoelectric crystal generally used is made of "quartz," but crystals made from tourmaline and Rochelle salt also display similar properties. Although tourmaline crystals are costlier, Rochelle salt crystals do not match the performance of the quartz crystal. Hence, for the purpose of clock, the quartz crystal is the one that is almost invariably used (Figure 5.7).

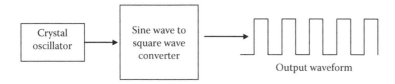

FIGURE 5.7
Crystal oscillator clock.

Since the waveform generated by the crystal oscillator is a sine wave, a sine wave to square wave converter is used to get a proper square wave clock.

These crystals have a very low temperature coefficient, causing very little variation in the clock frequencies with environmental conditions. Another factor that affects the stability of these clocks is "ageing." With time, the stability of the clock decreases. However, as an overall performance, a very high stability over long periods can be obtained if the factors affecting frequency stability are kept under control (such as temperature, humidity, dust). Even the effect of ageing can be minimized by proper treatment and environment.

Accuracies of the order of one part in 100 megacycles are easily possible with proper environmental conditioning. These clocks are able to generate frequencies up to hundreds of megacycles per second.

These clocks are used in all the communication systems today as the "node clock" or equipment clock. The accuracy requirements of the node clocks of the order of ±4.6 to ±50 ppm are easily met by these clocks at the required frequencies (tens to hundreds of megahertz).

Still higher stability requirements are called for, however, in clocks that are used for ultrahigh-frequency applications. Applications such as "network synchronization" or "plesiochronous operations" demand very high degrees of accuracy and stability. (The terms are explained in chapters on PDH and synchronization.) The crystal oscillator clocks fall short of meeting those requirements, making way for the use of "atomic clocks" for such applications.

5.4.3 Atomic Clocks

The atomic clocks are so far the most accurate clocks. The accuracy levels available are up to ±1 part in 10^{13} to 10^{14}, which works out to nearly 1 s in 1 million years. It has been a human endeavor ever since the start of civilization to calculate time more and more precisely. Crystal oscillators were a big step in that direction; however, the accuracy levels available after the invention of atomic clocks are far better than their crystal oscillator counterparts.

The first atomic clock as an experimental model was demonstrated in 1945, which had an even poorer accuracy than the crystal clock, but it proved the concept. A proper atomic clock made from cesium metal was finally made in the year 1955.

The atomic clock is made by locking the crystal oscillator clock to the atomic transitions. The electrons in the material of the atomic clock (say cesium 133) are excited externally to change their energy level state. The resonance frequency of this transition is the frequency of the atomic oscillator or atomic clock. This frequency for the element cesium 133 is 9,192,631,770 Hz. In fact, the standard of time, i.e., the "second," is now defined by this atomic constant. In other words, a "second" is the time that the cesium 133 atoms take to make 9,192,631,770 transitions.

The present technology atomic clocks are of the following types:

(i)   Cesium beam atomic clock

(ii) Hydrogen atomic clock

(iii) Rubidium atomic clock

Of these, the cesium beam clock is the most accurate one. Although the hydrogen clock has better short-term accuracy than the cesium beam clock, it has poorer long-term accuracy. The accuracy levels of cesium 133 clocks, as mentioned earlier, are up to 1 part in 10^{13} to 10^{14} (there is a little amount of variation due to the minor differences in the manufacturing process). They are used in all the prime applications and primary standards. For example, the navigation systems, Global Positioning System (GPS) and a primary reference clock (PRC) in long-distance communication links, all use the cesium beam clocks. The cost of the cesium beam clocks has also come down drastically over the years due to improvement in the technology of manufacturing and increasing volumes of production.

The rubidium clock, while small and costs low, has a poorer accuracy (in the range of ±1 part in 10^{12}). Due to its low cost and easy portability, the rubidium clock is more popular in applications where the accuracy requirement is not the highest, for example, the synchronization supply unit (SSU) of synchronous network of long-distance telecommunication links.

Our scientists, however, are not happy with the accuracy of 1 s in 1 million years. They are, hence, researching to make future clocks of the order of accuracies of 1 s in 1 billion years, which is the lifetime of the universe itself.

The GPS also deploys several cesium beam clocks in every satellite and is able to provide accurate timing information to the GPS receivers, the accuracy of which is comparable with the local cesium beam clocks. We will see in the chapter on synchronization that the synchronous networks use a clock derived from GPS, in addition to cesium beam-based primary and rubidium-based secondary clocks, to maintain the network-wide synchronization.

5.5 Clock Accuracy Required in Communication Systems

In modern digital communication systems, various data rates that have been defined as standard data rates, have to maintain their accuracies (or bit rates or clock speeds) within tolerable defined limits.

Sr. no.	Data rate (nominal; Mbps)	Nomenclature	Tolerance limits (ppm)
1	2	E1	±50
2	8	E2	±30
3	34	E3	±20
4	140	E4	±15

FIGURE 5.8
Various data rates of PDH and their node clock accuracy tolerance limits.

We will see in the chapter on plesiochronous digital hierarchy (PDH) the basic data rates defined by International Telecommunication Union (ITU-T), along with their accuracies in detail. Their required accuracies are given in Figure 5.8.

We can see that all these accuracies can be easily met using crystal oscillator clocks, and hence they are the ones used for these PDH node clocks. These accuracies are good enough for the node clocks of the PDH systems, but the clocks required for synchronization (the concept of synchronization in PDH systems will be clear, when you read the chapter on synchronization), which are deployed at the boundary of the network of a particular operator, are required to be of highest accuracies. These synchronization clocks are of cesium- or rubidium-type atomic clocks.

In the case of synchronous digital hierarchy (SDH) networks, the tributary data rates and their tolerances remain the same as those of the PDH, as the basic tributaries to SDH network are mostly PDH tributaries such as E1, E3, etc. However, for the purposes of network synchronization, the ITU-T defines three types of clock accuracies. They are called PRC, SSU, and synchronization equipment clock (SEC). Their levels of required accuracies and commonly used type of clock are indicated in Figure 5.9.

As can be seen from Figure 5.9, although the specified requirement of accuracy for a PRC is ±1 part in 10^{11}, which can be met even by a rubidium clock,

Sr. no.	Type of application in SDH	Accuracy required	Per ITU-T recommendation number	Type of clock deployed
1	PRC	± 1 part in 10^{11}	G.811	Cesium beam atomic clock
2	SSU	± 1 part in 10^8	G.812	Rubidium atomic clock or GPS clock
3	SEC	± 4.6 ppm	G.813	Crystal oscillator clock

FIGURE 5.9
The types of clocks defined by ITU-T for SDH network synchronization.

a cesium beam clock is invariably deployed by the network operators. This is due to affordable costs of cesium beam clocks and very little requirement. The whole network of an individual network operator can be synchronized with the help of a few PRCs only.

The second-level accuracy requirement of ±1 part in 10^8 is that of SSU. As a large number of SSUs are required in the network of an operator for complete synchronization, a rubidium atomic clock is deployed, which provides better accuracy than desired and is of low cost and small size. One more reason for the deployment of rubidium clock for SSUs is that the rubidium clock has excellent hold-over accuracy (of the order of ±2 × 10^{-11} per day and ±1 × 10^{-10} per month). The holdover accuracy is the ability of the clock to maintain the accuracy for a substantial period in the event of a failure of its master clock (see Section 7.8 for more details).

The SEC clock (or the node clock) has been defined to have an accuracy of better than ±4.6 ppm. For this purpose, the inexpensive crystal clocks are enough and they are the ones that are invariably deployed.

Review Questions

1. What do you understand by clock in reference to telecommunication? How does it compare with the wall clock that we know?
2. What is the significance of the clock in digital communication?
3. Explain how the clock performs the triggering function. Why are rising or falling edges used for triggering?
4. What role does the clock play in correct reception of digital signal and how?
5. Why is clock required at the transmitter?
6. What are the types of waveforms used for clock signals? What do you understand by "duty cycle?"
7. What are the main types of clocks?
8. How is an astable multivibrator used for generating the clock signal?
9. What is the frequency range of a multivibrator clock and what factors depend on it?
10. On what factors depends the accuracy of a clock and how can it be improved?
11. Why is crystal oscillator clock preferred over multivibrator clock?
12. What are the levels of accuracies obtained in crystal oscillator clocks and why?

13. Which are the types of clocks used for network synchronization and plesiochronous operations and why?
14. What are the level of accuracies achieved in atomic clocks and how?
15. What are the salient differences between cesium beam and rubidium clocks?
16. What are the accuracies required for various levels of PDH and what types of clocks are used to obtain them?
17. What do you understand by the terms PRC, SSU, and SEC? What are the accuracy requirements and what types of clocks are deployed?
18. What is the role of cost in the selection of a particular type of clock for an application? Explain with example.

Critical Thinking Questions

1. What will be the repercussions of triggering with the middle of the square wave clock pulse rather than edges?
2. Try to do the triggering function using other types of flip-flops or logic gates.
3. How can you improve the accuracy of multivibrator clocks?
4. What will be the implications on the communication systems if the accuracies of 1 s in 1 billion years are achieved in future clocks?

Bibliography

1. T.C. Bartee, *Digital Computer Fundamentals*, Tata McGraw-Hill, 1977.
2. J. Millman and H. Taub, *Pulse Digital and Switching Waveforms*, Tata McGraw-Hill, 2005.
3. J.C. Bellamy, *Digital Telephony*, John Wiley & Sons, Singapore, 2003.
4. F.E. Terman, *Electronic and Radio Engineering*, Fourth Edition, McGraw-Hill Kogakusha, Tokyo, Japan, 1955.
5. ITU-T Recommendation G.811, *Timing Characteristics of Primary Reference Clocks*, International Telecommunication Union, Geneva, Switzerland.
6. ITU-T Recommendation G.812, *Timing Requirements of Slave Clocks Suitable for Use as Node Clocks in Synchronization Networks*, International Telecommunication Union, Geneva, Switzerland.
7. ITU-T Recommendations G.813, *Timing Characteristics of SDH Equipment Slave Clocks (SEC)*, International Telecommunication Union, Geneva, Switzerland.

6

Signal Impairments, Error Detection, and Correction

The digital signal, when constructed, is in the form of well-defined pulses per the chosen line coding or digital modulation scheme. However, before this signal reaches the intended receiver, it undergoes a substantial amount of abuses, which reduce its size and distort its shape. The received pulses hardly look like pulses anymore. In fact, they look like two-dimensional profiles of "natural hillocks" when viewed on the oscilloscope. They have highly distorted shapes and much smaller amplitudes as compared with the original transmitted pulses. The transmitted and received pulses may look more or less like the ones shown in Figure 6.1.

The factors contributing to this pitiable condition of the pulses are not only the transmission media but also the various segments of the transmitting equipment. The biggest blow that the transmission media gives to the pulse is the attenuation. The received pulse amplitude may be thousands of times smaller than the amplitude of the transmitted pulse. The amount of attenuation depends upon the type of media, receiver thresholds, etc. (For example, the received power may be in the range of 100 db below [10^{-10} times] the transmitted power, in case of a radio link.) Transmission media, such as copper wire pair, coaxial cable, radio, or optical fiber, all contribute to noise and cause distortion to the pulse.

The transmitting equipment segments are not far behind the media when it comes to distorting the square shape of the pulse. The biggest effect is that of band-limiting filters, which restrict the bandwidth of the signal, chopping off the sharp edges and creating spurious waveforms at their output.

We will see about all these factors in detail in this chapter.

The digital signal suffers from the same set of impairing phenomenon that plagues analog signals (as seen in Chapter 2), and in addition to those, it has other factors affecting its correct detection. The most prominent of which are intersymbol interference (ISI) caused by ringing in the pulse waveform and channel-induced impairments, jitter in the bit stream due to various factors in the transmission media, and equipment inaccuracies, etc. All these factors shall be discussed in this chapter.

The impairments to the pulses cause "bit errors" in the received signal; therefore, there has to be a mechanism to detect these bit errors, so that they can be either corrected or the blocks of the bits containing errors may be

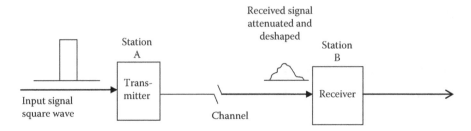

FIGURE 6.1
The distortion and deshaping of the pulse due to the channel.

retransmitted. We shall discuss in this chapter some of the error detection and error correction mechanisms.

6.1 Types of Signal Impairments

The impairment of digital signals can be broadly divided into the following four categories:

   (i) Attenuation
  (ii) Distortion
 (iii) ISI
 (iv) Jitter and wander

Let us discuss them briefly.

6.2 Attenuation

This is the biggest loss that the signal concedes while traveling from the transmitter to the receiver. Transmission media keep on absorbing the signal as it propagates, and there is currently no way to get rid of this factor. Different media have high or low losses (attenuation), and many a times, the choice of media is based upon this factor (for example, the losses in coaxial cable are less than those in copper wire pair), but the ultimate choice of media is governed by many other factors.

Along with the problems of analog signal, we have also discussed the effect of attenuation on the signal in detail. (The reader may refer to Section 2.2.1.)

Besides the transmission media, attenuation is also caused by many more factors if we look at it from the point of view of total signal power generated and the signal power received at the receiver, rather than considering received signal power against the power launched on the media alone. These factors are

(i) Attenuation due to antenna: In case the media is radio waves, the signal has to be radiated through antenna, which does not radiate all the power it receives. The antenna efficiency may be anywhere in the range of 50% to 80%. If the antenna is fitted with a radome (protective cover), the efficiency is further reduced, as the radome also causes some attenuation. Also, if the line of sight is not properly established, lesser signal will be received at the antenna.

(ii) Attenuation due to limited bandwidth: Majority of the systems operate on limited bandwidth for each channel. The band-limiting filters in the transmitter remove many frequency components from the signal, converting the square pulse into a different type of waveform, stripping off substantial energy from the pulse, and thus reducing the transmitted power.

6.3 Distortion

During the propagation, the signal gets distorted (see Figure 6.1) due to a number of factors. The important ones are

(i) Noise

(ii) Interference

(iii) Amplitude distortion due to frequency dependent amplification/ attenuation

(iv) Phase or delay distortion due to frequency dependent velocity of propagation.

(v) Harmonic distortion

(vi) Intermodulation distortion

We have discussed all these issues in detail in Chapter 2 in connection with analog signals. They all equally apply to digital signals as well because, after all the digital or analog signals, all signals are voltage and current waveforms. The only advantage the digital systems have is that they do not have to reconstruct the transmitted waveform as it is in the receiver; instead, they have to just find out whether 1 or 0 was transmitted.

6.4 Intersymbol Interference

Each pulse representing a 1 or a 0 is called a symbol in digital transmission. (Although a symbol may carry more than one bit [see Section 4.2.7], to develop an understanding of ISI, we will consider one symbol as one bit.) These symbols are transmitted in a sequence and are required to be detected distinctively at the receiver. However, due to various factors such as noise, distortion, jitter, and the shape of the transmitted waveform, the shape of the symbol received at the receiver input gets distorted to a great extent, including the spreading of the symbol in time domain. Thus, the symbols start interfering with each other. The phenomenon is called ISI. For any practical system, the ISI cannot be completely eliminated, but it has to be restricted within acceptable limits to get the desired performance of the link. Figure 6.2 illustrates the ISI phenomenon.

The transmitted signal shown in Figure 6.2 is an NRZ code where each 1 is represented by a positive pulse and 0 is represented by a negative pulse. Due to the factors stated above, the received pulses are distorted in shape and spread out in the time domain to overlap with the next or previous symbols for some portion of themselves. The receiver determines whether the received symbol is a 1 or a 0 depending upon whether the positive threshold is crossed by the pulse or the negative threshold is crossed, at the instance of rising edge of the clock. The amplitude of the spread over portions of the symbol can either add to the amplitude of the previous or next symbol or subtract from it depending upon the polarity of the next or the previous symbol. As can be seen, the first three pulses or symbols will get correctly

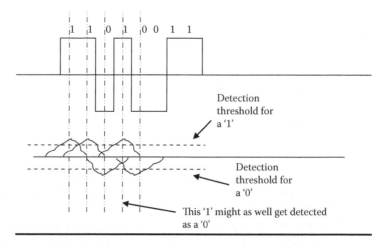

FIGURE 6.2
Wrong detection of a bit due to ISI.

detected as 1, 1, and 0, but the fourth symbol could be detected as either a 1 or a 0 (instead of a correct 1) because, even though the positive threshold is certainly crossed, the negative threshold might also be crossed because of the combined amplitudes of the spread over portions of the previous and the next pulses at the clocking instance of the present pulse.

The distorted shape of the received pulses gives rise to increased ISI as seen above. Hence, all the factors listed above in Section 6.3, which are responsible for causing distortion, are also responsible for ISI. However, the phase distortion (or delay distortion) is of specific importance with respect to ISI. In phase distortion phenomenon, the transmission medium allows the components of the signal pertaining to different frequencies to travel through it at slightly different velocities. This results in different frequency components reaching the receiver at slightly different timings, resulting in spreading of the pulse.

The shape of the transmitted pulses and the accuracy of the timing (clock) pulses to transmit and detect these pulses play a major role in increasing or decreasing the ISI. The better the timing accuracy, the lesser are the chances of ISI. The effect of pulse shape and the timing accuracy is explained in the next section.

6.4.1 Effect of Pulse Shape on ISI

The shape of the pulses actually transmitted in the baseband transmission system (pulse transmission system) is never a perfect square pulse. This is because the square pulse consists of components of infinite number of frequencies, requiring infinite bandwidth for transmission. However, in the practical communication systems, the bandwidth is not only almost always "limited," but bandwidth efficiency is also considered as a major factor in the performance parameters of the communication system. Thus, the bandwidth of the square pulses is truncated before transmission, which results in the change in shape of the pulse altogether. Figure 6.3 shows the input pulse to the band pass filter and the shape of the output pulse after passing through the band pass filter.

Before we discuss the shape of the output pulse and its implications on ISI, let us first see why we have chosen to have a pulse of width T as the input pulse and a band pass filter of bandwidth $1/2T$. Harry Nyquist established that in digital baseband transmission, the maximum symbol rate without ISI can be twice the available bandwidth. There will be $1/T$ symbols per second for a system with each symbol of width T seconds. Hence, the bandwidth required to transmit the $1/T$ symbols without ISI would be $1/2T$, or we can simply say

$$\text{Symbol rate} = 2 \times \text{bandwidth}$$

$$\text{or } Rs = 2BW$$

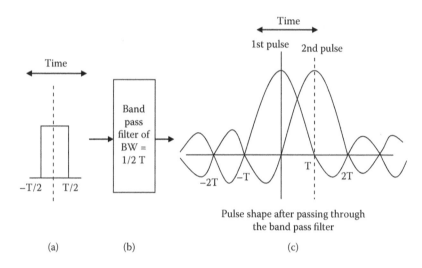

FIGURE 6.3
Shape of the square pulse before passing through a band pass filter (a) and after passing through a band pass filter (c). (b) Band pass filter.

where Rs is the symbol rate (also called Nyquist rate). (Do not confuse this with the Nyquist's sampling theorem, which states that any signal can be faithfully reconstructed from its samples if the sampling is done at twice the maximum frequency of the signal. This theorem is used for analog-to-digital [A/D] and digital-to-analog [D/A] conversion. The subject under discussion here is, however, the transmission of symbols without ISI and not A/D or D/A conversion.)

In Figure 6.3, take note of the shape of the square pulses after passing through the band pass filter. Besides the main lobe, which is widened to a width of 2T in shape, there are an infinite number of side lobes present, which are called "ringing" and are associated with the pulse. The ringing frequency is the same as that of the pulse. A second pulse following the first one also generates a similar pattern as shown. The whole sequence looks jumbled up, but Nyquist showed that the symbols (or pulses) can be recovered faithfully, despite the presence of so much ringing around them, if they are sampled (at the receiver) at the exact incidences where the adjacent symbols are zero crossing (having zero amplitudes). Thus, a maximum symbol rate of twice the bandwidth is very much achievable.

In the practical systems, however, slight timing inaccuracies are unavoidable. The side lobes of the adjacent symbols become relevant as soon as the zero crossing is violated in the sampling instances at the receiver, and gives rise to increased ISI in proportion to timing inaccuracy. Therefore, it is preferable to design the system with this knowledge. The solution is to allow a little larger bandwidth than the minimum required, and design the band pass filters with a specific frequency response so as to reduce the amplitudes of the side lobes and eventually end the side lobes in a finite time. A large variety of filter designs are available to tackle this issue, however, the one called "raised cosine response" filter is the most popular one. The response is called raised cosine because it is a function of COS^2 (cosine square) of the bandwidth (Figure 6.4).

Of these responses shown in Figure 6.4, the one with the bandwidth of $1/2T$ (the Nyquist BW) has amplitudes of the side lobes similar to those shown in Figure 6.3 for the ideal Nyquist filter. However, as the bandwidth is increased, the side lobes in the output of the raised cosine filter start reducing in amplitude. The second curve has a bandwidth 1.5 times the Nyquist bandwidth, and the third one has 2 times the Nyquist bandwidth. It can be seen that with the raised cosine filter in place and doubling of the bandwidth, the side lobes are quite suppressed in amplitude and they die out after some time instead of continuing indefinitely. This type of symbol can tolerate some amount of inaccuracies in the sample times at the receiver and still give close to "zero-ISI" performance.

In the whole link, what matters is the shape of the pulse/symbol at the input of the receiver detection circuitry and not at the transmitter output alone. Thus, the filters of transmitter and receiver are designed in such a way that they compliment each other, to present the best possible symbol at the detection circuitry of the receiver, taking into account the adverse effects of the transmitting medium.

One popular approach to reduce the ISI further is to make use of equalizers. The equalizer in the receiver estimates the pulse distortion and uses filters to clean the pulses to bring them to more acceptable shapes before they are presented to the detection circuitry.

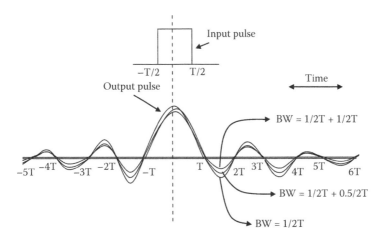

FIGURE 6.4
Shape of the square pulse after passing through the "raised cosine filter."

6.5 Jitter and Wander

In an ideal digital communication link, the phase of the received pulses should remain constant. When a multiplexed digital bit stream is acquired by a receiver, the pulses of the received signal do not exactly coincide with their ideal positions. The rising and falling edges of the received pulses should fall at equal distance with respect to the receiver clock. However, in practice, this seldom happens. The pulses arrive at the receiver with varying degree of phase differences. This difference in their phases is sometimes positive and sometimes negative. This variation in the phase difference from pulse to pulse makes them generate a shaky pattern on the oscilloscope giving the appearance of many pulses trying to coincide with each other, but not able to do so. Because of this shaky or jittery pattern they generate, they are said to be containing jitter. The pattern is shown in Figure 6.5.

In other words, the received pulses are "shaky" or "jittery." Thus, this phase difference is called jitter. It is the biggest nuisance to the correct detection of digital signals.

The phase difference in the received pulses and the consequent "jitter frequency" are illustrated in Figure 6.6. To accommodate one full cycle of jitter frequency, the phase difference shown in the figure is highly exaggerated.

The phase of the received pulses normally varies gradually with respect to the local clock. When this variation is fast, it is called "jitter," and when the variation is slow, it is called "wander." Normally, the variations with a frequency of more than 10 Hz (fast variations) are called jitter, and those with a frequency of less than 10 Hz (very slow variations) are called wander.

In this example (Figure 6.6), one cycle of jitter frequency is shown to contain only nine bits of the transmit signal, which is highly exaggerated for the purpose of understanding the principle of jitter frequency. Let us calculate what these values will be like in actual practice for the jitter or wander frequencies. For example, let us take a wander frequency of 10 Hz.

One cycle of this frequency = 0.1 s (1/10)

Thus, the number of bits of an E1 bit stream involved in generating one cycle of this wander frequency equal to 0.1 × 2,048,000 bits (an E1 bit stream is at 2048 kbps) or equal to 204,800 bits or 204.8 kilobits.

FIGURE 6.5
Received pulses with jitter as seen on the oscilloscope.

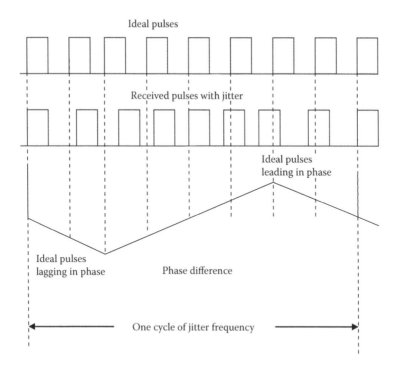

Ideal pulses

Received pulses with jitter

Ideal pulses
leading in phase

Ideal pulses
lagging in phase

Phase difference

One cycle of jitter frequency

FIGURE 6.6
Jitter in the received digital stream.

Hence, the total amount of wander will be spread over these many bits.

Both jitter and wander are potential dangers to the communication link, as they may not only harm the correct timing recovery but may also throw the whole system out of synchronization. Jitter also leads to increased ISI, which, in turn, increases the bit error rate (BER) aside from directly causing bit errors if not controlled properly.

The main causes of jitter and wander are the following:

(i) *Timing content pattern in the signal:* As we have seen in Chapter 4, various types of line codes have different patterns of timing contents. Other than the Manchester code, all codes have timing pulses that depend upon data because 1s and 0s are represented by such pulses, which may lead to absence of timing pulses for long intervals when there are a large number of repetitive sequences of 1s or 0s. These long intervals of absence of timing pulses give rise to increased jitter because locally generated clock pulses have to be inserted to fill these large gaps in the recovered clock (because the clock is required to have continuous pulses). Every time a pulse is externally inserted in the ongoing sequence, jitter is introduced because the phase of externally inserted pulse differs from the ongoing sequence due to implementation imperfections. (Whatever amount of accuracy is maintained in inserting the pulses, some variation in the phase of the inserted pulses is bound to come, with respect to the ongoing sequence, due to inherent tolerances of the electronic devices involved, noise, and other imperfections.)

(ii) *Instability of the transmit clock:* If the transmit clock is not very stable, it may introduce frequency variations in the transmitted signal, which will lead to variations in the phases of individual bits as compared with each other, leading to jitter and wander creation. Even "very stable" clocks are not "fully stable," thus some amount of jitter is always introduced by the clock instability. The effort should be to minimize it using as stable a clock as possible economically. Various types of clocks and their accuracies are discussed in Chapter 5. The stability of a clock is best at a specified temperature. Variations in the temperature cause deterioration in the accuracy/stability of the clock. The problem becomes more pronounced with the ageing of the clock, the older the clock the more the instability. Of all the causes of jitter and wander in the digital signal, the clock instability is the largest contributor.

(iii) *Noise, interference, and distortion:* The effect of noise, interference, and distortion on the analog signal has been discussed in detail in Chapter 2. The digital signal also gets affected by these factors in a similar way. The degradation causes the phases of the individual pulses to vary with respect to each other, causing jitter.

(iv) *Changes in the climatic conditions:* Variations in the ambient temperature during the course of 24 hours causes the transmission path length to vary slightly due to expansion and contraction of the metallic or OFC media. It also changes the properties of the media affecting the velocity of propagation of the signal. Both these factors affect the instances at which the pulses reach the receiver, giving rise to jitter and wander. The propagation delay in the metallic media (copper) is as much as 725 ps/km per 1°C increase in temperature, whereas in optical fiber media, it is much less at nearly 80 ps/km per 1°C increase in temperature. Owing to the large variation in the temperature during the course of 24 hours and large lengths of transmission media, there is a considerable variation in the propagation delay and consequently in the path length, giving rise to wander generation. (Although both jitter and wander are present, usually, the slow variation during 24 hours causes the frequency of variation in the phase of the received pulses to be much lower than the 10-Hz mark; thus, most of it falls in the category of wander.)

(v) *Jitter due to bit dropping and pointer movement:* Whenever timing adjustments take place due to reasons such as insertion of special timing pulses, removal of stuff bits, pointer adjustment, etc., jitter gets added to the signal because during these timing adjustments, the phase can never be matched perfectly in the practical systems as explained in (i) above. We will see in subsequent chapters on plesiochronous digital hierarchy (PDH) and synchronous digital hierarchy (SDH) that for equalizing the bit rates of E1 tributaries before multiplexing, additional bits are introduced in the tributaries. When these bits are removed at the demultiplexing stage, the clock has to have irregularity to some extent because some bits that were introduced as stuff bits for equalizing the bit rates of all tributaries are required to be removed. This irregularity of the clock for dropping the stuff bits results in introduction of jitter due to implementation imperfections during the readjustment of the clock. This type of jitter is called "waiting time jitter." During the demultiplexing process, to remove the "extra" or "stuff" bits from the multiplexed bit stream, all the bits are "written into" a memory element but only the required ones (minus the stuff bits) are "read out" to the output port. Thus, during the period of stuff bits, the equipment clock has to "wait" before giving the next "read" pulse, and the jitter generated during this process is called waiting time jitter.

In case of SDH, the bit rate equalization between tributaries is also achieved by means of a mechanism called pointer. Every time the pointer operation takes place, it introduces additional jitter due to similar reasons as stated above.

6.5.1 Control of Jitter and Wander

Jitter and wander are impairments that have to be lived with in the digital transmission systems. They cannot be completely eliminated, but they can be controlled within specified limits so as not to cause unacceptable system malfunctions.

The low-frequency jitter can be removed to a large extent using a special circuit called "phase locked loop," or PLL, which will be discussed shortly. (Note that the "frequency" being talked about here is not the frequency of the digital signal but the frequency of "variation" of the "phase" of received pulses with respect to the ideal pulses; see Section 6.5.)

However, the very low frequency jitter (i.e., wander) and the very high frequency jitter are beyond the capability of a PLL. In regenerators of digital signal, the outgoing clock is derived from the incoming clock. The jitter components of the incoming clock, which could not be removed by the PLL, get embedded in the outgoing clock. Thus, despite the deployment the PLL circuit on all the network elements (or nodes or regenerators as they are called) of a communication link, the wander and the very high frequency jitter are passed on from node to node and continue to accumulate. They may finally exceed values that can be easily tolerated by the input port of a subsequent node, resulting in slips (see Section 7.5 for a detailed discussion of slips), bit errors, or even loss of synchronization.

To remove this accumulated wander and jitter, a technique known as "elastic store implementation" (also called "slip buffer implementation") is used on a few nodes in the network at appropriate intervals. Elastic store is a buffer memory that is used to store the incoming bits of the digital signal at the receiver and transmit the output bits after completely retiming the pulses. (We will discuss the elastic store concept shortly.) This complete retiming of the pulses can remove all of the high frequency jitter. However, the wander, being of a low frequency, needs a very large amount of memory for elastic store, thus it may not be possible to eliminate the wander completely if it is beyond the limits specified by the ITU-T. The quickly varying jitter (high-frequency jitter), on the other hand, needs a small amount of memory for elastic store and thus can be easily tackled. The use of elastic store, however, introduces some amount of delay; thus, it cannot be used on every node or else the total delay in a long network will become very high, which will affect speech quality (like pauses in the speech in the international calls working on satellites). Thus, the elastic store is deployed on a few selected intermittent nodes and only when it is necessary from the point of view of removing the accumulated jitter and wander.

ITU-T has defined the limits of jitter and wander that can be allowed on the input port of a node in the network as a whole, irrespective of the number of repeater nodes preceding the said node. These limits specify the jitter at all the levels, i.e., very low frequency (wander), low-frequency jitter, and high-frequency jitter. The equipment is designed by the vendors so that they can handle the jitter and wander up to the limits specified by ITU-T (G.823 and G.825 recommendations for PDH and SDH systems, respectively). The equipment installed at the nodes with elastic store for removal of the accumulated jitter and wander have to be able to remove the jitter and wander completely, if they are within the limits defined by ITU-T, by completely retiming the regenerated digital signal.

Some other means of minimizing the jitter and wander are as follows:

(i) *Using scrambler:* The use of a scrambler (see Section 4.4.2 for scrambler principles) will minimize the pattern dependent jitter by removing long sequences of 1s and 0s.

(ii) *Minimizing pointer movement:* Minimizing pointer movement can reduce the pointer jitter. Pointer movement can be minimized in SDH (for pointer is described in Chapter 10 on SDH) by maintaining the tributary bit rate constant.

(iii) *Maintaining a proper environment for the system clock:* Maintaining a proper environment for the system clock, i.e., correct and constant temperature and humidity will minimize the so-called clock noise. ("Clock noise" is the variation in the clock frequency due to its instability. The instability increases with variation in temperature and ageing of the clock. The older the clock, the more unstable it is.)

(iv) *Using OFC as the transmission media:* Using OFC as the transmission media will minimize jitter due to path length changes and velocity of propagation changes.

All or some of these measures have to be deployed by the network operator to bring the jitter and wander within the limits specified by the ITU-T. If all these measures and deployment of PLL (which we will discuss in the next section) cannot bring the jitter below the specified limits, the number of repeaters in the network without the deployment of elastic store have to be limited, such that the jitter and wander are within specified limits. The elastic store then "cleans" the signal completely by retiming the transmitted pulses.

Let us now discuss the principles of jitter control through the mechanisms of PLL and elastic store.

6.5.1.1 Phase-Locked Loop

6.5.1.1.1 PLL Principle

The readers who have seen a traditional radio receiver will know how troublesome it is to properly tune the set—you tune a station and within a few minutes, the receiver is out of tuning, which creates more noise than music. This happens due to drift in the tuned frequency because of a change in the values of the tuning capacitor and inductor due to change in temperature and even due to the movement of the capacitor spindle due to vibrations. I have a similar radio. Somehow, my son loves radio more than TV, and listening again and again to more noise than music had always been frustrating. I could not throw the set away because it was expensive equipment with good deck and power stereo. I purchased a PLL FM receiver for just 80 rupees and used its earphone socket to feed the auxiliary input of the two-in-one stereo set, and it was perfectly tuned and remained tuned for as long as the set was on.

This is the magic of the PLL circuit, which is so effective and cheap. In fact, it is indispensable in most electronic equipment now. PLL is basically a feedback control system. In our club, we used to play a game where the participants were blindfolded and asked to break the earthen pitcher with a bamboo stick at a distance of 5–6 meters in a straight line. To watch them trying to hit the pitcher was real fun, as no one reached anywhere near the pitcher. But there were smart guys who could manage to keep a slit open for visibility from underneath the loose blindfold, and not only did they hit the pitcher but hit it precisely at the spot where it got shattered in no time. When someone was moving toward the pitcher with his eyes open, his brain received continuous feedback and the error in movement was continuously corrected. In the absence of open eyes, there was no feedback and the person landed up at wild distances from the object. This is how precisely a PLL works.

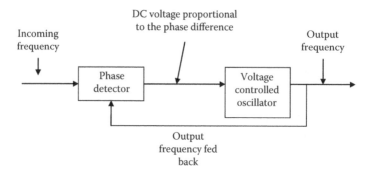

FIGURE 6.7
Principle or workings of PLL.

Figure 6.7 illustrates the basic arrangement that explains the workings of a PLL. A phase detector detects the phase difference between the incoming signal and the output of the local oscillator and produces a voltage proportional to the phase difference. The phase detector determines the difference in the timing of zero crossing of the two inputs and produces a voltage that is proportional to this difference. This voltage is applied to a voltage-controlled oscillator, the output frequency of which varies with this applied voltage. The higher the difference in the phase of incoming and outgoing signal, the higher the DC voltage produced and vice versa. This voltage gets applied to the oscillator and produces the positive or negative correction in the output frequency of the oscillator. By applying the correction, the phase difference is reduced and consequently less output voltage is produced for applying the correction. The cycle continues, and soon a condition is arrived, when the incoming and outgoing signals are completely in phase with each other.

As mentioned earlier, it is a wonderful technique available at a very low price, in the form of integrated circuits, and thus it is widely deployed.

Of course, the frequencies of the incoming signal should be close to that of a voltage controlled oscillator (local oscillator); otherwise, it could be out of the so-called "capture range" of the PLL circuit and the phase or frequency locking cannot be achieved.

6.5.1.1.2 PLL for Jitter Control

When the PLL is used to control jitter in the received signal, a low-pass filter is added between the phase detector and the voltage-controlled oscillator. The low-pass filter becomes necessary to eliminate the noise, which is at high frequency, and if not removed, may cause unpredictable output from the VCO. The modified arrangement is shown in Figure 6.8.

If the frequency of the incoming signal is slightly different from that of the local oscillator frequency due to jitter, there will be a phase difference in both of them with respect to each other. This phase difference is detected by a phase detector. The phase detector output is then passed through a low-pass

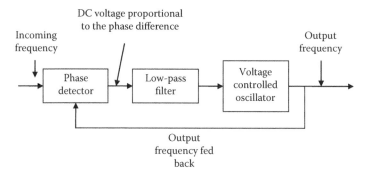

FIGURE 6.8
PLL for jitter control.

filter as mentioned above, and then applied to the VCO, to correct the frequency of the VCO in accordance with the phase difference. In other words the VCO output signal phase locks the incoming signal. The clock generated by the local oscillator (VCO) thus locks on to the timing of the incoming signal. This locking of timing of received bits with that of the local oscillator maintains the synchronization.

Note that the timings of the local oscillator adapt to the timings of the incoming signal bits and not the other way around, which means that it inherits the jitter of the incoming signal, although to a greatly reduced extent. Thus, if this recovered timing is used for the outgoing signal from the node (which is usually the case at an intermediate communication link node) the unfiltered jitter is passed on to the next node. The process repeats at a number of nodes, giving rise to accumulation of jitter, until it is removed by another means like elastic store.

Also, the jitter content of the incoming signal at lower and higher frequencies are passed on to the output signal due to the band pass nature of the low-pass filter. The wander, which is defined as the jitter at <10 Hz, is not fully removed by the PLL. Similarly, the very high frequency jitter is not removed by the PLL. The lower cutoff frequency of the low-pass filter is kept very low, so that it can remove the noise, interference, etc. from the incoming signal; however, it cannot be reduced below a certain value due to practical considerations. Thus, the wander and very high frequency jitter are passed on to the next repeater section.

The wander and very high frequency jitter thus keep on accumulating from repeater to repeater (node to node). The accumulated jitter and wander are finally removed by the use of elastic store at selected nodes. We will see the functioning of the elastic store in the next section.

6.5.1.2 Elastic Store

The "elastic store" or "slip buffer" implementation is an arrangement of buffer memory implanted at a few intermediate nodes in a synchronized telecommunication network to completely retime the bits of the received signal. This complete retiming facilitates the outgoing bits from the particular node to be pure and clean, i.e., free from any impairment whatsoever including jitter and wander.

All the incoming bits in a receiver are stored in a buffer memory by timing them through the recovered clock from the incoming bit stream itself. The outgoing bits from the node transmitter, however, are timed through an independent accurate clock whose frequency is the same as the average long-term frequency of the incoming bits. (Impairments such as jitter and wander create only short-term variations with respect to time. In a synchronized network, the total number of transmitted and received bits is the same on an overall average basis, although there could be short-term variations in the numbers due to jitter and wander.) The principle of operation of elastic store for this purpose is illustrated in Figure 6.9.

The incoming bits to the node receiver are stored in a buffer memory dedicated for this purpose. The incoming bits are written into this memory, the clock timings for which are the recovered clock timings from the incoming bit stream. This incoming bit stream contains jitter and wander. After the buffer memory is almost half-filled, the bits are read out to the transmitter as outgoing bits from the node. This reading out is done by timing the bits through a local clock with the same average frequency as that of the incoming signal. Since the read clock is totally independent of the incoming clock, the outgoing bits are totally jitter- and wander-free.

When the incoming bits are received at a slower rate as compared with the average rate, the buffer memory starts emptying. On the other hand, when the incoming bits arrive faster, the buffer memory starts filling more and more. The operation of the elastic store is on a first in/first out (FIFO) principle, which ensures that the last available bit is read out by the read clock, irrespective of its location in the buffer memory. The size of the buffer memory

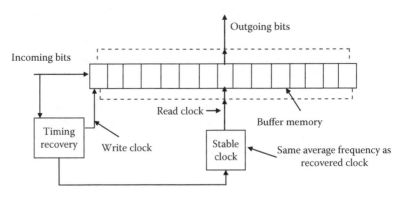

FIGURE 6.9
Principle of use of elastic store for jitter removal.

is kept such that it never "overflows" or "underflows" for the specified jitter and wander content in the incoming signal. While the high frequency jitter may need small buffer memory, the requirement may be much higher for the wander, which varies slowly, but varies to a much greater extent as discussed in the preceding sections.

An important point to note here is that the outgoing signal is delayed by a number of bits equal to half the length of the buffer memory on an average. The telecommunication networks could be very long, particularly for international calls, involving many operators and many countries. If every operator introduces a delay in the signal processing, the total delay may exceed the acceptable levels (like the delay in international calls made on satellite media cause speech pauses). Thus, the elastic store is not deployed at all the nodes, but it is deployed only at selected intermediate nodes such that the total jitter and wander are maintained within the permissible limits.

6.6 Eye Diagram

An "eye diagram," or an "eye pattern" as it is synonymously termed, is a wonderful tool for the qualitative assessment of the received signal quality in the digital communication, with respect to the impairments caused to it by various factors like noise, distortion, ISI, jitter, etc. The eye diagram is generated on an oscilloscope when a random pulse sequence is fed to the vertical input of the oscilloscope while the horizontal input is set as a time base of the same frequency as that of the pulse sequence. If an undistorted pulse sequence (locally generated) of square pulses, having positive and negative pulses representing 1s and 0s, is connected to the vertical input as stated above, the oscilloscope will display a box-like pattern. However, when an actual received pulse sequence is connected, the box degrades to look like an "eye," because of deshaping of pulses due to various impairments to the signal, as discussed in this chapter so far. Both the "box" and the "eye" pattern are shown in Figure 6.10.

The eye diagram or the eye pattern as shown in Figure 6.10 is generated because the successive pulses, distorted in shape, keep hitting the oscilloscope screen per their individual shapes. The cathode ray tube retains the display during successive hits due to its persistence. Many pulse shapes combine in the eye diagram to display the maximum and minimum limits of many factors.

The vertical opening of the eye indicates the margin available for clear detection of the signal from the noise or the "noise margin." The more the vertical opening, the more the margin and the better it is. The lesser the

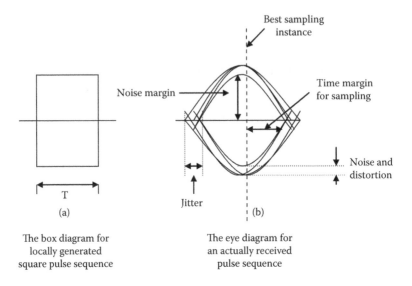

FIGURE 6.10
Eye diagram generated by a random pulse sequence on an oscilloscope.

opening, the more difficult it will be to detect the pulses from the noise. The horizontal opening of the eye gives the time margin for clear detection by sampling at the correct instance. The optimum sampling instance is in the center, but as we know, there are problems due to various impairments, thus the horizontal opening indicates the permissible deviation of the sampling instances. Again, the more the opening, the better it is. The top and bottom clusters of the eye indicate the noise and distortion in the received signal; obviously, the lesser they are, the better it is. Similarly, the left and right clusters of the eye indicate the time variance of the pulse or jitter. Again, the more the opening, the lesser is the jitter, and because all these deshaping factors of the eye increase the ISI, the more the opening of the eye, the lesser is the ISI.

6.7 Error Detection

Despite adoption of all the measures needed to avoid errors in the detection of the pulses or bits in the received bit stream, errors do occur. Thus, it is necessary to detect the errors in the received data and take necessary corrective action. Error detection is also necessary from the point of view of knowing the effectiveness of the error avoidance measures implemented in the system and investigating or strengthening them if necessary.

The methods used for detecting the errors in the received data (or bit stream) are (i) use of error detecting line codes, (ii) inserting parity bits, and (iii) cyclic redundancy check (CRC).

6.7.1 Use of Error Detecting Line Codes

Of many coding techniques that have been developed to code the digital signal for transmission, some provide inherent error detection capabilities. For example, the AMI, HDB-3, and CMI codes described in Sections 4.2.4 and 4.2.5 provide for error detection in the transmitted bit stream. An example of AMI code is illustrated in Figure 6.11.

As can be seen in Figure 6.11, any bit error that is caused by an extra pulse (a 1) where there is no pulse (a 0), or any missing pulse (missing 1) will give rise to a violation of the AMI rule. The AMI rule says that each 1 will have an opposite polarity as compared with its preceding or succeeding 1s. Thus, if any bit change takes place (in the media due to the various impairments) as mentioned above, there will be two consecutive pulses of the same polarity, which can be easily detected, indicating an error.

Of course if two consequent 0s are replaced by spurious pulses of 1s with opposite polarities than they will not be detected, but such happenings are very rare. Normally, the error will be of one odd bit or a burst of a large number of bits (due to impulse noise of lightening and so forth).

6.7.2 Inserting Parity Bits

One of the simplest and most popular methods of checking the "in circuit" performance of a telecommunication link is "parity check." Since this check is generally performed by bit interleaving of a particular set of bits, it is popularly called "bit-interleaved parity" (BIP).

Let us see how it is implemented.

The whole of the data of a frame is divided into a number of blocks. The number of blocks will be equal to the number of bits performing the parity check. Consider a data stream of 64 bits divided into 8 blocks for an eight-bit parity check.

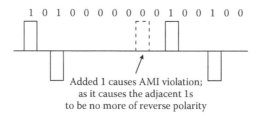

1 0 1 0 0 0 0 0 0 0 1 0 0 1 0 0

Added 1 causes AMI violation;
as it causes the adjacent 1s
to be no more of reverse polarity

FIGURE 6.11
Error detection through polarity violation in AMI coding.

Block number	1	2	3	4	5	6	7	8
	1	0	1	0	1	1	0	1
	1	1	0	1	0	0	1	0
	0	1	1	0	1	1	1	0
	1	0	1	1	1	1	0	1
	0	0	0	1	1	0	1	1
	1	1	0	1	0	1	0	0
	0	1	1	0	0	1	1	0
	1	1	0	0	0	0	1	1
Parity bits	1	1	0	0	0	1	1	0

FIGURE 6.12
Checking of errors through parity bits.

As can be seen in Figure 6.12, the 1s in each column are added, and if the total of the number of 1s in the column is an odd number, then a 1 is placed in the result row, i.e., the parity bit. Similarly, if the number of 1s is even, then a 0 is placed as the parity bit. This ensures that the number of 1s in a column including parity bit is "even." Thus, this is called "even parity." Conversely, if we wish to ensure the number of 1s to be odd in every column including parity bit, then it will be called an "odd parity." In this example, we can see that the total number of 1s in block number 1 are 5, i.e., an "odd" number, thus, a 1 is placed on the parity bit to make it "even." Similarly, the total number of 1s in block number 3 are 4, which is "even," thus a 0 is placed on the parity bit to keep it "even" and so on.

The parity bits so generated are transmitted along with the rest of the bits of the frame. At the receiver, calculations are performed again in the same way as are done at the transmitter. The parity bits received at the receiver from the transmitter and those generated locally are matched at the receiver. If both of them match correctly for all the bits independently for each parity bit, there is no bit error in the transmission, and if there is a mismatch, then there are blocks (a block is a set of a defined number of bits) with errors, in as many numbers as the number of mismatched parity bits.

Although the technique is very simple and straightforward, its accuracy is very limited. Suppose there are two bit errors in the block of data being checked by the parity bit. If both of these two errors are positive or negative, the parity will still detect an error, but if the bit errors are one positive and another negative, they will cancel each other and parity will show no error. Similarly, there could be a large number of bit errors, but as long as they balance themselves, no error will be detected by parity check. Also the number of bit errors is not known; all that is known is that there is some bit error. Then why is this method deployed in all modern systems? The answer is that despite its above-mentioned limitations, the error indications generated by the parity check are a good measure of transmission errors. More importantly, these indications enable performance monitoring without taking the circuit out of service.

In practical situations, we hardly ever need to know the exact number of bit errors happening in the system. All we are interested in is the general performance of the circuit. We need to know whether the errors are within the tolerable limits for a particular application or for a particular customer. The actual number of bit errors (BER) also tells us the same story. Thus, for making a decision about the action to be taken, parity check gives a reasonably good indication of circuit performance.

6.7.3 Through CRC

One of the most important procedures for error detection is CRC. A code word derived from a calculated manipulation of the data is transmitted along with the data. This code helps in checking the accuracy of the transmitted data. The code bits are redundant as far as the data are concerned. Because of this redundancy of the checking code and because its algorithm is based on cyclic codes, the procedure is called "CRC."

To generate the CRC code (or CRC check sum as it is popularly called), data are manipulated in a predefined manner through logical devices in such a way that a fixed number of data bits generate the CRC code. The CRC code so generated is also of a fixed number of bits, but much smaller than the data block. For example, in E1 transmission, a block of data of 2048 bits is logically manipulated to generate a 4-bit CRC check sum. This computation of the 4-bit CRC is done by using a predefined logical algorithm. The principle is illustrated in Figure 6.12.

This CRC check sum is then transmitted along with the data with clear identification of its location. At the receiver, the CRC check sum is separated from the data. The rest of the data are used to generate the CRC check sum again by deploying the same logic as that at the transmitter. This locally generated CRC word is then compared with the received CRC word. If the data bits of the block of data received have undergone any corruption in the channel leading to one or more bit errors, the locally generated CRC will not match with the received CRC, indicating an errored reception. Although there is a distant possibility that the CRC word (or CRC code or CRC check sum) may get corrupted in such a way that it gives the same result as the locally generated CRC from the errored bits of the data, this does not seem to pose a problem in practice because of its very low probability.

The limitation of CRC is that it only informs that there are bit errors in the received data, it does not tell us about the number of errors. Despite this limitation, it finds use in most modern communication systems.

A higher number of bits for CRC words, such as CRC 8, CRC 16, CRC 32, etc., are used for lengthier data blocks (Figure 6.13).

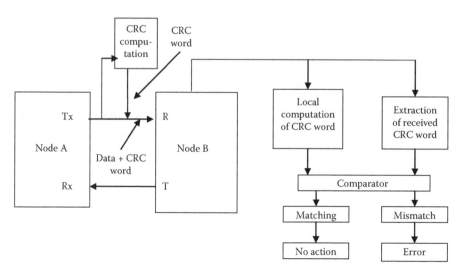

FIGURE 6.13
Error detection through CRC.

6.8 Error Correction

The detection of errors in the received data calls for either retransmission of the errored data frames or rejection of the complete set of data for the period concerned and repairing the circuit after shutting it down for some time in data communication applications. In speech circuits, it may cause intermittent inaudibility or circuit outage if the errors are numerous. Hence, if it is possible to detect and "correct" the errors in the received data, there will be a considerable improvement in the overall circuit performance.

Error correction can be performed on the received data by either (i) repetition of packets containing errored bits or (ii) forward error correction (FEC) or carrying out the error correction in the received signal itself.

6.8.1 Repetition of Packets Containing Errored Bits (Automatic Repeat Request)

In data communication systems, the data is divided into packets of a fixed number of bits or bytes. (For example, in an ATM packet, called a "cell," there are 53 bytes, which contain 48 bytes of data and 5 bytes of overhead.) Whenever bit errors take place, the errored bits are in some or other packets. On detecting an errored bit, the receiver immediately informs the transmitter for retransmission of the entire packet that contained an errored bit. The packet is retransmitted, and the packet containing an errored bit is simply

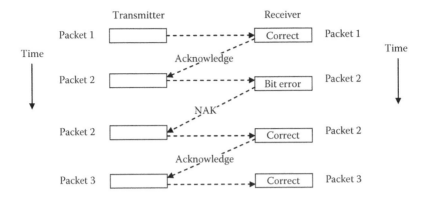

FIGURE 6.14
Retransmission process for packets containing bit errors.

discarded at the receiver. The procedure is illustrated in Figure 6.14 for one of the several protocols in use.

The transmitter transmits a packet and the receiver acknowledges it. The transmitter transmits the next packet only after the acknowledgment from the receiver that the packet had been received correctly (without any error). In case of a bit error, the receiver sends a different signal (negative acknowledgment, or NAK) to the transmitter indicating a bit error in the received packet, and the transmitter retransmits the same packet instead of transmitting the next packet. The process is called "automatic repeat request" (ARQ).

6.8.2 Forward Error Correction

The procedure discussed in the previous section (that of retransmission of data packets containing errored bits) cannot be adopted in real-time systems such as voice and video. In real-time systems, therefore, we have to devise a technique by which we can carry out the error correction in the received data itself, or simply live with the errored bits if they are not too many to cause appreciable quality degradation. Such a correction procedure is called FEC (in contrast with the kind of backward error correction by means of retransmission of packets).

The process involves inclusion of some redundant bits in the signal bit stream, which are to be used for error correction. Besides inserting extra bits, there has to be a mechanism to identify such bits within the whole bit stream and use them to correct the errored bits. The penalty to be paid is a higher number of bits than actually required for the transmission of data alone. Thus, reduced efficiency and higher bandwidths have to be acceptable.

The above-mentioned process of inserting extra bits and using them for error correction calls for the use of special line codes for the digital signal.

Such line codes are called structured sequences, which are further categorized in three categories called "block coding," "convolution coding," and
"turbo coding."

The block coding makes use of blocks of a fixed number of bits, say m,
to be transmitted as blocks of n bits, where $n > m$. For example, blocks of 3
bits each may be transmitted as blocks of 5 bits each. Since a 3-bit code will
generate $2^3 = 8$ codes and a 5-bit code will generate $2^5 = 32$ codes, only 8 of
the 32 codes are used for transmission. The selection may be done through
a look-up table. The error detection is done per the parity check procedure
explained earlier in this chapter. However, a complex analytical procedure
is adopted for determining the errored bit and replacing it with a proper bit.
The bandwidth used is 2/3 times more than what will be required without
block coding, and the coding efficiency is only 3/5 = 60%.

The convolution coding differs in the process of generation of code. Instead
of using a look-up table for determining an appropriate code, a logic circuit
is used to generate a sequence of higher bit rate, say 2 bits per data bit. The
bandwidth and the efficiency of coding are poor in this case as well.

Due to the disadvantage of higher bandwidth requirement and poor coding
efficiency, these FEC codes are used only where the channel is "noisy" and gives
unacceptable error rates with the use of normal codes, and also where the conventional methods of improving the error rate such as increasing the transmit
power, reducing repeater distances, etc. are either not feasible or prove unafford-
able. In normal long-distance communication systems on radio, optical fiber,
etc., these codes are rarely deployed. In case of optical fiber cable (OFC), where
the bandwidth is not a major issue, the network operator would still not like
to use these codes as they reduce the coding efficiency. The percentage of coding efficiency lost is actually the percentage of additional traffic lost that would
have been feasible otherwise, leading to revenue loss in the same percentage.
Moreover, in the long-haul backbone SDH systems used today on OFC, the error
rates are very low, and obtaining the error rates within the margins defined by
ITU-T is not usually a problem (we will see the acceptable error rates as defined
by ITU-T shortly in this chapter).

Notwithstanding the above argument, however, the FEC has made a big
comeback in the latest backbone technology, the optical transport networks
(OTN) standards. The deployment of FEC reduces the error rates substantially, thereby increasing the transmission length for the same acceptable
error rate limits for the same optical launched power. Higher transmission
distances means less expenditure on repeaters (less numbers).

Out of the many types of block codes developed, the "Reed–Solomon"
code is the most popular one. It is used in "digital video broadcast" (DVB)
and music CDs (for correcting disturbed tracks) and, most importantly, in
the latest OTN standards (RS 255/239 code). Another important code, the
Golay code, is used in deep space communication. The examples of uses of
convolution coding are satellite communications, military applications, GSM
mobile devices, and voice band modems such as V.32 and V.34, etc.

6.9 Link Performance

We have seen that the digital signal undergoes many kinds of degradations during its course of generation, processing in the transmitter, traveling
through the channel, processing in the receiver, and finally getting delivered
to the user. The square pulses lose their shape in the filters before leaving
the transmitter because of noise that gets added in the transmitter processing and in the transmission media, due to attenuation that takes place in the
antenna and channel and because of distortion due to various factors in the
channel. What ultimately needs to be ensured is that not only is the signal
received correctly (i.e., with acceptable and predefined error rates), but there
is also enough margin for the system parameters to tolerate reasonable variations in the functional and environmental conditions. A generalized communication link is shown in Figure 6.15.

For satisfactory reception of the signal as stated above, what matters finally
is the following:

(i) Whether the received signal strength is sufficient to be comfortably detected by the receiver. In other words, whether it is above the
receiver threshold (below which receiver cannot detect the presence
of a signal) and not only it is above the threshold but sufficiently above
the threshold (to have enough safety margin) for accommodating the
usual variations due to changes in the environmental conditions (such
as variations in the temperature, increased losses due to small defects,
etc.). The loss of signal, as we have already seen, is called attenuation.

(ii) The second factor is signal-to-noise ratio (SNR). This is the ratio of
signal power to noise power at the receiver input. The receiver has a
limited capability to distinguish signal from noise for an error free
(or defined acceptable error percentage) reception. If the SNR is less
than this designed value, the received signal will have more errors
than are acceptable.

Both these factors are expressed in decibels; thus, before discussing these
factors further, let us see what decibel is. Decibel (db) is a convenient unit
to represent ratios. For example, the ratio of two quantities A and B can be
expressed in decibels as

$$db = 10 \log A/B$$

Thus, if $A = 10$ and $B = 1$, the ratio (call it C) in decibels is 10 log 10/1 or
$C = 10 \times 1 = 10$ db.

If $A = 100$ and $B = 1$, C will be 10 log 100/1 or $C = 10 \times 2 = 20$ db

If $A = 1000$ and $B = 1$, C will be 10 log 1000/1 or $C = 10 \times 3 = 30$ db

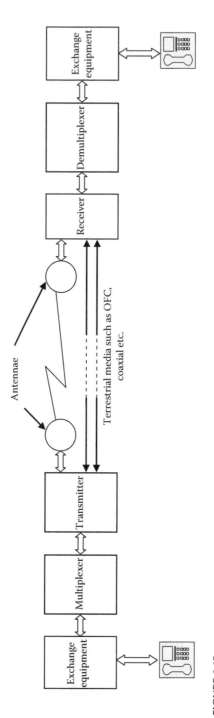

FIGURE 6.15
Full communication link with its components.

The above values of A and B have been specially chosen for illustration, as they will be very useful in visualizing various power levels when the reader comes across such relationships any time. Note that a ratio of 1000 is represented by 30. On extending it further, we can see that a ratio of 10^6 will be equal to 60 db and so on. This provides a phenomenal ease of plotting the graphs, which otherwise would have not been practicable on normally used paper sizes with original values like 10^6, 10^{12}, etc.

Although decibel represents a ratio only, it is used for representing absolute values of power as well by adding a subscript, with w (for watt) or m (for mill watt). Thus,

$$0 \text{ db}_w = 1 \text{ W}$$

$$10 \text{ db}_w = 10 \text{ W}$$

$$20 \text{ db}_w = 100 \text{ W}$$

$$30 \text{ db}_w = 1000 \text{ W}$$

and similarly

$$0 \text{ db}_m = 1 \text{ mW}$$

$$10 \text{ db}_m = 10 \text{ mW}$$

$$20 \text{ db}_m = 100 \text{ mW}$$

$$30 \text{ db}_m = 1000 \text{ mW or } 1 \text{ W}$$

In case $A = 10$ and $B = 100$ in the above equation, log 10/100 will be equal to –1; thus, C will be –10. It is very common to encounter such negative values when dealing in decibels.

SNR depends upon many factors as stated above, and it varies from system to system. Universal curves for SNR are shown in Figure 6.16.

As can be seen in Figure 6.16a, the SNR degrades as the length of the link increases because of addition of noise in the channel. Figure 6.16b shows the error rate variation with SNR. The higher the SNR, the lower is the error rate. These plots are true for all types of communication links, be it on copper wires, radio, or optical fiber.

Depending on the requirement and feasibility, SNR can be improved by increasing the transmitted power, reducing the link length, or switching over to a low-loss media. Improvement in the link can also be made by improving the receiver design to make it function satisfactorily with lower SNR. Normally, the links are designed to function with a margin. Let us say if the acceptable SNR to the receiver is 20 db, then the link's designed SNR should be 25 or 30 db for it to function with specified or better error rate.

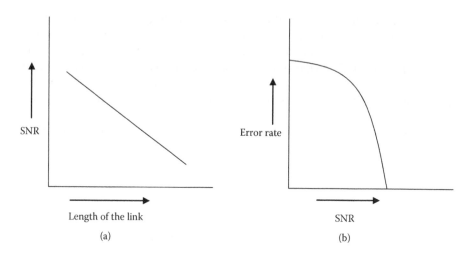

FIGURE 6.16
(a) Variation SNR with length of the link. (b) Variation of error rate as a function of SNR.

As far as attenuation is concerned, the designed signal strength at the receiver should be sufficiently above the receiver threshold for its satisfactory performance. For example, in optical fiber links operating at 1310 nm, the attenuation per kilometer is about 0.38 db. Thus, if the transmitted power is 0 db$_m$ (1 mW) and the receiver threshold is −30 db$_m$ (0.001 mW), the length of the link should be such that a margin of 8–10 db is available above receiver threshold. (The splices [optical fiber joints] also contribute to losses, and as such, some margin has to be allowed for these losses as well.) The link length can be calculated as follows:

Maximum losses that can be permitted = Trans power − receiver threshold
$$= -0 \text{ db}_m - (-30 \text{ db}_m)$$
$$= 30 \text{ db}_m$$

Margin for unforeseen system problems = 10 db$_m$
+ Margin for splice losses   = 3 db$_m$

Total margin = 13 db$_m$

Total losses that can be permitted in the media = Maximum losses that can be permitted − total margin = 30 − 13 = 17 db$_m$

Hence, the length of the link = 17/0.38 (the media losses being 0.38 db/km) = 44.73, or rounded off to 45 km.

This length of 45 km will, however, be subject to the SNR being within acceptable limits as stated in the previous paragraph.

6.10 Required Link Performance

To ensure that the telecommunication links perform to the acceptable levels, the limits of the acceptable levels of performance need to be defined so that all the manufacturers can design their equipment accordingly, and all the network operators can integrate the equipment adhering to those limits. Accordingly, the limits of acceptable levels have been defined by the ITU-T in their recommendation number G.821 and G.826.

All the factors that affect the performance of the digital communication link, as seen in the previous sections in this chapter, ultimately result in bit errors, and the higher the number of problems, the higher the number of bit errors and vice versa. Thus, the number of bit errors defines the quality or the performance of the link.

For the purpose of defining the acceptable levels of bit errors, ITU-T has considered a hypothetical link of 27,500 km length. This link is supposed to cross international borders. Of the total number of bit errors permitted in the end-to-end connection of 27,500 km long link, the percentages of contribution of national and international portions that should not be exceeded have also been defined by ITU-T. (ITU-T recommendation G.821 defines the acceptable bit error rates for the connections of 64 kbps or multiples of 64 kbps, up to 30 numbers of 64 kbps connections, i.e., up to an E1 containing 30 voice and two framing/signaling channels, also called ISDN/PRI connection. ISDN/PRI stands for International Subscriber Digital Network/Primary Rate Interface. The performance parameters for higher bit rate connections of PDH or SDH hierarchies are defined in G.826 and G.828 recommendations of ITU-T.)

The acceptable bit error rates/performance parameters are calculated by measuring the bit errors in a fixed interval of time, which may be a month or so. During this measurement interval, the total number of bit errors is measured for each individual "second" of time. The individual seconds with bit errors are defined as follows in G.821:

(i) *Errored seconds (ES):* These are the seconds found having one or more bit errors.
(ii) *Severely errored seconds (SES):* These are the seconds found having bit error rate of more than or equal to 1×10^{-3}.

The error performance parameters or the acceptable bit error rates are defined as follows with the help of the above two events.

(i) *Errored seconds ratio (ESR):* This is the ratio of the errored seconds (ES) to the total seconds in the measurement interval (a month or so). The acceptable limit of ESR is less than or equal to 0.08.

$$ESR \leq 0.08$$

(ii) *Severely errored seconds ratio (SESR)*: This is the ratio of the severely errored seconds to the total seconds in the measurement interval (a month or so). The acceptable limit of SESR is less than or equal to 0.002.

$$SESR \leq 0.002$$

(Another parameter called degraded minutes [DM] was initially introduced by ITU-T, which stood for those minutes in which the BER was greater than 10^{-6}. However, this was subsequently dropped, as there were hardly any cases reported having degraded minutes.)

As stated earlier, the above parameters are for a point-to-point connection at 64 kbps speeds or multiples thereof up to 30×64 kbps. For higher bit rates, the error performance parameters have been defined in G.826. Many will be discussed in Chapter 11 when we study the SDH performance monitoring. However, for complete details, interested readers may refer to the ITU-T recommendation G.826.

Review Questions

1. Depict through a schematic diagram how the digital signal gets affected due to transmission media and processing equipment.

2. Compare the possible impairments of digital signal with analog signal.

3. What are the common types of impairments caused to the digital signal between transmission and reception?

4. What is attenuation and why is it inevitable?

5. Other than transmission media, what are the factors causing attenuation?

6. What do you understand about distortion? What are the factors that cause distortion to the digital signal?

7. Explain the phenomenon of ISI through illustration. What is its significance?

8. What are the most significant factors affecting the ISI and why?

9. How does the shape of a square pulse change after passing through a band pass filter?

10. What are Nyquist rate and Nyquist's sampling theorem?

11. Why do received signals cause side lobes and how are they minimized?

12. What is "raised cosine" filter? How does it improve ISI?

13. Can an equalizer help in reducing ISI? If yes, how?

14. What is jitter? How is it different from wander?

15. Explain jitter frequency with illustration.
16. What are the main causes of jitter and wander? Briefly explain each.
17. Which one is the most predominant cause of jitter in digital signal and why?
18. Explain "waiting time jitter."
19. How are jitter and wander controlled? Are they required to be completely eliminated for practical systems?
20. What are the means of controlling jitter? Briefly explain.
21. What is PLL? Explain the principle or workings of PLL with the help of a schematic diagram.
22. Why PLL is so popular?
23. How is the PLL used for jitter control?
24. What is the purpose of elastic store in a communication network?
25. Explain the principle functioning of elastic store.
26. What is an "eye diagram," and what is its significance?
27. Why is error detection required? What are the methods?
28. Can the errors be detected by using a particular code? If so how?
29. What is parity? Where is it used and how?
30. What are the limitations of parity check as an error detection mechanism?
31. What do you understand about BER?

Critical Thinking Questions

1. Illustrate the phenomenon of ISI with the help of a code different from the one explained in this chapter.
2. Compare the effect of all the factors causing distortion, pulse shape, and timing accuracy on ISI with the approximate percentage in which they will affect ISI.
3. Can the ISI be completely eliminated by increasing the bandwidth many folds? Discuss.
4. Can the jitter cause transmitter and receiver to go out of synchronization? If yes, how?
5. What are the impairments that cannot be seen on the eye diagram and why?
6. Design your own error-detecting code.

Bibliography

1. D. Roddy and J. Coolen, *Electronic Communication*, Fourth Edition, Prentice-Hall, India, New Delhi, 2001.
2. T.C. Bartee, *Digital Computer Fundamentals*, Tata McGraw-Hill, 1977.
3. W.H. Hayt, J.E. Kemmerly, and S.M. Durbin, *Engineering Circuit Analysis*, Sixth Edition, Tata McGraw-Hill, 2008.
4. J.G. Proakis and D.G. Manolakis, *Digital Signal Processing Principles, Algorithms, and Applications*, Third Edition, Prentice-Hall, India, New Delhi, 2005.
5. P.V. Shreekanth, *Digital Microwave Communication Systems with Selected Topics in Mobile Communications*, University Press, 2003.
6. H. Taub and D.L. Schilling, *Principles of Communication Systems*, Second Edition, Tata McGraw-Hill, 2005.
7. M.E. Van Valkenburg, *Network Analysis*, Third Edition, Prentice-Hall, India, New Delhi, 1974.
8. B. Sklar, *Digital Communications "Fundamentals and Applications,"* Pearson Education, 2002.
9. J.C. Bellamy, *Digital Telephony*, John Wiley and Sons, Singapore, 2003.
10. J.D. Ryder, *Network Lines and Fields*, Prentice-Hall, 1975.
11. S.B. Wicker, *Error Control Systems for Digital Communication and Storage*, Prentice-Hall, 1995.
12. S. Ramo, J.R. Whinnery, and T. Van Duzer, *Fields and Waves in Communication Electronics*, John Wiley and Sons, 1994.
13. G. Kennedy and B. Davis, *Electronic Communication Systems*, Fourth Edition, Tata McGraw-Hill, New Delhi, 2005.
14. S. Haykin, *An Introduction to Analog and Digital Communications*, John Wiley and Sons (Asia), 1989.
15. B.P. Lathi, *Modern Digital and Analog Communication Systems*, Oxford University Press, Oxford, UK, 1998.
16. S. Salivahanan, A. Vallavaraj, and C. Gnanapriya, *Digital Signal Processing*, Tata McGraw-Hill, 2000.
17. ITU-T Recommendation G.113, *Transmission Impairments*, International Telecommunication Union, Geneva, Switzerland.
18. ITU-T Recommendation G.102, *Transmission Performance Objectives and Recommendations*, International Telecommunication Union, Geneva, Switzerland.
19. ITU-T Recommendation G.120, *Transmission Characteristics of National Networks*, International Telecommunication Union, Geneva, Switzerland.
20. ITU-T Recommendation G.821, *Error Performance of an International Digital Connection Operating at a Bit Rate below the Primary Rate and Forming Part of an Integrated Services Digital Network*, International Telecommunication Union, Geneva, Switzerland.
21. ITU-T Recommendation G.826, *End-to-End Error Performance Parameters and Objectives for International, Constant Bit-Rate Digital Paths and Connections*, International Telecommunication Union, Geneva, Switzerland.

7

Synchronization

Synchronization is perhaps the most important, most complex, and most confusing about subject of digital communication. This is a subject in which a real concept building is required if one has to understand or maintain the digital communication systems. The need for synchronization arose with the advent of digital communication itself. The digital communication signal consists of a stream of digits or bits, which are either 1s or 0s, i.e., a pulse or no pulse. When this pulse train leaves the transmitter of one telecommunication node and reaches the receiver of another telecommunication node, the receiver needs to know the location of each bit in the pulse (or bit) train precisely. As all the bits of the received signal look alike, if a mistake is made in identifying a bit, the whole sequence of bits, when converted to analog signal, will produce only garbage. Let us take an arbitrary example, where each bit represents a letter (Figures 7.1 and 7.2).

The transmitted sentence was "Will you tell a story," and if the receiver placed the second bit in the position of first bit, the received sentence becomes "illy out ella s torya." You can see that the meaning of the message is completely gone. Although this is not an example of a practical system, and is exaggerated, it most appropriately demonstrates the effect of loss of synchronization in a digital communication system.

Thus, a bit-by-bit correspondence has to be maintained between each and every pair of transmitter and receiver for a practical communication system to be feasible, and the process of achieving this correspondence is called synchronization.

To maintain the bit-by-bit correspondence or to achieve the synchronization, we need to detect the incoming signal pulses to the receiver correctly. These pulses are called "symbols" because they are, more often than not, not simple pulses but are different types of pulse combinations generated by different types of line codes (see Chapter 4 for details of line coding). In case of carrier communication systems the line codes are modulated with different types of modulation schemes giving rise to very different types of transmitted symbols. Thus, if we are able to detect these symbols correctly we can achieve the desired synchronization, which we may call "symbol synchronization."

The exact phase of the received digital signal is not of much importance as a little bit of variation in the phase will not affect the correct detection of the symbol. However, in case the digital modulation deployed is of the phase shift keying (PSK) type, the detection of the correct phase of the signal becomes important, without which the symbol cannot be detected correctly. In this

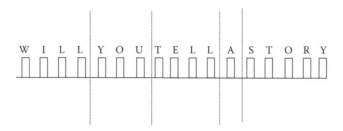

FIGURE 7.1
Transmitted sequence.

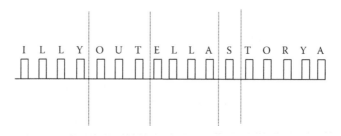

FIGURE 7.2
Received sequence. A mismatch of 1 bit in the transmitted and received signal produces garbage.

case, the detection has to be the so called "coherent detection," in which the phase correspondence has to be ensured. We will see about these variations in subsequent sections. After the symbols are detected correctly, the timings of the bits have to be matched with the timings of the receiver clock to achieve continuous synchronization between the transmitted and the received bits. This process is called timing recovery. Further, to be able to use the received bit stream for extracting the transmitted information, the bit stream is required to be framed. The framing process is the one that finally enables us to achieve the bit-by-bit correspondence between the received and the transmitted signal.

7.1 Synchronization Process

Thus, the synchronization process involves the following steps:

(i) Correct detection of symbols
(ii) Timing recovery and clock synchronization
(iii) Framing

Let us see them in brief.

7.1.1 Correct Detection of Symbols

To correctly detect the symbols from the received signal, the following steps are involved:

(i) Demodulation
(ii) Filtering
(iii) Amplification
(iv) Equalization and pulse shaping
(v) Decision making

These steps are illustrated in Figure 7.3.

7.1.1.1 Demodulation

When the digital base band signal is modulated, such as in the case of the transmission on microwave radio, optical fiber, etc., it needs to be demodulated first at the receiver to get back the envelop or the base band signal before further processing. In the case of pulse transmission, however, this step is not required as the signal is already in the base band form.

The demodulation process will be different for different types of modulations used, such as frequency shift keying (FSK), amplitude shift keying (ASK), phase shift keying (PSK), etc. The demodulators for each of these types of modulations are discussed in Chapter 4 on line coding and digital modulation.

The process of FSK detection is sometimes called frequency synchronization, and that of PSK detection is similarly called phase synchronization. Let us see their difference in brief.

7.1.1.1.1 Frequency Synchronization

Let us take a case of FSK modulation. The requirement of the receiver in this case is to be able to clearly identify the frequency of the received signal, so as

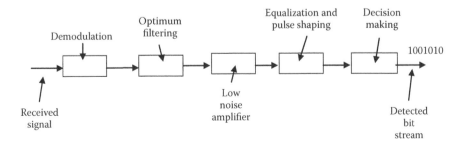

FIGURE 7.3
Steps involved in detection of symbols.

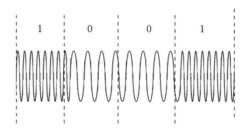

FIGURE 7.4
FSK modulation, having different frequencies representing 1s and 0s.

to determine whether the particular frequency of the received signal at the given moment is the one pertaining to a 1 or a 0. The phase of the received signal with respect to the transmitted signal is not very important. Even if a cycle or two of the transmitted signal are missed in the received signal, it will not make a difference, as the decision about whether the signal is a 1 or a 0 is based on its frequency and not phase. For example, in case of a 34-Mbps system on microwave radio of 7-GHz frequency, we will have nearly 200 cycles of sinusoidal carrier signal for transmission of each bit. The FSK modulation scheme will change this number slightly for the 1s and 0s, but the overall number of cycles will remain nearly in the same range. Thus, if the received signal has a phase difference with respect to the transmitted signal, it will not affect the decision of the receiver about the bit being a 1 or a 0, as long as the frequency is correctly identified. This is called "frequency synchronization." Figure 7.4 illustrates the frequencies of 1s and 0s.

The precision and accuracy required in the receiver circuit are much less as compared to the phase detection discussed next. This leads to simpler designs of the receiver, resulting in lower costs.

7.1.1.1.2 Phase Synchronization

In case the modulation scheme deployed by the transmitter is PSK, the receiver will necessarily have to detect the phase of the incoming signal to be able to make a correct decision about the signal bit being a 1 or a 0. This is owing to the fact that the PSK modulation merely changes the phase of the carrier signal by a predefined amount to differentiate a 1 from a 0 or vice versa. This phase difference could be 180, 90 degrees, etc., depending upon the specific PSK scheme (see Section 4.3.3 for PSK details). The PSK symbols are shown in Figure 7.5.

The detection in this case has to be much more precise; otherwise, it may lead to an incorrect decision. This type of detection where the phase of the received signal is detected correctly is called "coherent detection." Coherent detection is the most precise one among all the categories of detections. It needs to deploy more complex and precision circuitry; hence, it is obviously costlier. The PSK has an advantage over FSK in that it requires lower

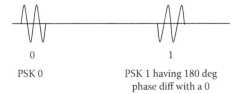

0   1

PSK 0   PSK 1 having 180 deg
                             phase diff with a 0

FIGURE 7.5
PSK symbols identified by phase difference.

signal-to-noise ratio of up to 4–5 dB for the same error probability as that of FSK. (Both types of synchronizations discussed above deploy a circuit called "phase locked loop" (PLL) for synchronizing the receiver with the transmitter. While the frequency synchronization deploys a simple PLL, called a first-order PLL, the phase synchronization deploys a much more accurate PLL, called second-order PLL. The principles of functioning of PLL have been explained in Section 6.5.1.1.)

7.1.1.2 Filtering

The received waveform is fed into an optimum filter, which is actually a two stage filter. The first stage is a normal filter that removes the unwanted frequency components from the signal. The second stage is a matched filter, which is designed specifically to suit the type of received waveform; this filter smoothens the signal further and maximizes the signal-to-noise ratio.

7.1.1.3 Amplification

In long transmission sections, the amplitude of the received signal is very low, necessitating amplification before further processing. Since the process of amplification amplifies the noise besides amplifying the signal, the amplifiers used are special low noise amplifiers to keep the signal-to-noise ratio from deteriorating.

7.1.1.4 Equalization and Pulse Shaping

After amplification, the signal is fed into an equalizer with a pulse shaper. This is a circuit designed with the knowledge of the characteristics of the transmission medium. The distortion caused by the transmission medium is more or less deterministic (meaning, its characteristics are known). Thus, the equalizer is designed to have inverse characteristics of that of transmission medium in as far as the distortion is concerned, so that the signal undergoes inverse of the distortion caused due to media, with the net result that the effect of distortion due to media is more or less nullified. This process helps in minimizing the channel-induced intersymbol interference (ISI).

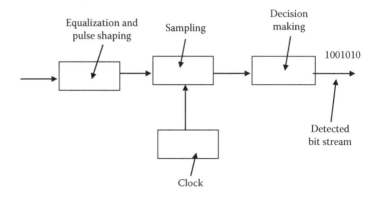

FIGURE 7.6
Process of decision making.

7.1.1.5 Decision Making

After the previous four steps, the signal is sampled at the clock frequency, and depending upon its amplitude and the type of code employed, its value is determined as a 1 or 0. The process is shown in Figure 7.6.

The clock, however, may not have exactly the same frequency as that of the received signal, which is almost always the case, and thus it is necessary to match the timings of this clock with the frequency of the received signal. This matching is performed by deriving a clock signal from the received signal. The process is called "clock recovery" or "timing recovery" and is discussed in the next section.

7.1.2 Timing Recovery and Clock Synchronization

Timing recovery or clock recovery implies the reconstruction of the transmitter clock accurately at the receiver. This is inevitable for achieving the synchronization between the transmitted and received bit streams.

As seen in the previous section, the better shaped pulses after the equalization process are sampled with the help of a local clock, to determine their amplitudes at specific clock intervals, which are supposed to fall in the middle of the pulses. The decision making circuit then decides whether the signal pulse is a 1 or a 0, and the bit stream is reconstructed.

However, the matters are not that simple. If the clock does not precisely match the transmitter clock, after a few samples, the sampling clock pulses may not fall in the middle of the received pulses, with some of the pulses not getting sampled or some getting sampled twice. Unfortunately, the practical clocks, however accurate they are, can never have exactly the same timings. Thus, the receiver clock needs to be synchronized to the transmitted clock. For this purpose, the timing or clock has to be recovered from the received bit stream, and the receiver clock has to be synchronized with it.

To recover the timing from the received signal, the received signal is fed into a zero crossing detector after the pulses are refined by the optimum filter and the equalizer. Before feeding to the zero crossing detector, the pulses are reshaped to improve the zero crossing performance. The zero crossing detector captures the timings of the zero crossing of the signal pulses and drives an oscillator. The output of the oscillator is converted to square wave to generate the clock pulses at the recovered timings. The arrangement is shown in Figure 7.7. The clock generated from the recovered timings is called "line clock."

Although the line clock can be used for sampling theoretically, in practice, a local oscillator is required for two reasons:

(i) The incoming signal may not have a zero crossing after every bit/ symbol, as many of the line codes have no voltage level for indicating zeros. It is in fact the sampling process that determines whether the bit is a 1 if the voltage present is positive (or negative or either, as the case may be; refer to Chapter 4 for a discussion of line codes, or a 0 if no voltage is present at the sampling instant. Thus, the sampling process requires a clock pulse at the defined periodicity of pulses or at the bit rate.

(ii) The equipment at any particular location may have to run on their own clock in case a communication from the node supplying the clock or the timing information fails due to some reason.

The local clock so required has to have good enough accuracy to perform satisfactorily in the absence of the timing pulses from the source transmitter. We will discuss these aspects in detail subsequently in this chapter. For the time being, it is enough to understand that a good local clock is inevitable in any receiver.

This local clock is the one that is used for sampling shown in Figure 7.7. To maintain the accuracy of detection, this local clock has to be synchronized

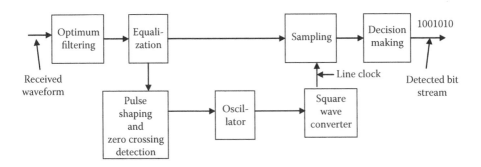

FIGURE 7.7
Timing recovery from the received signal.

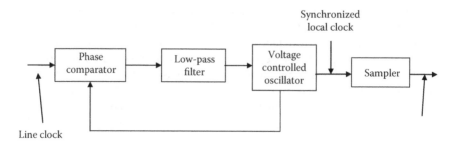

FIGURE 7.8
Synchronization of line clock with local clock using PLL.

to the line clock. This synchronization is done by means of a PLL (for a discussion on PLL, see Section 6.5.1.1). For this purpose, the local clock uses a "voltage-controlled oscillator" (VCO). The line clock and the local clock are both fed to a phase detector, whose output is a voltage that is proportional to the phase difference in the phases of the two clocks. This voltage, when applied to the VCO, corrects the phase of the local clock to bridge the difference, which leads to minimizing the phase difference. Successive pulses finally lead to a situation when the local clock is in "phase lock" with the line clock. This is the situation of complete synchronization of the transmitter and receiver. Figure 7.8 illustrates the scheme. (Instead of recovering the clock from the received signal, the synchronization can be achieved by transmitting the clock signal separately and then synchronizing the receiver clock with it. This, however, adds heavily to system costs and hence remains only a theory in digital systems, as practically all systems use recovered clock from data [bit] stream. The system has not been discussed here; however, essentially, it uses the same principles of bit timing coincidences.)

7.1.3 Framing

With the clock recovery and detection of 1s and 0s, we are able to correctly reconstruct the received bit stream. However, this bit stream will be of little use if we do not know where the stream begins and ends. A wrong counting of bits may lead to a situation similar to the one shown in Figure 7.1.

The job of the demultiplexer in the receiver is to separate all the individual channels from a multiplexed bit stream, so that they can be fed to the respective user circuits. Any mistake in identification will result into a wrong connection or a garbage reproduction. Now, how does the receiver know as to which bit pertains to which channel because all the bits look alike, and rightly so because they are so designed. To help the receiver identify various channels, the concept of framing has been devised. In fact, this is a fundamental building block of any digital communication system, without which no practical digital communication system can work.

The continuous digital bit stream is broken into small blocks of a definite length called "packets." With each of these packets, a certain number of bits are added with a predefined and fixed pattern in the beginning of the packet (in some cases, they may be at the end too). The blocks so formed are known as "frames." The extra bits that are so added are called "framing bits." Thus, the entire bit stream is broken into frames of fixed sizes. When this bit stream is received by a receiver, it looks for the framing bits, which have a fixed predefined pattern, say, 1111010000. When the receiver comes across this pattern in the bit stream it starts identifying the bits of the complete frame per the predefined length of the frame. After completing the counting of first frame it identifies the framing bits of the second frame, only to reconfirm the desired bit pattern. Thus, counting and identifying the individual channel bits from their respective positions next to the framing bits enables the receiver to assign a particular bit to a particular channel, and thus identify and separate the individual channels.

Let us consider a digital bit stream from which the bits are being continuously received. Let us say it is a 2-Mbps stream consisting of 32 numbers of 64-kbps voice channels. (A 2.048-Mbps bit stream is the first level of digital communication in the European system and is popularly called an E1 or a 2-Mbps bit stream. It is also popularly referred to as a PCM. Although PCM stands for "pulse code modulation," the E1 system was called by that name due to the first-time use of PCM technology.)

In the case of a 2.048-kbps signal, the period for one sample of the voice signal, i.e., 1/8000 s, or 125 µs, is taken as a frame. Thus, the bit stream flows at the rate of 8000 frames/s.

If the receiver of the bit stream is able to identify the beginning of the frame, then it can count and find which is the first bit of the frame, second bit, and so on, and it will be able to identify various channels with the help of the fixed definitions per the standards and separate different channels.

This identification of the beginning of the frame is done by the receiver by observing a predetermined pattern of bits in the incoming stream as stated above. Let us say that a particular sequence of bits (see Figure 7.9) has been

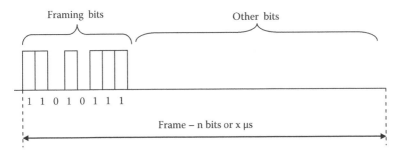

FIGURE 7.9
Framing bits, in a frame of n bits or x microseconds.

inserted in each and every frame during transmission of data. The receiver will continuously look for this sequence, say, 11010110, in the incoming bit stream, and wherever it matches with this sequence, it will take it as the beginning of the frame. The frame identification will be repeated for a few frames to confirm that the bit pattern is actually arriving from the framing bits and not from the data bits, as the possibility of receiving the same bit pattern in data bits cannot be ruled out.

Thus, the bit to bit correspondence is ensured in the received data and the synchronization is achieved.

7.1.4 Use of Multiframes

The above scheme of framing results in a wastage of many bits in each frame, which cannot be used for the data. This additional burden on the bit stream is called overhead. (The overhead may include many more bits for many other purposes. In fact, all the bits carried in a frame other than the data bits are called overhead bits. Many a times the words "bit stream" and "data stream" are used synonymously; however, whenever the subject is of defining the duties of various bits separately, data bits are meant for "user data," while the other bits such as framing, overhead, etc. are mentioned separately.)

The framing bits as described in the previous section are always required for achieving the synchronization of the digital signal, and they are required to be provided in every frame. However, there are some functions other than the signal synchronization, such as signaling of individual speech channels and so forth, which are not required to be performed in every frame (i.e., 8000 times every second). Such functions may be performed a little more slowly, say, a few times every second. These functions are performed by including the concerned function bits once after many frames, say, after 16, 32, 64 frames, etc. Such blocks of frames are called multiframes.

These multiframes are also required to be "framed" for the same reason of identification of individual bits. Obviously, the framing bits for multiframes are distributed in many frames (multiframes), for example, the final bit of each frame may carry one bit of a particular sequence for the purpose of framing. Thus, for the bit sequence of Figure 7.9 (if it is sent one bit in each frame) to be received, the receiver will have to look for a minimum of 8 frames. Although this process results in certain wastage of information contained in a few initial frames, which have to be passed before the beginning of the frame could be located, the savings on the number of bits on all the subsequent frames (which may run into billions) is a much larger gain (see Figure 7.10). However, both types of systems are in use for different applications.

The bits used for the purpose of frame identification are called framing bits. The type of framing shown in Figure 7.10 is called multiframe framing.

The above discussions bring out the general principles of framing. In actual practice, there are hundreds of framing patterns, practically a different

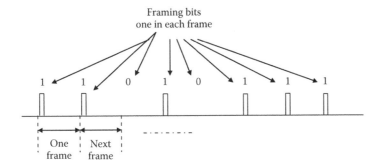

FIGURE 7.10
Framing using one bit in each frame: multiframe framing.

pattern for each application. But none of them are difficult to understand if the above principles are understood. (It is recommended that further contents of this chapter should be read after reading the chapters on PDH and SDH for better appreciation.)

7.2 Synchronous vs. Asynchronous Systems

We have learned that the digital communication cannot work without synchronization. However, some systems are called asynchronous systems, and others are called synchronous systems, depending upon the process of achieving the synchronization.

7.2.1 Asynchronous Systems

In case the transmission of digital signal is in frames of bits that are not continuous (such as data communication), the synchronization can be easily achieved. For example, let us say the receiver clock is of stratum 4 accuracy (in North American standards, the accuracy of clocks is defined as "stratum 1" to "stratum 4", with stratum 1 being the most accurate and stratum 4 being the least), which is defined to have an accuracy of ±32 ppm, i.e., ±32 parts in 10^6. If our data frames are transmitted one by one with some gap in between, the synchronization will be required only for one frame at a time, and with every frame, there will be a new beginning of synchronization, with the intervening gap absorbing the difference in the clocks rates. Frames are usually of small size, e.g., an ATM frame is 53 bytes long, which is equal to $53 \times 8 = 424$ bits. A typical asynchronous network is shown in Figure 7.11.

Now if both transmitter and receiver are of ±32 ppm accuracy, in the worst case, they will have a variation of $32 + 32 = 64$ bits in 10^6 bits.

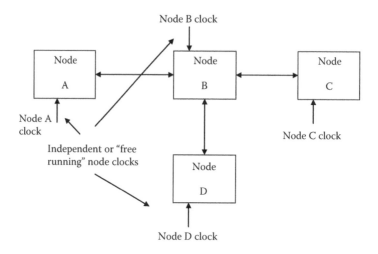

FIGURE 7.11
An asynchronous network in which each node is working on its own "independent" or "free running" clock.

Thus, in 424 bits, they will have a variation of $64 \times 424/10^6 = 0.027$ bit at most. Which is negligible because $[100 - 0.027] = 99.97\%$ of bit width of the last bit of the frame is available for the receiver to recognize. Thus, data transmission can take place without any problem. Generally, the frames or groups of the frames are not very long, leading to satisfactory data reception with reasonably accurate clocks; however, better clock accuracies are needed for large frame length protocols.

This methodology is very popular in data transmission systems with discrete data packets being transmitted. But the problem comes when the signal transmitted has continuous bits, such as that of voice communication TDM signals (or even the data being carried on TDM signals). The problem is more serious if the data rates are high, e.g., E2 rate of 8 Mbps or E3 rates of 34 Mbps. In such a case of continuous bits, the difference in the frequencies of the free running node clocks eventually leads to loss of data, which is called "slips" (free running node clocks, slips, etc., are described in detail subsequently in this chapter). Then the question is how do the systems get synchronized? Obviously, they work by improving the clock accuracy and tolerating the slips or controlling them in such a manner that they do not create many problems. This will be clear as you read through the chapter.

Such asynchronous systems have the advantage of not requiring line synchronization, but they suffer from poor accuracy. In the past, the provision of synchronization through highly accurate clocks and their distribution in the network was an extremely costly affair, but with the passage of time, the highly accurate timing sources such as atomic clocks and timing through

GPS have become quite affordable. Thus, the present trend is to go for synchronized systems normally. Asynchronous systems using highly accurate clocks for their nodes (to minimize clock mismatches) continue to be in use presently (called plesiochronous networks described subsequently in this chapter) but are on their way out, being replaced by synchronous networks called synchronous digital hierarchy (SDH) due to numerous advantages of the synchronous networks, which you will learn in the chapter on SDH.

There is a school of thought that started advocating the asynchronous networks again because of their simplicity, not requiring synchronization, and the development of protocols that can carry real-time services such as voice and video on data networks. However, most experts still recommend (and the industry uses) synchronization because of accuracy of transmission, effective use of bandwidth, and the present low cost of synchronization devices. (This is mentioned here for the purpose of completeness. Some of the terms may not be clear now but they will be as you read through the subsequent sections.)

7.2.2 Synchronous Systems

Let us have a look at Figure 7.12. The clock recovered by the timing recovery mechanism from the incoming bit stream from node A can be used to drive the local node clock through PLL. Thus, a bit-by-bit correspondence or synchronization is maintained. The clock of node A is called "slave clock." This process is called master–slave synchronization of the nodes, and the system comprising A and B nodes is called a synchronous system.

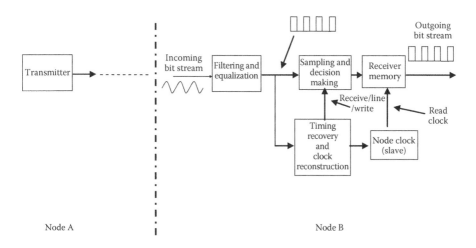

FIGURE 7.12
Synchronous system with master–slave synchronization.

Note that there is a difference in a synchronous system and a synchronous network. Let us see what is implied by a "synchronous network."

7.3 Synchronous Networks

While discussing synchronous systems, we used the words *system* and *network* synonymously. Although, in general, the meaning of both the words may be taken as the same, let us define them separately for clarity of understanding of concepts.

Let us call a single location of equipment as a "node" (for which there is no confusion), any two nodes functioning together as a "system," and more than two nodes functioning together as a "network."

It is necessary to understand the difference because in a large asynchronous network, it may be possible that a few nodes are operating synchronously or even a particular node may operate synchronously with one node and asynchronously with another node.

As opposed to the "asynchronous" networks, the local clocks of "all" the nodes in a "synchronous" network are driven from a master clock. Figure 7.13 illustrates the principle. (This is only an illustration of the principle; in practical systems, the approach is a little different, which will be clear after reading Section 7.8.)

When we say "network," it generally means a large number of nodes spread over wide geographical areas.

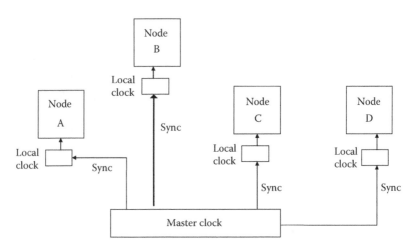

FIGURE 7.13
Local clocks of all the nodes are synchronized with a master clock, in a synchronous network.

Synchronous networks are generally deployed using SDH, which will be discussed in detail in Chapter 10.

7.4 Plesiochronous Networks

We have seen that no digital communication system can work without synchronization. The systems that depend only on framing and whose nodes do not receive timing from a master clock are called asynchronous systems, and those systems and networks that derive their timings from a master clock are called synchronous systems or networks.

The plesiochronous networks fall in the category of asynchronous networks because all the nodes in a plesiochronous network operate independently on their own clock in "free running" mode. The asynchronous networks shown in Figure 7.11 and redrawn here as Figure 7.14 also represents a plesiochronous network.

The word *plesiochronous* is from the Greek word *plesio* meaning "nearly the same" and *chronous* means "time." Thus, the meaning of *plesiochronous* is "nearly the same time," which is taken for "nearly the same frequency." It means that the nodes in a plesiochronous network operate at nearly the same frequency. The network nodes by this definition use very high accuracy clocks. At higher rates of plesiochronous transmission such as 34 Mbps or higher, the node clock accuracy is recommended to be of the order of ± 1 part in 10^{11}.

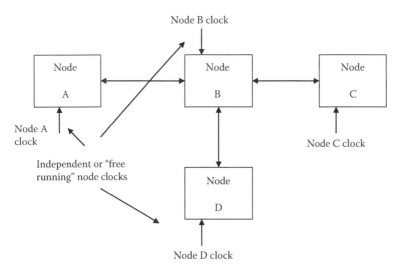

FIGURE 7.14
Asynchronous/plesiochronous network in which each node works on its own "independent" or "free running" clock.

Clock accuracies have been defined by the strata orders such as strata 1, 2, 3, 4, etc. in the North American standards for SONET applications, whereas ITU-T has defined the clock accuracies in terms of primary reference clock (PRC), synchronization supply unit (SSU), synchronization equipment clock (SEC), etc. in their recommendations of G.811, G.812, and G.813 for SDH applications. The details of clock accuracies are dealt with in Section 5.5.

This philosophy of plesiochronous network was considered the best at that time because of the following reasons:

1. The cost of network-wide synchronization, by distributing the timings from a master clock, was considered too high because the systems were always justified on the basis of the cost of small networks.

2. The need for synchronized networks was not really a pressing need because although the long-distance lines are connected through digital transmission, the interexchange connectivity remained analog through E&M signaling or loop signaling, thereby easily tolerating the slips. (The problem with the slips became intolerable only when the fully electronic exchanges were interconnected at the digital level, i.e., T1 and E1 connectivity.)

3. The cost of highly accurate clock sources to be deployed in individual nodes was consistently coming down.

These factors made plesiochronous networks the ultimate choice of the time, and based on plesiochronous digital hierarchy (PDH), which will be discussed in detail in Chapter 8, if became popular all over the world, although in different formats in different countries.

However, as stated above in item 2, the occurrence of slips in the interconnectivity of the digital exchanges gave rise to unacceptable quality standards, as expectations for the quality of the services were consistently growing. The telecommunication operators then chose to build their own synchronization systems within the PDH hierarchies. This limitation and the other limitations of the PDH networks about extraction and insertion of tributaries (in PDH networks, adding or dropping of a tributary involves a complete demultiplexing and remultiplexing of all the stages; this will be detailed in Chapter 8), the growing demands of higher and higher capacities, stabilization of the cost of the highly accurate clocks (the decreasing cost was a good reason for its acceptance, which stood no more), deployment of GPS satellites (for timing signals), and the part introduction of synchronization by the telecommunication operators as mentioned above were the reasons that paved the way for a new synchronized system of networks called SDH. (In North America, there is an equivalent standard known as SONET, i.e., "Synchronous Optical Networks." In fact, SONET was the first standard developed that paved the way for SDH.) Thus, plesiochronous networks are almost obsolete. Whenever they still exist, they are being replaced by SDH.

7.5 Slips

When the clock frequency of the incoming data stream and that of the local clock do not match perfectly, there are slips. Let us say the incoming signal "lags" the local clock by 1/100 of a bit during each clock pulse. Thus, the incoming signal will lag by 1 bit after 100 bits of a local clock. In due course of time, this lag causes some bits to be missed by the receiver, which are called slips. Now if the frequency of local clock was 100 bps, then there is 1 slip/s. Although in actual practice, the number of slips is much less owing to the high accuracy of clocks; nevertheless, they are a big problem, as they can throw entire data out of synchronization if not tackled properly.

This was an example of negative slips, but if the incoming bits were "leading" instead of lagging, it would have resulted in positive slips. Slips also occur due to the impairment of the transmitted signal in the transmission media such as jitter and wander, but the numbers are much smaller as compared to those due to poor accuracy of free running node clocks.

You may wonder that if the frequency (or bit rate) of incoming and outgoing signals are not the same, how is the synchronization maintained? Synchronization is achieved by temporarily storing the incoming bits in a buffer memory and then transmitting them out one by one, maintaining the bit by bit sequence properly (see Figure 7.15).

The buffer memory is filled to half of its capacity if the bit rates of the incoming signal and outgoing signal are the same. If the incoming signal is faster, more memory is filled and the outgoing location is accordingly shifted toward the right. When the incoming bit rate is less than outgoing, less than half the memory is filled, and the outgoing location is shifted leftward. Thus, although a little delay is introduced due to serial storage in memory, bit-by-bit correspondence or the sequence is maintained.

If the bit rates are so different, the memory is either completely filled or completely emptied. This is explained in the next section.

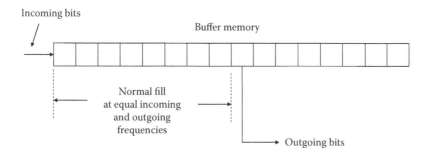

FIGURE 7.15
Use of a buffer memory to maintain synchronization.

7.6 Elastic Store and Controlled Slips

In asynchronous or plesiochronous networks, the incoming bits to a network element (node) are stored in a buffer memory, called "elastic store," temporarily for a period of one or two frames, and then they are clocked out using local clock. The arrangement helps in maintaining synchronization, by allowing slips to occur in a controlled manner. (This type of elastic store is also used in synchronous networks but only at a few selected nodes. This storage eliminates jitter and wander, making the outgoing signal clean; see Section 6.5.1.2 for a discussion on jitter removal using elastic store.)

When network elements use a small buffer memory, which cannot accommodate up to two complete frames of data (in case of E1, it is $256 \times 2 = 512$ bit long), uncontrolled slips occur. Thus, for slips to occur in a controlled manner, an elastic store of more than two frame lengths is deployed. This is explained as follows.

Let us consider a case where the buffer memory or elastic store used can accommodate only two-thirds of a frame, and the arrangement is to clock in/out half of it at a time, i.e., one-third of a frame at a time. (The data is "clocked in" in the buffer by the incoming clock and "clocked out" from the buffer by the local clock.) Thus, the buffer is divided into two parts: on an incoming clock, one-third of the frame is clocked in, and on the second pulse of incoming clock, the second one-third of the frame is clocked in, and between these two "clocking-in" operations, the data from the first part, i.e., the first one-third of the frame, is "clocked out." (The "clocking in" and "clocking out" clocks are not to be confused with the "write clock" and "read clock," which are supposed to be at the frequency of the bit stream. Clocking in and out of elastic store is an internal arrangement and the frequency used is a fraction of the bit stream frequency, depending upon the length of the elastic store.) (Figure 7.16). Normally, pulse 1 of incoming clock "clocks in" the data, and pulse 1 of local clock follows and "clocks out" the data.

If the local clock is faster, the pulse number 2 of the local clock appears before the pulse number 2 of the incoming clock, causing retransmission of the one-third frame F1 part I. The receiver on the other end will, however, be expecting part II, thus looking for framing bits after the next part, i.e., part III of F1 frame. Now since one-third of the frame has been retransmitted, the bit sequence is actually lost. The synchronization gets lost and the receiver and the transmitter have to begin the process of frame alignment altogether again, which is a time-consuming procedure and the result is a loss of a large number of frames, as the resynchronization may take several seconds.

To avoid this problem, the width of elastic store is increased to two frame lengths, and a technique called controlled slips is used. A complete frame is either retransmitted or lost, but synchronization is maintained. The loss of one frame as such is much smaller a loss, compared to the loss of synchronization, which results in a loss of thousands of frames (Figure 7.17).

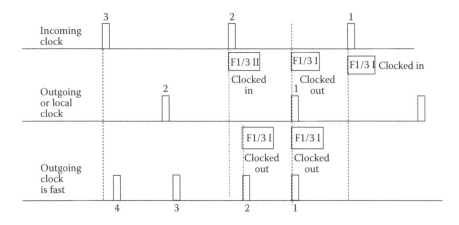

FIGURE 7.16
Uncontrolled slips.

We can see in Figure 7.17a and b that when the incoming and outgoing clock rates are the same, the frames are clocked in serially. When F1 and F2 are clocked in, the outgoing clock "clocks out" F1 after some time, then the next pulse of incoming clock "clocks in," F3 and F2 is moved to the outgoing position and F1 is deleted from the buffer.

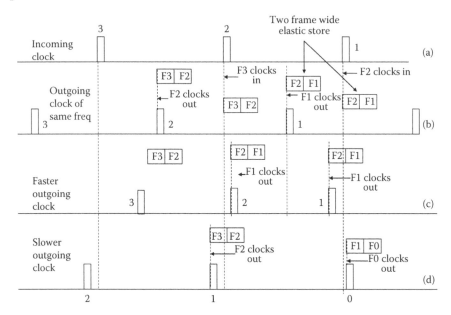

FIGURE 7.17
"Controlled slips" operation allows the loss of a complete frame at a time. (a, b) Process when the incoming and outgoing clock rates are same. (c) Process when the outgoing clock is faster than the incoming clock, and (d) process when the outgoing clock is slower than the incoming clock.

When the rate of outgoing clock is faster than incoming clock over a period, pulse 2 of the outgoing clock occurs before pulse 2 of the incoming clock, leading to "clocking out" of the frame 1, which was already "clocked out" by pulse 1, thus leading to retransmission of one frame (Figure 7.17c).

When the rate of outgoing clock is slower than incoming clock (Figure 7.17d) over a period, pulse 1 of the outgoing clock lags behind the pulse 2 of the incoming clock. Thus, pulse 1 of the outgoing clock "clocks out F2," whereas pulse 0 of the outgoing clock had "clocked out F0." Thus, frame F1 is not transmitted at all, leading to a loss of one frame.

In actual practice, the above steps of retransmission and loss of frame are deliberately performed when the said pulses are actually about 16 bits away from their position required for these actions. The frame is retransmitted or omitted, and the pulses brought back to their normal position, until the next such operation takes places and so on.

7.7 Line Synchronization vs. Networks Synchronization

Presumably, and it must be clear, line synchronization and network synchronization are different subjects. However, in view of a large number of jargons used, it would be appropriate to separately explain their meaning clearly.

The arrangement shown in Figure 7.13 is an example of line synchronization, i.e., synchronization on a particular transmission line with defined nodes. It is also called "node synchronization" or "data synchronization."

This type of synchronization arrangement may exist in many discrete applications, such as connecting two isolated digital exchanges through digital trunk line connections (i.e., an E1 or a T1 connection). It could also be a part of a big asynchronous network where these nodes are synchronized to each other, and the rest of the nodes connected to these nodes are not synchronized. (E1 is a composite signal format comprised of 30 voice + 2 signaling channels of 64 kbps per European standards, and a T1 is a 24-channel composite signal per American standards.)

On the other hand, network synchronization talks about synchronization of the whole network or at least that of a particular telecommunication operator. This envisages the distribution of timing pulses throughout the network from one common "master clock." Thus, at any given time, all the bits flowing in all directions are all synchronized.

However, in actual practice, it is not possible to connect all the nodes of the network to a single master clock simply because of the large number of nodes and their geographically diverse locations. Thus, the timing distribution is done by the master clock with the assistance of secondary timing units called SSU or stand-alone synchronization elements (SASE). Figure 7.18 illustrates the principle.

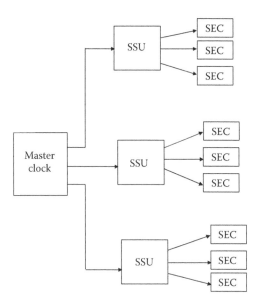

FIGURE 7.18
Timing distribution in network synchronization.

The master clock supplies the timing to the SSU, which in turn supplies timings to the actual nodes and the local clocks at the nodes, which are called "SEC." All these clocks have been defined in ITU-T recommendations G.811, G.812, and G.813. (For details of the clock accuracies, please see Section 5.5.)

7.8 Types of Network Synchronization

Depending upon the method of implementation, synchronization may be categorized as follows:

(i) Master–slave synchronization
   (a) Using primary and secondary reference clocks
   (b) Timing distribution through GPS
(ii) Mutual synchronization
(iii) Plesiochronous synchronization

We have read about plesiochronous operation (see Sections 7.1 and 7.4) in detail.

Let us have a brief look at the principle of master–slave synchronization and mutual synchronization.

7.8.1 Master–Slave Synchronization

7.8.1.1 Using Primary and Secondary Reference Clocks

The process explained in Section 7.7 through Figure 7.18 is nothing but master–slave synchronization. It is also called "hierarchical master–slave synchronization" because it uses a hierarchy of clocks, i.e., master clock, SSU, SEC, etc.

This is by and large the most dominant form of synchronization in all the modern networks such as interconnection of digital exchanges, SDH, etc. More details about its implementation will be discussed subsequently in this chapter under the heading "network synchronization engineering."

7.8.1.2 By Timing Distribution through GPS

The basic arrangement remains the same as discussed in previous sections. The only difference is that the master clock is replaced by a timing source that receives the timing information from GPS satellites (Global Positioning System), called GPS/PRC.

GPS is a cluster of low-orbit satellites maintained by the U.S. Department of Defense. The primary purpose of these satellites is to facilitate identification of the position of any object on the earth. It has become a very popular system today worldwide and has found numerous applications.

The GPS satellites house a number of very highly accurate timing sources (of PRC accuracy) based on the cesium beam technology. The timing signals in the satellite are also regularly monitored and synchronized by the ground-based deployment of a cluster of PRC clocks, which are of highest accuracy. Besides cesium beam sources, hydrogen-based sources are also used. (The accuracy of cesium beam PRC is generally better than that specified by ITU-T G.811, i.e., $\pm 1 \times 10^{-11}$. Generally, it is in the range of $\pm 2 \times 10^{-12}$, but it is possible to find accuracies up to a few parts in 10^{13} or 10^{14}. With the same specifications, the actual accuracy varies in this range because of manufacturing tolerances going in favor or against. It is only a matter of picking up the proper GPS clock from the lot.)

Thus, the accuracy of the GPS clock is as good or as bad as that of autonomous PRC.

However, in practical deployment, a mix of the two is preferred. The cesium beam PRC can be easily afforded by the big network operators. Thus, a few PRCs backed by a few GPS clocks are deployed, both of which take charge of each other's outages.

7.8.2 Mutual Synchronization

In this method of synchronization, all the nodes of the network have highly accurate clocks that keep exchanging the timing information with each other. Finally, the network synchronizes on a common average frequency (see Figure 7.19).

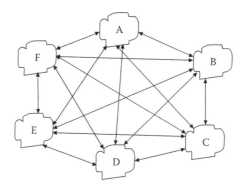

FIGURE 7.19
Mutual synchronization.

The synchronized frequency is highly stable, and the network is not affected if one of the nodes or its clock fails. This type of synchronization is considered the best from the point of view of reliability. However, it is also the most complex one to implement; thus, it has not found much practical use and there are a very few systems in the world that are working on this philosophy.

It is the master–slave synchronization which has found the most practical implementation.

7.9 Pseudosynchronization

All the modern TDM networks are now adopting SDH and thus implementing the "master–slave" synchronization, in which all the nodes of the "entire" network are synchronized to one master clock. But what happens when there has to be a communication to or from another operator's network? The other operator has a similar system of master and slave clocks. If these two networks have to synchronize with each other, one of the operators has to accept the clock of the other operator as a master clock, but that is again not possible because there are some operators connected at the other end of his network. So whom will he be taking as his master clock supplier? Obviously, it is not possible for him to accept any other operator's clock as a master clock (the only possible exception could be an operator whose network is connected to only one other operator).

Then there are international boundaries where again the networks change. Thus, it is not possible to operate all the networks of the world on single master clock timings.

Thus, the only solution is to go for plesiochronous connectivity between operators. Since all the networks within themselves will operate at very accurate master clocks (generally, cesium beam clocks with an accuracy of $\pm 1 \times 10^{-11}$ or better), the number of slips are restricted to bare minimum (see Figure 7.20).

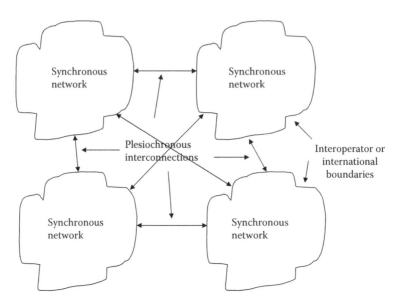

FIGURE 7.20
Pseudosynchronization in international networks.

For this purpose, ITU-T has defined the maximum number of slips that can be allowed on such international connection in recommendation G.822, "Controlled Slip Rate Objectives for International Digital Connection."

7.9.1 Permitted Number of Slips in Pseudosynchronization Network

According to the G.822 standard, only 1 slip in 70 days is allowed (per the accuracy of the PRC clock defined in G.811; for clocks details, see Section 5.5; slip calculations are explained in the next section in this chapter) for every plesiochronous communication. An international connection has been defined by ITU-T G.801. Its length is assumed to be 27,500 km and is assumed to have 13 plesiochronous connections.

For this international connection, the number of permissible slips for each network will be 70/12 (12 connections for 13 nodes) = 1 slip in 5.8 days. However, it is also perceived by the ITU-T that there may be some nodes that are a part of the same synchronization network, and in actual practice, there may not be as many independent nodes. Thus, the total number of international connections have been presumed to be 4 for the purpose of this definition, which leads to 70/4 = 1 slip in 17.5 days. However, the worst-case limit on number of slips due to plesiochronous operation alone continues to be 1 slip in 5.8 days.

Performance category	Mean slip rate	Period of time (total time 1 year)
(a) Rate-1	<5 slips in 24 h	>98.9%
(b)	>5 slips in 24 h <30 slips in 1 h	<1.0%
(c)	>30 slips in 1 h	<0.1%

FIGURE 7.21
Required slip rates for 64 kbps international connection.

Further, it has been realized by ITU-T that in practical networks, the number of slips may actually exceed due to various reasons such as configuration of the international digital networks, national timing control arrangement, jitter and wander, operational characteristics of various types of switches and transmission links, and temporary disturbances such as network rearrangements, protection switching, errors, etc. Considering all these, the required rates have been relaxed by ITU-T per Figure 7.21. The measurement is to be done over a period of 1 year to conclude the rate.

Although these are only recommendations of ITU-T, good manufacturers of equipment try to give a better performance than these recommendations. The operators have to choose proper equipment and strategies to meet these requirements to survive at the international level.

Thus, plesiochronous operation is the only practice that provides a practical means to make an international connection feasible. We can hope for a day when all the telecommunication operators of the world will synchronize with a common master clock and there will be no slips.

7.9.2 Slip Rate Calculations

In the preceding section, we have seen the maximum allowable number of slips under various conditions in a digital communication. How are these rates, such as 1 slip in 70 days and so forth, arrived at? Let us see.

The PRC clocks, per G.811, has to have an accuracy of ± 1 part in 10^{11} parts (also written as $\pm 1 \times 10^{-11}$). With this limit, how long will it take to produce 1 slip? Before that, we should know what is meant by 1 slip. We have seen in Section 7.5 that the difference in clock rates of transmitter and receiver nodes produces slips, i.e., if the transmitter is sending the bits at 100 bits/s and the receiver's local clock is running at 101 bits/s, there will be 1 slip/s.

We have also seen the concept of "controlled slips" in Section 7.6. "Controlled slips" means that slips are counted in number of frames or 1 slip means the slip of one frame and not 1 bit. Whenever we talk of number of slips, it is always the number of "frame slips." The number of bits per slip may vary depending upon the frame length, i.e., number of bits in one frame. For this purpose, the frame is also taken as the standard frame of 125 µs. Recall the sampling theorem in Chapter 3, where it was mentioned that the sampling rate for a voice signal is standardized at 8000 samples/s. These 8000 samples/s work out to 125 µs between each sample. This 125 µs is further time division multiplexed to give a 30-channel PCM (E1), and a much higher order of multiplexing is carried out in SDH/SONET. However, the frame length continues to be 125 µs in higher-order multiplexing (excluding PDH), and thus there are 8000 frames/s. Thus, when we talk of slips in SDH systems, it is in terms of this frame of 125 µs.

Let us now calculate the slip rates:

(i) For a PRC clock with $\pm 1 \times 10^{-11}$ accuracy, the worst case difference will be 2 parts in 10^{11}, when the connecting nodes/networks have a perfect mismatch, or in terms of frame slips, there will be 2 frame slips in 10^{11} frames.

Or there will be 1 slip in $10^{11}/2$ frames.

As there are a total of 8000 frames in 1 s, there will be $(8000)/10^{11}/2$ slips/s.

Or there will be 1 slip in $10^{11}/2 \times 8000 = 6{,}250{,}000$ s $= 72.33$ days (nominal slip rate per ITU-T is 1 slip in 70 days).

(ii) Similarly, the frequency accuracy of a SDH equipment slave clock per G.813 is ± 4.6 ppm, i.e., ± 4.6 parts in 10^6 parts.

Or $(4.6 + 4.6) = 9.2$ slips in 10^6 frames (the worst case).

Or we can say there will be 1 slip in $10^6/9.2$ frames.

Since there are 8000 frames in 1 s, there will be

$10^{(6)}/9.2$ frames in $1/8000 \times 10^6/9.2 = 13.5$ s

Or there will be 1 slip in 13.5 s.

Similar calculations can be done for any given accuracy of the clock.

7.10 Synchronization Network Engineering (Planning a Synchronous Network)

We have so far seen many aspects of synchronization. Now let us have a look at how to put the various pieces of information together to plan the synchronization scheme for a practical network, which is called "synchronization network engineering." The subject is too vast to be contained in this text;

however, we will try to see in brief the principles involved. (For a detailed study, you may refer to ITU-T recommendation G.803.)

7.10.1 Network Synchronization Areas

The entire network of the operator may be driven either by a single master clock or it may be divided into a number of "synchronization areas." The choice will basically depend upon the size of the network. It will be desirable to break the network into "synchronization areas" if it is too big geographically and has a large number of nodes. Both the options have their advantages and disadvantages. The big network will avoid "plesiochronous" working between synchronization areas (see Figure 7.20; synchronization networks A–D may be considered as "synchronization areas" of one network), thus minimizing the total number of slips in the network. On the other hand, the smaller "synchronization areas" connected together with plesiochronous operation will have a greater number of slips but will be easy to manage (the distribution of clock will be simplified), but they will have longer "synchronization trails," which means a larger accumulation of jitter and wander (the accumulation of jitter and wander is proportional to the number of nodes in the synchronization chain), thus more number of slips.

Hence, the operator has to make a choice between the two philosophies based on his circumstances.

7.10.2 Synchronization Chain

If operating conditions are ideal, there is no limit to the number of nodes, which can be synchronized to a master clock. However, the environment limits this. ITU-T G.803 defines a synchronization chain for this purpose (Figure 7.22).

Although the maximum number of SSUs permitted is restricted to 10 and the number of SECs between two SSUs is restricted to 20, the total number of SECs in a synchronization chain is not allowed to increase beyond 60. (Master clock SSUs and SECs are explained in the next section.)

7.10.3 Master Clock SSUs and SECs

7.10.3.1 Master Clock (PRC)

The master clock of the synchronized network is the clock in which the timing signals of all the clocks of the network are synchronized. It should have an accuracy of ± 1 part in 10^{11} or better; per G.811, this is called a PRC (in the United States, it is called stratum 1 accuracy). In practice, "cesium beam" atomic clocks are deployed for this purpose, which generally have a better accuracy than this requirement. The master clock has to continuously verify its timing signal with Universal Time Coordinated (UTC) using a

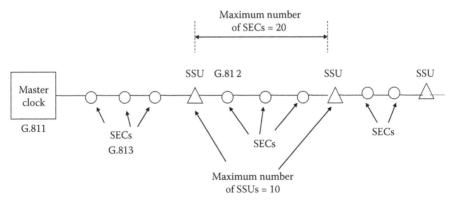

FIGURE 7.22
Limits recommended by ITU-T for maximum number of SSUs, SECs, and total number of nodes.

UTC-based navigation system such as GPS or LORAN-C. (UTC is the time scale maintained by the "Bureau international des poids measures" [BIPM] and the International Earth Rotation Series [IERS], which foresees the basis of a coordinated dissemination of standard frequencies and time signal. We have already discussed GPS in Section 7.10. LORAN-C is a land-based navigation system that was developed for maritime navigation. It is operated by the U.S. Coast Guard. It also uses the highest timing standards of cesium beam atomic clocks).

Since all the nodes in the network will be synchronized to the master clock, whatever the accuracy of the master clock is, the network will stay synchronized with it. Then why should it be of the highest accuracy?

The reasons are the following:

(i) The wander and jitter have to be limited, or else they will accumulate beyond manageable proportion in the long synchronization chain.

(ii) The higher the accuracy, the lesser the slip rate at the plesiochronous boundaries with other networks. Thus, to keep the slip rate within permissible limits per ITU-T G.822, the accuracy has to be of the defined order.

(iii) Due to an interruption in the media link or a defect in the equipment or due to maintenance outage, a particular node may not be able to receive the timings from the master source for some period of time. During this period, the node clock has to maintain its functioning on its own clock; thus, the local clock has to be of acceptable accuracy to be able to sustain the functioning for a reasonable period of time.

7.10.3.2 Synchronization Supply Unit

This is a slave clock, to be driven by the master clock. The accuracy of this clock is defined by ITU-T G.812 as 1 part in 10^8 or better.

SSU is designed to be a part of the timing distribution hierarchy (see Figure 7.23).

As can be seen in Figure 7.23, SSU acts as a kind of an agent of master clock in distribution of its timing information. It receives the timing information from the master clock and distributes it among all the SECs connected to it and to all the local utilities at its physical locations, such as PBX, local node, etc. The term SSU is given by ITU-T and is used by all the countries except the United States and Japan. In the United States and Japan, it is called building integrated timing supply (BITS), and it supplies timing to all the utilities in the building in addition to the slave nodes.

The most important feature of an SSU is its "holdover accuracy." In the normal course, the SSU receives timings from the PRC; however, in case of a failure in the link, when it is disconnected with PRC, it should be able to maintain a very high accuracy for a substantial period of time in its "free running mode" (when it stops receiving timing information from PRC, it runs on its own called free running mode). In fact, it should be able to maintain the accuracy of the PRC for a few days. (Normally, the network is so designed that SSU will have at least two reference sources [PRCs] connected to it from different directions. Thus, in case of a link failure from one direction, the other reference is selected, either manually or automatically. The

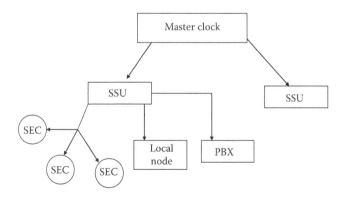

FIGURE 7.23
Role of SSU in a synchronous network.

manual selection can be local or through remote operation from a "network management center." The actual mode of switch over is decided based on local conditions such as staffing [round-the-clock or only one shift] and the PRC autonomy period of the clock; we will shortly see what is meant by the PRC autonomy period.)

The worst-case holdover performance that is guaranteed by the manufacturers is in the range of 9 days for atomic clock-based SSU and about 21 h for a crystal oscillator-based SSU, during which period, there will be no degradation in the performance of the network. These figures are for the SSUs maintained at a constant specified temperature. For a 10°C variation in temperature, the figures will drift to 33 h and 8.3 h, respectively.

The practical implication of this holdover accuracy is the requirement of time for repair of the PRC link. Thus, with the above accuracy, as long as the link is repaired within 9 days, there is no fear of loss of synchronization.

This period is known as the PRC autonomy period of an SSU (see Figure 7.24). Thus, depending upon the ability of second or third references source (of PRC) to SSU, the methodology of switch over to the standby reference in case of failure of main reference (manual or automatic, local or remote, etc.) and the expected time to repair the failed link, the type of SSU may be selected. The area where second reference is not available should be provided with better holdover performance SSU.

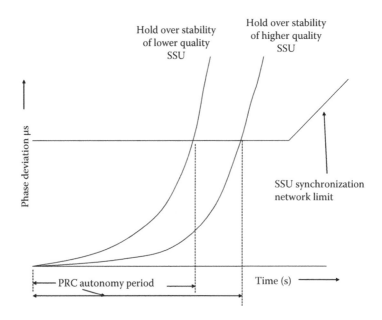

FIGURE 7.24
PRC autonomy period of SSU.

SSU could be a part of the equipment such as PBX and therefore others installed in the same premises. When it is not a part of any other equipment and is meant only for synchronization network, it is known as SASE.

7.10.3.3 SDH Equipment Slave Clocks

Figure 7.23 shows the position of SECs in the synchronization hierarchy. Since the present application of synchronized network is mostly in SDH, the clock has been so named. It is the local clock of a "network element" (NE) or a "node," which is actually a regenerator or add/drop station of a synchronous network.

This clock is of the least accuracy in the hierarchy. As defined in ITU-T G.813, its free running accuracy is ±4.6 ppm (±4.6 × 10^{-6}). The poor accuracy does not really matter because the clock almost always remains "locked" (synchronized) to the master clock through SSU. If there is a failure of a link, it will generally receive the timings from another SSU from another direction (Figure 7.25).

Node 2 SEC receives the timings information from SSU-B instead of the normal SSU-A when the link between nodes 1 and 2 fails. The switchover to the standby reference is generally automatic (and thus, fast) in case of SEC because their "free running" and "holdover" accuracies are very poor and thus large changeover periods are unacceptable. The changeover is done on preprogrammed logic or is based on SSM to be seen later in this chapter) in case there is more than one standby reference available.

7.10.4 Typical Synchronous Network

A typical synchronous network of an operator using only one synchronization area is shown in Figure 7.26 (For an operator with more than one synchronization area, this will constitute one of such areas).

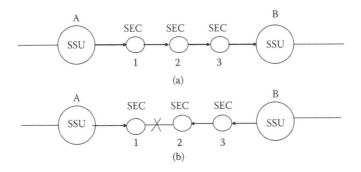

FIGURE 7.25
(a) Normal reference chain of synchronization. (b) Failure case.

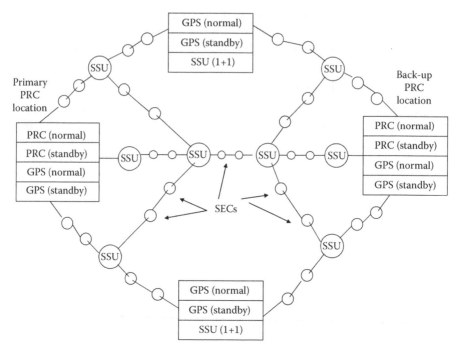

FIGURE 7.26
Typical synchronous network of an operator.

Referring to Figure 7.26, the PRCs (autonomous clock, generally cesium beam based) are provided at more than one strategic location, and at each location, they are located at a separate geographic location from other systems to avoid the possibility of network shutdown in case of a local disturbance such as fire, earthquake, etc. Within the PRC location, it is provided with a standby to take care of the network in case of a problem in one PRC. In addition, two numbers of GPS-based PRCs have also been provided as standby sources of timing reference and also for continuous monitoring of UTC. One number of GPS receiver would have been enough, but for reliability purposes, two numbers have been shown, which can be easily afforded due to their low cost.

At a few locations, it is again desirable to provide a pair of GPS receivers along with SSU to act as standby timing sources in case of problems with primary and backup PRC locations. These SSUs could have 1 + 1 rubidium oscillators, while the other SSUs may have 1 rubidium + 1 crystal oscillator. (A rubidium oscillator has a much better accuracy than a crystal oscillator, but is more expensive.)

With these standbys and backups, it is expected that the network synchronization will have the desired robustness. Although each standby or backup provision adds to cost, the costs involved are more than justified when avoiding the collapse of network synchronization, which may eventually result in complete disruption of traffic and is of paramount concern.

As can be seen, the SSUs have been provided with more than one (and as many more as easily possible, keeping the geography of the area and interconnections possibilities) reference, so that in the event of a failure of PRC references from one side, it is able to quickly switch over to one of the other sides and maintain synchronization.

7.10.5 Synchronization Principles and Timing Loops

While planning the synchronization network, the following principles should be followed:

(i) A tree topology should be adopted in the distribution of synchronization, with PRC on top. No part of the network should remain isolated and there should not be any internal timing loops (Figure 7.27). ("Timing loop" is not to be confused with "loop timing," which is a method of master–slave synchronization. In fact, the arrangement shown in Figure 7.27 is an example of loop timing when it connects between two EPBXs.)

(ii) The tree branches should be as short as possible because the longer the branches (i.e., more NEs and SSUs), the larger is the accumulation of jitter and wander.

(iii) No clock should trace itself to a clock of lower accuracy. In case no clock of higher or equal accuracy is available in case of a failure of normal reference, the clock should go to its holdover mode. (This will avoid formation of timing loops, which will be explained shortly).

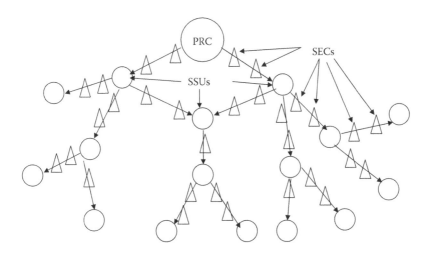

FIGURE 7.27
Tree topology to be followed. The SSUs trace their timings to PRC, and SECs trace their timing to the SSUs.

7.10.5.1 Timing Loops

A timing loop is formed when a clock traces its timing back to itself (Figure 7.28).

In Figure 7.28, SEC A receives timings from SSU, but a second reference has been provided from SEC C. In case of disruption of link to SSU, SEC A starts receiving timings from SEC C, which receives its timing from SEC B, which in turn receives its timing from SEC A; thus SEC A traces its timing back to itself. This leads to the formation of the so-called timing loop among nodes A, B, and C.

Now, what is the significance of timing loops? When a timing loop is formed, the inaccuracies of the clocks pile up because at no stage is any connection applied in the form of synchronization from a better clock.

The drift finally builds up to such a level that it throws the entire network or a portion of it out of synchronization. Thus, while planning the network, care has to be taken that not only the timing loops are eliminated in the distribution of timings but the probability of accidental creation of timing loops are also minimized, if not completely eliminated. Accidental creation of timing loops can take place if during disruption of services an engineer programs the node's reference timing from another direction without verifying the timing flow.

Accidental creation of a timing loop is illustrated in Figure 7.29. Figure 7.29a shows the normal timing arrangement with the arrows representing the direction of supply of timing interaction. Figure 7.29b shows a disruption

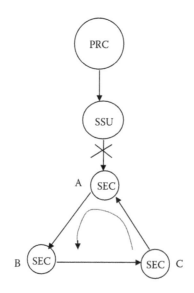

FIGURE 7.28
Potential timing loop.

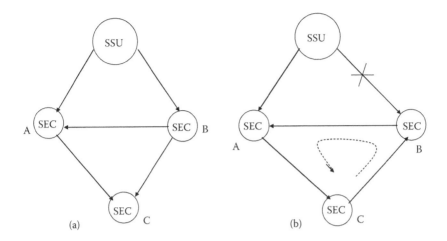

FIGURE 7.29
Accidental formation of a timing loop. (a) Normal timing directions; (b) timings programmed after a fault in the link between SSU and node B leading to formation of a timing loop.

in the link between SSU and node B. During this disruption, the site engineer programs node B to receive timing from node C. This creates a timing loop ACBA.

What makes matters worse is that there are no alarms about the timing loops in the network management system; thus, timing loops are detected only with the disruption of traffic.

Although there are no means to completely prevent such accidental timing loops, "frequency offset guarding" can be provided in some SSUs to tackle this problem partially.

The clocks in the timing loop will start drifting in frequency in an uncontrolled manner. If one of the clocks in the loop is equipped with this mechanism, it can detect the drift in frequency at an early stage and disqualify its current reference to break the loop.

7.10.6 Synchronization Status Message

SECs will generally have two or more timing references (see Figure 7.30).

In case of disruption in its normal reference, it will have to pick up timing reference from another direction if there is only one more reference aside from the selected one. But if there are more then one standby references, then which one would be selected? Naturally, the one that is better, that is, more accurate and stable, should be selected.

The selection at the SEC level is generally automatic because it has to be done very quickly, as the free running accuracy of SEC is very poor ($\pm4.6 \times 10^{-6}$).

The information about the quality of the standby reference is conveyed to the SEC by means of SSM. This message is conveyed in the overhead bits of

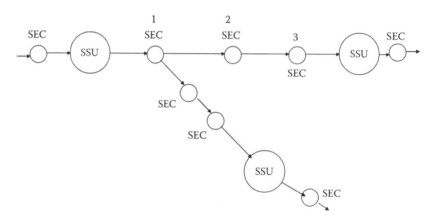

FIGURE 7.30
SECs generally have two or more references for timing.

the SDH transmission (see Section 11.5.1.4). However, SSM conveys the quality of the reference as it is at the reference node because SSM is formed at that
node. It has no information on the impairment of the clock due to transmission link (wander and jitter) between the two nodes, but still it is considered
better than switching over to one of the references without any knowledge
about their quality.

Review Questions

1. What do you understand about synchronization? What is its significance in digital communication? Explain with an example.
2. What is the difference between a symbol and a bit? Explain symbol
   synchronization.
3. What do you understand about coherent detection and why do we
   need it?
4. List the steps involved in the synchronization process.
5. What do you understand by correct detection of symbols? What are
   the steps involved in correct detection? Illustrate with an example.
6. Differentiate clearly between a base band signal and a modulated
   signal.
7. What do you understand about modulation and demodulation?
8. Explain frequency synchronization with an illustration.

9. What do you understand by phase synchronization? How precisely is frequency synchronization different from phase synchronization?

10. Why do we need coherent detection in case of phase synchronization?

11. What is the role of filtering in the process of synchronization?

12. Why is low-noise amplification deployed in the receiver?

13. Equalization process reduces distortion—is it true? Explain.

14. What is the need of clock recovery in the process of synchronization?

15. Explain the process of timing recovery or clock recovery with the help of a schematic diagram.

16. When we use a recovered clock for synchronization, do we still need a local oscillator? If yes, why?

17. What do you understand about zero crossing detection?

18. How are the line clock and local clock synchronized with the help of a PLL?

19. Explain the concept of framing emphasizing its role in synchronization of a digital signal.

20. What is the difference between a packet and a frame?

21. What do you understand about overhead bits? Why are they necessary?

22. What is the need of multiframes? Explain with an example.

23. Is synchronization required in asynchronous systems? Explain the working principle of an asynchronous network.

24. Bring out the salient differences between synchronous and asynchronous systems.

25. Higher clock accuracies are called for in synchronous systems—why?

26. What are master and slave clocks? What is master–slave synchronization?

27. Explain the concept of synchronous networks as against synchronous systems.

28. What do you understand about the term "plesiochronous?" What are plesiochronous networks?

29. Make a table comparing the salient features of asynchronous, synchronous, and plesiochronous networks.

30. What were the reasons for the popularity of the plesiochronous networks?

31. How and why did SDH come into existence?

32. What do you understand about slips? What is their significance in synchronization?

33. How is the problem of slips tackled?

34. Explain the principles of controlled slips.

35. What is elastic store and how are slips controlled with its help? What is the minimum size of elastic store required for slips to occur in a controlled manner?

36. Work out the pros and cons of controlled and uncontrolled slips.

37. What do you understand about line synchronization, node synchronization, and data synchronization?

38. How is line synchronization different from network synchronization?

39. Explain the mechanism of clock distribution in a synchronous network per ITU-T standards with the help of a diagram.

40. What are the expected accuracies of different types of clocks of a timing distribution network?

41. What are the popular types of network synchronization?

42. Enumerate and explain the types of master–slave synchronization.

43. What are the merits and demerits of mutual synchronization of a network?

44. What is pseudosynchronization? Why is it required? Please explain.

45. What are the permitted limits on the number of slips on international connections and why? What will be the implications of increasing or decreasing them?

46. What are frame slips as opposed to bit slips? What will be the permitted slip rates if bit slips are calculated instead of frame slips?

47. Calculate slip rates for the defined accuracies of PRC, SSU, and SEC clocks.

48. What do you understand about synchronization network engineering?

49. What are the criteria for choosing synchronization areas?

50. What are the elements of a synchronization chain and what are their limits?

51. What are the properties of a master clock and why are they so?

52. Compare the salient features of SSU with those of a master clock.

53. What is holdover performance of a clock? What is its importance?

54. SEC is required to be of lower accuracy than SSU—why? What are the accuracy levels required for SEC?

55. What do you understand about SSM?

56. Explain a typical synchronous network of an operator with the help of a schematic diagram.

57. What are synchronization principles in order to avoid timing loops?

58. Are timing loops desirable? How can they can be tackled?

59. Can the timing loops be created accidentally? If so, how?

Critical Thinking Questions

1. Can you think of any real-life example of synchronization requirement and the repercussions if it is not maintained?

2. Suggest alternatives to timing recovery from the received signal. How will they work?

3. How many frames are there in an E1 bit stream and why? Consider working with different frame numbers and work out the implications.

4. Have you heard of any other type of synchronization that is not discussed in this chapter? Recall and compare it with any of the types discussed here.

5. What will be the effect of low-pass filter characteristics on the performance of PLL?

6. Compare the desired properties of PRC, SSU, and SEC clocks.

Bibliography

1. J.G. Proakis and D.G. Manolakis, *Digital Signal Processing Principles, Algorithms and Applications*, Third Edition, Prentice-Hall, India, New Delhi, 2005.

2. B.P. Lathi, *Modern Digital and Analog Communication Systems*, Oxford University Press, Oxford, UK, 1998.

3. NIIT, *Advanced Digital Communication Systems*, Prentice-Hall, India, New Delhi, 2005.

4. H. Taub and D.L. Schilling, *Principles of Communication Systems*, Second Edition, Tata McGraw-Hill, 2005.

5. J. Millman and H. Taub, *Pulse Digital and Switching Waveforms*, Tata McGraw-Hill, 2005.

6. P. Moulton and J. Moulton, *The Telecommunications Survival Guide, Understanding and Applying Telecommunication Technologies to Save Money and Develop New Business*, Pearson Education, 2001.

7. P.V. Shreekanth, *Digital Microwave Communication Systems with Selected Topics in Mobile Communications*, University Press, 2003.

8. S. Haykin, *An Introduction to Analog and Digital Communications*, John Wiley and Sons Asia, 1989.

9. L.E. Frenzel, *Principles of Electronic Communication Systems*, Tata McGraw-Hill, 2008.

10. D. Roddy and J. Coolen, *Electronic Communication*, Fourth Edition, Prentice-Hall, India, New Delhi, 2001.

11. B. Sklar, *Digital Communications, Fundamentals and Applications*, Pearson Education, 2002.

12. ITU-T Recommendation G.811, *Timing Characteristics of Primary Reference Clocks*, International Telecommunication Union, Geneva, Switzerland.
13. ITU-T Recommendation G.812, *Timing Requirements of Slave Clocks Suitable for Use as Node Clocks in Synchronization Networks*, International Telecommunication Union, Geneva, Switzerland.
14. ITU-T Recommendation G.813, *Timing Characteristics of SDH Equipment Slave Clocks (SEC)*, International Telecommunication Union, Geneva, Switzerland.
15. ITU-T Recommendation G.822, *Controlled Slip Rate Objectives on An International Digital Connection*, International Telecommunication Union, Geneva, Switzerland.
16. ITU-T Recommendation G.803, *Architecture of Transport Networks Based on SDH*, International Telecommunication Union, Geneva, Switzerland.

8

Plesiochronous Digital Hierarchy

We have seen the methods of construction of digital signals from an analog voice signal and the principles of time division multiplexing in the previous chapters. Let us now have a look at the process through which these signals are actually used in the practical communication systems to enable the media to carry a large number of channels.

The 64-kbps digital signal is never used as it is for the communication link, for the simple reason that it is too costly to use two pairs of copper wires over long distances for only one channel. Thus, a large number of 64-kbps signals are multiplexed into one physical channel, and sent across the distances, for much better utilization of physical resources such as copper cables or any other media.

The time division multiplexing began with the multiplexing of 24 channels. Besides time division multiplexing, a "pulse code modulation" (PCM) technique was used to convert the analog signals to digital signals. Since this was the first modern digital communication system and used PCM techniques, the system was popularly called a PCM. Thus, the PCM was the first digital communication system with 24 voice channels multiplexed together, and it employed PCM/TDM technology.

This PCM subsequently formed the first level of hierarchical system of multiplexing. The system was called "plesiochronous digital hierarchy" (PDH). *Plesio* is a Greek word that means "nearly the same" and *knonous* is another Greek word meaning "time." Thus, the meaning of *plesiochronous* is "nearly the same time." The nomenclature *plesiochronous* is related to the means of achieving the synchronization; however, since the multiplexing hierarchy was developed to work with this system, it was called PDH.

The first system was introduced by Bell Labs in 1961 and was called a "T1" or "DS1" (digital signal 1) consisting of 24 multiplexed channels. Subsequently, the PCM standards defined by International Telecommunication Union (ITU) specified a system of 32 channels (30 speech channels and 2 signaling channels) called an "E1." Further developments facilitated multiplexing many more voice channels on one physical channel. The multiplexing hierarchy developed is the PDH, as brought out in the previous paragraph. There are four levels in the PDH: I, II, III, and IV. Level I is popularly known as PCM; however, in PDH, its proper terminology is T1 or E1, as stated earlier. Likewise, level II is called T2 or E2; level III is called T3 or E3; level IV is called T4 or E4. While T1, T2, etc. standards are used in North America and Japan, the rest of the world uses the E1, E2, etc., per ITU–Telecommunication

Standardization Sector (ITU-T) standards. (ITU, which was earlier known as Committee Consultative International for Telegraphy and Telephony, or CCITT, is an international body responsible for developing telecommunication standards.)

In this chapter, we will see the details of each of these levels, and the hierarchy as a whole including the synchronization aspects. (We will discuss only the "E1," "E2," etc., per the ITU-T standards, which will serve the purpose of understanding of the concepts. The "T1," "T2," etc., multiplexing procedures work on similar (although not exactly same) principles and can be understood easily.)

8.1 Pulse Code Modulation

PCM stands for "pulse code modulation" alone; however, the communication systems developed using the 64-kbps PCM and time division multiplexing of 32 channels are popularly known as PCM. Each of these 32 channels is called a time slot (TS) because it occupies a particular TS in each of the 8000 frames (of 125 μs each) every second. (Each TS of PCM is of 8 bits or 1 byte in length.)

In actual systems, two TSs are reserved for signaling and alarm purposes. Thus, a practical PCM system consists of 30 voice channels and two operation and maintenance (O&M) channels (Figure 8.1).

TSs 0 and 16 are used for synchronization, signaling, alarms, error detection, etc., and the rest are used for pulse code modulated voice channels.

The 125-μs segment containing 32 TSs shown in Figure 8.1 is called an "E1 frame." (The concept of "frame" is explained in detail in Section 7.1.3.)

The rate of the bit stream E1 so formed is thus

E1 = 64 kbps × 32 = 2.048 Mbps (as each channel is of 64 kbps).

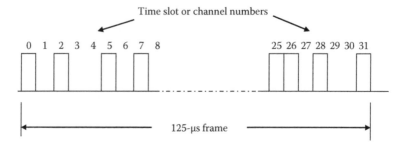

FIGURE 8.1
The 32 TSs making up a PCM or E1 frame.

This 2.048-Mbps bit stream is popularly known as a 2-Mbps bit stream. Now let us have a look at the multiplexing process in a PCM.

8.1.1 Multiplexer

The multiplexer looks at each of the 32 signals in sequence within a time frame of 125 μs and queues them up as a common bit stream of 2.048 Mbps. The process is called "interleaving." Two types of interleaving are popular, "bit interleaving" and "byte interleaving." They are illustrated in Figures 8.2 and 8.3.

Only a few channels have been shown for the purpose of clarity.

Let us see now the E1 multiplexer (Figure 8.4).

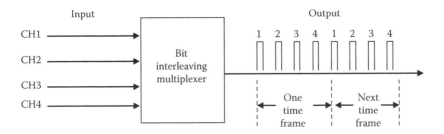

FIGURE 8.2
The bit interleaving multiplexer polls on the first bit of each channel in sequence and places a corresponding bit, reduced in duration to a quarter, serially one after the other, then goes for a second round of polling and so on.

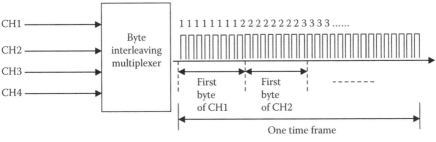

FIGURE 8.3
The byte interleaving multiplexer polls on the first byte of each channel in sequence and places a corresponding byte, reduced in duration to a quarter, serially one after the other, then goes for a second round of polling and so on.

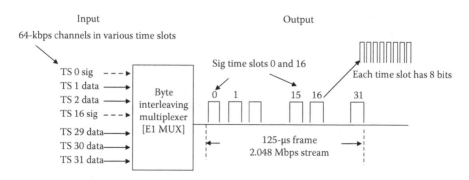

FIGURE 8.4
The E1 MUX showing the byte interleaving of 30 voice channels and 2 signaling channels of 64 kbps each. (Please note that the E1 channels or TSs are numbered from 0 to 31 and not 1 to 32.)

8.1.2 PCM/E1/2MB Multiplexing Structure

As brought out earlier, while North America and Japan went for a basic PCM system of 24 channels at a rate of 1.544 Mbps, called a DS1 or a T1, ITU-T recommended the use of the E1 instead, which is a 32-channel 2.048-Mbps system and was adopted by the rest of the world.

The E1 or the PCM consists of 30 voice channels of 64 kbps and 2 channels of 64 kbps for signaling, framing, etc. Since voice channel bits are available in the form of bytes (each sample is an 8-bit byte), byte interleaving has been adopted in this case as the multiplexing methodology.

This means that 1 byte of each channel is allotted one TS in the 125-µs frame. Also the signaling/framing and others are accomplished by bytes, in TSs 0 and 16 (see Figure 8.5).

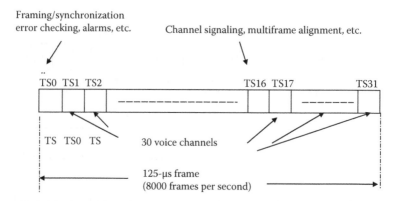

FIGURE 8.5
PCM/E1 multiplexing structure.

Bit rate = 30 × 8 bits for voice channel (30 bytes)

+ 2 × 8 bits for signaling/framing channel per frame (2 bytes).

= 256 bits per frame

Thus, the bit rate per second = 256 bits × 8000 frames/s

= 2048 kbps

= 2.048 Mbps

Now let us have a look at the details of the functions of TSs 0 and 16 and the frame alignment process of an E1.

8.1.3 Time Slot 0

The TS 0 in the above frame consists of 1 byte. Various bits of this byte have been assigned different functions as follows (Figure 8.6).

As you can see, the assignment of different bits is different for even and odd frames. To understand this, let us first understand the concept of multiframe in an E1. We will come back to Figure 8.6 to discuss further details.

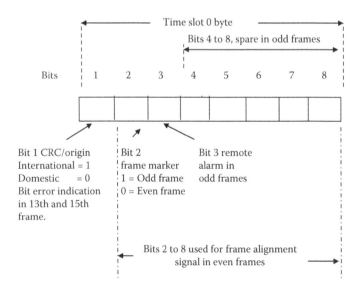

FIGURE 8.6
Assignment of various bits of TS 0 in E1 frame.

8.1.4 Multiframe

PCM needs a four bit signaling for voice "call processing." To accommodate this four bit signaling of each channel (for 30 channels, 4 × 30 = 120 bits are required, which will be quite a large burden on a single frame, and as such cannot obviously be accommodated), a concept of multiframe has been deployed in E1 multiplexing. The four bit signaling has been catered for two channels in one frame (8 bits of the TS), thus making a total of 15 frames, and one frame has been added for the synchronization of the multiframe itself. Thus, the complete signaling and synchronization takes place within a multiframe of 16 frames. (It may appear too long a wait before the frame synchronization is achieved, but it hardly takes 16/8000 = 0.002 s.) This signaling and multiframe alignment is performed in TS 16, the details of which are dealt with in the next section. The multiframe is shown in Figure 8.7.

We have seen the assignment of functions of different bits of TS 0 in Section 8.1.3 above. Let us now have a closer look at the duties of various bits therein.

8.1.4.1 Time Slot 0 Details

8.1.4.1.1 Bit 1

Bit 1 of TS 0 has been reserved for marking the origin of the bit stream. If its value is 1, then the E1 is of an international origin, and if it is 0, then E1 is of domestic origin. However, this usage is changed wherever "error detection is deployed using CRC." (Cyclic redundancy check [CRC] is a very powerful and popular error-detecting mechanism in digital transmission. For a discussion on CRC, please see Section 6.7.) The bit is then utilized for CRC in all "even" frames of a multiframe. This bit in odd frames 13 and 15 is used for "CRC error messages." When used for the CRC, the bit in remaining odd

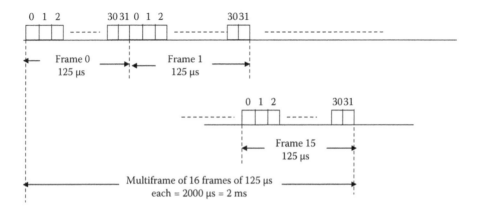

FIGURE 8.7
A 16-frame multiframe of E1.

frames (other than 13 and 15) remains 0 and does not mean anything. (In some systems, it is used for a 6-bit CRC sequence, which is utilized for interworking with the nodes that do not deploy the CRC procedure.)

8.1.4.1.2 *Bit 2*

Bit 2 is used to mark the frame as "even" or "odd" frame. If bit 2 is 0, then the whole frame is an even frame, and if bit 2 is 1, then the frame is an odd one. This information about even and odd frames is used for further utilization of TS 0 and TS 16 bits in the formation of a multiframe. By marking the frame, the bit also makes a part of the 7-bits (bits 2–8) frame alignment signal (FAS). (FAS is a bit pattern used for locating the beginning of the frame by the receiver. The receiver looks for this pattern in the received bit stream continuously, until it finds it. As soon as the FAS is located, the frame is said to be aligned.)

8.1.4.1.3 *Bit 3*

Bit 3 of odd frames of a multiframe is used for remote fault alarm indication. A 0 at bit 3 indicates no alarm on the remote (next) node, and a 1 indicates a fault alarm. Bit 3 of even frames is a part of the 7-bit (bits 2–8) FAS.

8.1.4.1.4 *Bits 4 to 8*

Bits 4 to 8 of even frames of a multiframe along with bits 2 and 3 make a 7-bit FAS. The pattern defined for this FAS is 0011011. The receiver looks for this pattern on alternate frames to synchronize with the transmitter to mark the beginning of frames. These bits in odd frames are not used and are kept as spare bits that are to be used by the network operator for country-specific applications.

8.1.5 Time Slot 16

The eight bits of TS 16 are mainly used for signaling over a multiframe of 16 frames. The first frame is used for multiframe synchronization and alarms, whereas the rest 15 frames are used for providing 4 bit signaling (known as channel-associated signaling, or CAS) to two channels each (15 frames × 2 = 30 channels); see Figure 8.8. (When we need to talk to someone using the telephone, we need to communicate our intention to the exchange. We do this by lifting the receiver of our telephone set [or taking the telephone "off" the hook, called "off hook" condition]. In other words, we communicate to the exchange that we have an intention to talk to someone. The exchange responds by extending a dial tone to us, communicating that now you are allowed to dial the telephone number of the desired party. These communications such as ringing tone to the called party, ring back tone to us, busy tone, etc. are called signals, and the process is called signaling. The technology of signaling has acquired many shapes during the process of development of communication systems. Initially, the signaling information

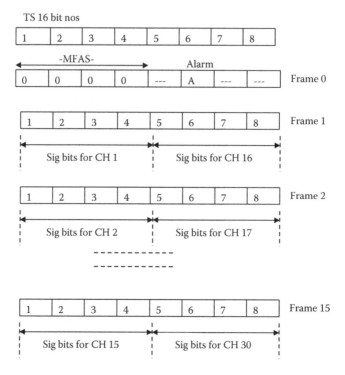

FIGURE 8.8
TS 16-bit assignment.

was transmitted on the voice channel itself and was accordingly called "in-channel signaling." Further developments of switching systems [exchanges], particularly the stored program-controlled exchanges found it to be much more convenient and efficient to dedicate a separate channel for carrying the signaling information of all voice channels. The process was called "common channel signaling" [CCS]. The signaling information in the E1 standard is carried in one particular channel, for all the voice channels. Hence it is a form of CCS. However, since there are 4 dedicated bits in the 16-frame multiframe for carrying the signaling information of each channel, the signaling is also called CAS. Thus, the signaling in an E1 is an example of channel-associated CCS.)

Note that channels are counted from 1 to 30, whereas TSs are counted from 0 to 31 and 16 frames of a multiframe are from 0 to 15. (This is very important to remember because many times during further reading you may get confused on this.)

The first 4 bits of the first frame of the multiframe are 0000 and are used for multiframe alignment. Bit 6 is used for remote alarm for multiframe defects, and the rest of the bits, i.e., 5, 7, and 8, are spare bits, which are presently unassigned and their application is to be specified in the future.

8.2 Higher-Order Multiplexing/Hierarchical Multiplexing

We have seen how the basic digital communication system works in the form of PCM or E1. In this PDH level 1 (E1), we could accommodate 30 voice channels on a single transmission medium, thus the telephone calls became more accessible and cheaper. This further boosted the demand of services, leading to technologies that could accommodate even more numbers of channels on a single transmission medium. Higher-order or higher-rate multiplexers, therefore, emerged to accommodate more numbers of channels in the multiplexer. The higher-order multiplexing is carried out in a hierarchical manner multiplying the number of channels at each stage of multiplexing by a factor of 4. The bit rates are arranged in a hierarchy of 2, 8, 34, 140, etc. nominal bit rates. (The exact bit rates are not integral multiples of 1 Mbps, but for the ease of reference, they are called so, e.g., E1 is called 2 Mbps, although the exact bit rate is 2.048 Mbps.) The hierarchy is known as PDH as already seen in the beginning of this chapter.

The PCM that we have seen in the previous section is also a part of the PDH system and is known as the PDH level I or E1.

Four E1s at the bit rate of 2.048 Mbps each combine to constitute an E2 of 8.448 Mbps, 4 E2s combine to constitute an E3 of 34.368 Mbps, and 4 E3s combine to form an E4 of 139.264 Mbps bit rate. (More than E4 rates are rarely used and are not standardized by ITU-T; wherever they are used, they are proprietary standards of the manufacturer; Figure 8.9.)

Note that the outgoing bit rates are not the exact multiples of the incoming bit rates, but they are more.

For example, $4 \times E1 = 4 \times 2.048 = 8.192$ Mbps, whereas the E2 rate is 8.448 Mbps. The rates are kept higher deliberately for a purpose. A large number of "overhead" bits are added for maintenance/management purposes. (Overhead in digital transmission has the same meaning as the word implies. It means the number of "extra" bits, which are required by the system, to properly carry and receive the "data" bits. Thus, the transmitted digital signal stream consists of "overhead" bits and "data" bits. The data bits [bits are called data bits irrespective of whether they carry the bits for voice communication or for data communication] are also called the "payload" bits.)

The overhead bits in this case include the bits for framing, national use, alarm, justification opportunity, justification control, etc. (We will see about "justification" in detail in a subsequent section.)

The E1 bit stream is treated as a payload at the input of an E2 multiplexer. The multiplexer adds the necessary overhead bits like the ones mentioned above to be able to carry and deliver the E1s properly to the receiver. At the receiver, the overheads are removed and the E1s are restored in their original shapes. Similarly, the E2s are treated as payloads by the E3 multiplexer. These E2s include the E1s and the E2 overheads, but as far as the E3 is concerned, each E2 is only a payload at its input. The E3 multiplexer adds its own

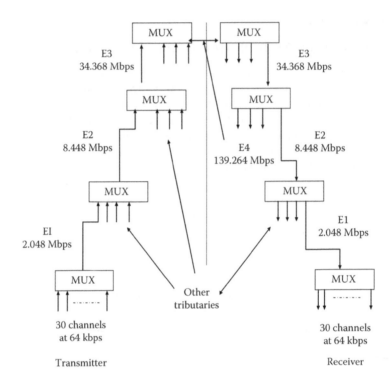

FIGURE 8.9
The PDH multiplexing hierarchies E1 to E4.

overheads to the E2s. The E3 demultiplexer at the receiver restores the E2s in their original shapes, then E2 payload is again demultiplexed to get the E1s, after removal of the E2 overheads. Similar is the case of multiplexing the E3s into E4 and demultiplexing of E4 into E3s.

Let us have a look at the details of E2 multiplexing now.

8.2.1 E2 Multiplexing

The PDH level II or E2 bit stream can be formed by time division multiplexing of four E1 bit streams as shown in Figure 8.10 below.

But there is a problem. If all the E1s are originated at the same place, with the same set of equipment, and without involvement of any transmission media, then these bit rates are exactly the same for each of the E1 (2.048 Mbps); otherwise, there could be a little variation in their bit rates with respect to each other. In the case of variation in bit rates, the multiplexed output may look like as shown in Figure 8.11 below.

It can be seen that in the output E2 bit stream, although the first set of four bits are correctly multiplexed in the proper sequence, i.e., the first bit of A channel, then first bit of B channel and so on, but in the second set, the bit

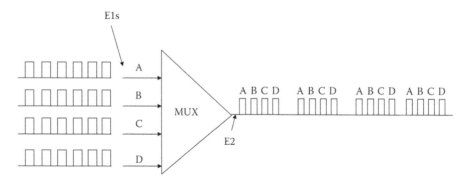

FIGURE 8.10
Time division multiplexing of 4 E1s to form an E2 bit stream.

belonging to B channel is uncertain, and in the third set, the bit sequence has become 3 2 3 3, meaning that the second bit of the third set has been picked up from the second bit of B channel instead of the third bit, as in the case of A, C, and D channels. Thus, the proper sequencing has been lost, and the channel cannot be identified by the demultiplexer by the position of bits in the bit stream. (The difference in the bit rate has been exaggerated for the purpose of illustration. In actual practice, the difference in various E1 rates will be a few parts per million only; however, it is enough to give us the same result. Also, although the bits have been shown separated from each other, in most cases, there are no spaces between bits; thus, the wrong bit could be picked up by the multiplexer in the second bit of B channel itself.)

Thus, to get a properly multiplexed E2 bit stream from four E1s of different bit rates, we will have to find some mechanism to make the bit rates of all the E1s equal to each other, which could be meaningfully demultiplexed later into original E1s.

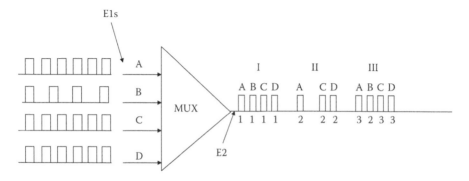

FIGURE 8.11
Time division multiplexing of 4 E1s whose bit rates are not the same. For clarity, only one E1 is shown to have a different bit rate.

Fortunately, there is a technique available called "bit stuffing" (in North American standards) or "justification" (in European standards per ITU-T) to solve the problem. However, before looking into this technique, let us know why the bit rates of different E1s are different.

8.2.2 E1 Bit Rate Variation

When we receive four E1 streams, from four different nodes or functions, some may originate from digital exchanges, some from other E1 multiplexers over long transmission lines and so on. Each of these bit streams is supposed to have a bit rate of 2.048 Mbps. However, in actual practice, the bit rates are never exactly 2.048 Mbps, but each of these E1s has a slightly different bit rate. The reasons for this difference are mainly twofold: first, the clocks of the nodes generating the E1 streams may be of poor accuracy, and second, because they are received via different types of media such as copper wires, coaxial cable, radio, or optical cable, or may be on different lengths of similar media. Transmission media may cause severe impairments to cause the bit rate variation. (For a detailed discussion on this, refer to Sections 6.1 to 6.5.)

Now, let us see how to overcome this problem with the help of the technique of "bit stuffing" or "justification."

8.2.3 Justification or Bit Stuffing

The purpose of justification or bit stuffing is to remove the difference in the bit rates of all the four incoming E1s. Thus, if a particular E1's bit rate is slower, we can "stuff" additional bits into the stream, or if it is faster, we can remove a few bits, to make the rates equal. The process is called "justification" on the same analogy as text adjustment in a document where we stuff additional space or reduce some space to justify the document so that it gives a neat appearance (see Figure 8.12).

As shown in Figure 8.12 (for illustrative purpose only, 3 E1s have been taken and the difference in the bit rates is highly exaggerated), the first E1 has the normal bit rate, while the second E1's bit rate is slower than normal, and that of the third E1 is faster than normal.

When the normal E1 is able to deliver 8 bits in time T, the second E1 is able to deliver only 7 bits and the third E1 is able to deliver 9 bits during the same period. Thus, if we add one bit to the second E1 during time T, and drop one bit from the third E1 within the same time duration, we would achieve the objective of equal bit rates. The added bit is called a "stuff bit" and its addition is called "positive justification" (or positive stuffing) and the dropping of the bit is called "negative justification" (or negative stuffing).

So, we are able to achieve equal bit rates for all the incoming E1s (incoming to the E2 multiplexer). But now there are new problems. Notice in Figure 8.12b that after adding a stuff bit in the bit stream between bits 5 and 6 (it could have been anywhere else too), the spacing between bit 5 and stuff bit and

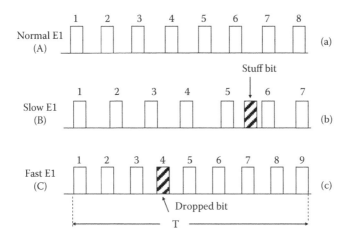

FIGURE 8.12
Bit stuffing or justification. Three E1s are shown; the first E1 is of normal bit rate (a), the second E1 is slow in bit rate (b), and the third E1 is faster (c).

between stuff bit and bit 6 are less than the normal spacing, thus although there are an equal number of bits during time T in each E1, the multiplexer will again get confused, unless the bits are equally spaced too. In Figure 8.12(c), you can see that a similar problem of nonuniform spacing between bits is created if bit 4 (or any other bit) is dropped. Moreover, dropping a bit from a stream will throw the original multiplexing of 64-kbps channels into E1 out of gear, which means, when the E1 will be finally demultiplexed into 64-kbps channels, it will not be possible to identify the individual channels correctly. Thus, the bits cannot be arbitrarily dropped.

To overcome these problems, a little different practical approach is adopted. The bit rate of the E2 multiplexer is kept higher than that required for multiplexing 4 E1s of normal bit rate, so as to accommodate the highest rate of faster incoming bit stream, and in slower incoming bit stream, some extra bits are inserted as "stuff bits" or "justification bits." To understand the concept, let us consider the Figure 8.12 again.

If we boost the bit rates of all the E1s to be equal to that of Figure 8.12(c), before multiplexing, then we will have 9 bits in each E1, during the time T. Thus, the E1 of Figure 8.12a will have one extra bit and that of Figure 8.12b will have two extra bits and that of Figure 8.12c will have no extra bits. So if we are able to insert these extra bits, identify them, and remove them from the bit stream at the time of separating the E1s again (demultiplexing E2 into 4 E1s), we will be able to achieve our objective of properly multiplexing and demultiplexing the E1s of different bit rates. Figure 8.13 illustrates the concept.

As you can see in Figure 8.13, the four incoming E1 bit streams (or E1 tributaries) have different bit rates, having 8, 7, 9, and 7 bits in period T,

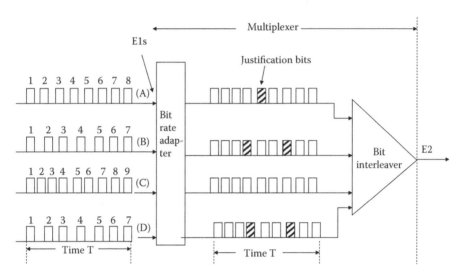

FIGURE 8.13
The bit rate adaptation unit in the E2 multiplexer makes the bit rates of each incoming bit stream (also called tributary) "the same" before bit interleaving.

respectively. They are sent to a bit rate adaptation unit in the multiplexer, which boosts the bit rate of slower tributaries, to make them all equal to the fastest one, i.e., tributary C. This is done by artificially introducing additional bits in tributaries A, B, and D. These additional bits, called "stuff bits" or "justification bits," are 1 in number for tributary A and 2 each for tributaries B and D. (For illustration purpose, we have shown any arbitrary bits as justification bits, but we will soon see that their locations are fixed, so that they can be identified and removed later.)

The example above shows up to two justification bits for a set of 7 bits, but how many justification bits are required in actual practice? Let us see it in the next section.

8.2.4 Number of Justification Bits

The actual number of justification or stuff bits required will depend upon the extent of variation in the bit rates of E1 tributaries. The variation in the bit rates could be due to the clock inaccuracies of the E1 nodes or due to various impairments in the transmission media. (For a detailed discussion on impairments in transmission media and clock, please refer to Section 6.1 to 6.5.)

The E1 nodes are required to use a local clock of a defined accuracy per ITU-T G. 732 standard, which specifies the bit rate as 2048 kbps ± 50 ppm (parts per million). Thus, the number of bits, which can be either less or more as compared to a normal E1 rate (2.048 Mbps), will be

$$(2.048 \times 10^6) \times (\pm) (50 \times 10^{-6}) = \pm 102 \text{ bits/s.}$$

Hence, if we keep the bit rate of each E1 within the above limits before interleaving, we can achieve the objective of accommodating all the variations in the E1 bit rates due to clock inaccuracies. Thus, we need to adopt the bit rate as follows:

$$\text{Required bit rate} = 2.048 \text{ Mbps} \pm 102 \text{ bits/s}$$
$$= 2.048 \text{ Mbps} + 102 \text{ bits/s}$$
$$(-102 \text{ will get accommodated automatically})$$
$$= 2{,}048{,}102 \text{ bits/s}$$

However, in practice, the bit stuffing has to be done in a regular and pre-defined manner so that the stuff bits can be easily identified, therefore, the stuff bits are included in every frame so that they appear at predefined locations in the bit stream. The frame length adopted in E2 is 848 bits. This frame of 848 bits carries the bits belonging to 4 E1s (the details will be seen later in this chapter). Thus, if one bit of justification has to be provided at the least for each E1, there will be four justification bits per frame and will mean that we will have

$$\text{Number of justification bits per E1 per second} = (4/848) \times 2.048 \times 10^6$$
$$= 9660 \text{ bits/s}$$

(Here, just for the purpose of simplification we have presumed that the rest of the 844 bits are bits from tributary E1s, although there are actually a number of other overhead bits added during the E2 multiplexing process. We will see the details soon.)

However, note that our requirement for meeting the clock inaccuracies was only 102 bits/s. Though we have to keep some margin for transmission media impairments, it cannot be so much to need overall 9660 bit/s padding. Then why do we not add stuff bits in a few frames only every second, rather than in every frame? That can be done, but is not done, to keep the system simple because four bits per E2 frame is not a big penalty, considering that we have to add a large number of bits per frame in any case for overheads such as frame alignment, alarms, etc.

8.2.5 Justification Control

We have just seen that the bit rate of each E1 is boosted before multiplexing it into an E2 stream to accommodate the variations in individual E1 bit rates with respect to each other. The bit rate is made sufficiently high so that it is equal to or more than the bit rate of the fastest E1. Thus, for the faster bit rate E1, there may not be any justification bits in addition to tributary bits.

These justification bits are added before multiplexing in the transmitter, and are removed after demultiplexing in the receiver to get the E1 back into its original shape.

But how does the receiver know that the incoming E2 stream contains justification bits or not?

This information is conveyed to the receiver by means of "justification control bits." The justification control bits are carried in the E2 bit stream at fixed locations in each frame. To understand the concept, let us have a look at Figure 8.14, which shows an arbitrary bit stream (for illustration only), showing one justification bit and one justification control bit. These bits are predefined by the design at location number 18 and location number of 10 in each frame.

When receiver demultiplexes this bit stream, it looks at bit 10 to determine whether the bit 18 is a justification bit or a data bit. If the value of bit 10 is 1, it will take bit 18 as a justification bit and reject it from the demultiplexed E1 bit stream, and if it finds the value of bit 10 as 0, it will take bit 18 as a data bit and will demultiplex the E1 without rejecting the bit 18. (Hence, the position number 18, or whichever it is, is called a "justification opportunity" bit, meaning that you can use it for justification if you so desire.)

But there could be a problem. What happens if the justification control bit gets corrupt in the transmission media and a 1 becomes 0 or vise versa? The prediction about bit 18 will go wrong and the received E1 bit stream may not make any sense. To solve the problem we can add one more control bit. When both of them are 1, the bit 18 is a justification bit and when both of them are 0, it will be a data bit.

But there is still a problem, if one bit gets corrupt and becomes 0 from 1 and the other remains 1, then what will we infer? Receiver will be at a loss to take the decision about bit 18 being a justification bit or not. So we add one more control bit to have a total of 3 control bits and presume that not more than one control bit will get corrupted at any given time and take a decision about the fate of bit 18 on the basis of majority voting. If two out of three control bits are 1, the bit 18 is a justification bit, and if two of them are 0, it is a data bit (see Figure 8.15). (It appears to be too much of a tax to use three bits

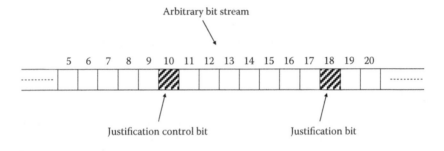

FIGURE 8.14
An arbitrary bit stream showing a justification bit at location 18 and a justification control bit at location 10.

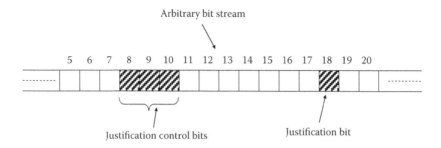

FIGURE 8.15
Bits 8–10, i.e., three bits used for justification control to decide whether bit 18 is a justification bit or not, on majority voting principles.

to verify the credentials of one bit, but remember that even one bit per frame is much more than enough to take care of the highest possible variations in the E1 tributaries as we have seen earlier in the previous section. Hence, a total of 4 bits per E1, one for justification and 3 for its control [total 16 bits], of 848 bits of a frame are not a big penalty.)

But what if two out of the three control bits get corrupted and give a wrong indication? Somehow, the experience shows and the systems have proven that this possibility is very rare, and our majority voting concept works well.

And let us now see exactly how these justification bits and justification control bits are incorporated in an actual E2 frame.

8.2.6 E2 Frame Structure

The E2 frame is 848 bits long in accordance with ITU-T standards G.742. (Although ITU-T has defined another standard G 745 also for E2 in which case the frame length is 1056 bits, the G 742 standard is more popular and is adopted by most manufacturers.)

For ease of representation, the frame is divided into four sets of 212 bits each (see Figure 8.16).

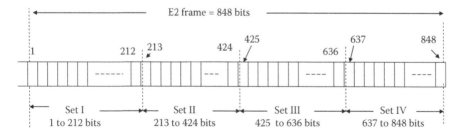

FIGURE 8.16
An E2 frame having 4 sets of 212 bits each.

8.2.6.1 Set I

The details of 212 bits of set I are as shown in Figure 8.17.

As can be seen, the first set has to carry the FAS word to facilitate synchronization by the receiver (see Chapter 7 for framing and synchronization). Thus, bits 1 to 10 carry the FAS word. The value of the FAS word as defined by ITU-T is 1111010000. This bit pattern has been chosen because it is statistically rarely probable to occur in the bit stream. Thus, whenever the receiver encounters this pattern of 1s and 0s, it takes it as the beginning of the E2 frame.

8.2.6.1.1 Bit 11

Bit 11 is used for extending fault alarm such as "loss of incoming signal," "failure of power supply," etc. to a remote multiplexing station for the purpose of maintenance and repairs.

8.2.6.1.2 Bit 12

Bit 12 is reserved for national use.

8.2.6.1.3 Bits 13 to 212

These 200 bits are used for data from tributaries. The data is bit interleaved, meaning that bit 13 is from tributary 1, 14 is from tributary 2, and so on and repeating the tributaries after every 4 bits.

There might be a question in the reader's mind about the choice of use of bit interleaving against byte interleaving. Let us have a look at it before proceeding to the details of set II.

8.2.6.1.4 The Choice of Bit Interleaving vs. Byte Interleaving

Section 8.1.1 provides an explanation of bit interleaving and byte interleaving.

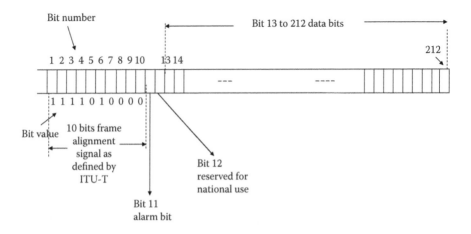

FIGURE 8.17
Set I made of first 212 bits of E2 frame.

Now the question is how we choose either of them for a particular application. We have seen (Section 8.1.1) that the PCM or PDH level 1 signal, i.e., E1, is created by time division multiplexing of 64-kbps signals through "byte interleaving," and in the preceding discussion, we have been talking about multiplexing the 4 E1s to form an E2 signal by "bit interleaving."

The reason for the choice is the frame structure. Each 125-μs frame of an E1 is divided into 32 TSs and each TS is equal to 1 byte. The duties of TSs 0 and 16 are to handle the framing, signaling, and other activities, whereas other 30 TSs carry digitized voice signals. Thus, if the 32 channels are multiplexed or interleaved in the transmitter, 1 byte of each channel at a time and demultiplexed at the receiver similarly, all the functions of TS 0 and 16 are achieved by utilizing individual bits subsequent to demultiplexing.

However, in case of an E2 frame, as we have seen in the preceding section, the functions have been assigned at bit levels, which cannot be broken or combined into TSs of 1 byte each. For example, it we take the first eight bits to make a byte, it will accommodate only 8 bits of the 10 bit long FAS. Two bits of FAS will go into the next byte, which will contain 2 FAS bits, 2 bits for alarm and national use, and 4 data bits. Similarly, in set II onward, bits 1 to 4 are justification control bits, 1 bit pertaining to each E1; thus, if we do byte multiplexing, it will put all four justification control bits in the same TS. Thus, since the functioning, i.e., framing, alarms justification, control, etc., are all at bit level, it is more convenient to use bit interleaving rather than byte interleaving. When we see synchronous digital hierarchy (SDH) (Chapter 10), we will find that SDH adopts byte interleaving, basically owing to the fact that stuffing/justification, control, etc. functions are all performed at the byte level in SDH, which makes it convenient for byte interleaving. And when byte interleaving can serve the purpose conveniently, there is no point in going for bit interleaving, as the implementation of byte interleaving requires 8 times fewer number of steps.

Let us now resume our discussion of further sets of an E2 frame.

8.2.6.2 Sets II and III

The details of the bits of sets II and III are as shown in Figure 8.18.

Sets II and III have similar distribution of the duties of various bits. The first four bits are justification control bits, one for each E1 tributary, and the remaining 208 bits are data bits.

As we have seen in Section 8.2.6, there are three justification control bits for each E1.

The first bit of sets II–IV is used for justification control of first E1, the second bit of each of these sets is used for the justification control of second E1, and so on up to fourth E1. Thus, if two of the bits out of the first bits of each of these three sets (total 3 bits) are 1, the justification opportunity bit (justification opportunity bit is a bit in a predefined position; this bit can either be declared as a data bit or a stuff bit, depending upon the declaration made by the justification

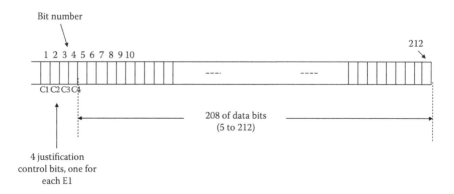

FIGURE 8.18
Details of sets II and III.

control bits) is either taken as a justification bit or it is taken as a data bit by the receiver depending upon the value of the justification control bits.

8.2.6.3 Set IV

The details of the bits of set IV are shown in Figure 8.19.

The first four bits are for justification control of the four E1s, respectively. The next four bits, i.e., bits 5–8, are the justification opportunity bits or as commonly called "justification bits," each pertaining to one E1 in sequence. The rest of the bits, i.e., from bits 9 to 212 (204 bits), are tributary data bits.

As explained earlier, the justification control bits (they are also known as justification "service" bits) C1, C2, C3, and C4 of each set indicate to the receiver whether bits 5, 6, 7, or 8 (as the case may be) are justification bits or data bits through majority voting.

Now let us combine all four sets to form the complete frame, and have a look at the overall bit rate of the E2 bit stream.

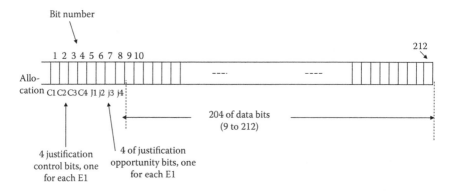

FIGURE 8.19
Details of bit allocations of set IV of an E2 frame.

8.2.7 E2 Bit Rate and Bit Rate Adaptation of E1s

If we summarize the preceding discussion of four sets of the E2 frame, we have the bit distribution as under (Figure 8.20).

Please note that the number of bits carrying data in each set is different. Set I has 200 data bits, as compared with 208 of them in sets II and III and 204 data bits in set IV. However, this distribution is only for the purpose of management, it has no consequence on transmission, as the entire E2 frame travels as one frame down the transmission media. Hence, we can say that there are a total of 820 data bits and 28 other bits (we may call them overhead bits) in an E2 frame of 848 bits.

Note that these 820 data bits are pure data bits (as against 4 numbers of justification opportunity bits, which could be either data bits or stuff/justification bits). This means that there will be 820 data bits in the frame if all the justification opportunity bits are "justification bits," and there will be 824 data bits in the frame if all the justification opportunity bits are "data bits." Thus, our minimum bit rate will be 820 bits/frame and maximum bit rate will be 824 bits/frame.

Let us see how this translates into bit rate per second.

The bits per second at the output of an E2 multiplexer have been defined as 8.448 Mbps; therefore, we can calculate the following parameters:

(i) The number of frames per second will be

= 8,448,000/848

= 9962.26415 frames/s

(ii) The duration of each frame will be

= 1 s/9962.26415

= 100.378788 μs

(Please notice that the frame duration here, and also in subsequent PDH level multiplexing, is not 125 μs, which was the case for E1, and will again be the case for SDH.)

Set no.	FAS bits	Alarm bits	National use bits	Justification control bits	Justification opportunity bits	Data bits	Total bits
1	10	1	1	0	0	200	212
2	0	0	0	4	0	208	212
3	0	0	0	4	0	208	212
4	0	0	0	4	4	204	212
Total	10	1	1	12	4	820	848

FIGURE 8.20
Distribution of bits in an E2 frame.

(iii) The minimum tributary bit rate per second (for E2 bit stream) will be

= minimum number of data bits per frame × number of frames per second

= 820 × 9962.26415

= 8,169,056.6 bits/s

Therefore, the minimum bit rate per second for each E1 tributary will be

= 8,169,056.6/4 bits/s

= 2,042,264.15 bits/s

= 2042.3 kbps (approximate)

= 2.042 Mbps (approximate)

(iv) The maximum tributary data per second (for E2 bit stream) will be

= maximum number of data bits per frame × number of frames per second.

= 824 × 9962.26415

= 8,208,905.66 bits/s

(This excludes all the overhead bits like framing bits, alarm bits, national use bit and justification control, and opportunity bits, which are presumed to all be used for the management of tributary data.)

Therefore, the maximum bit rates per second for each E1 tributary will be

= 8,208,905.66/4

= 2,052,226.41 bits/s

or = 2052.2 kbps (approximate)

or = 2.052 Mbps (approximate)

The parameters for an E2 bit stream are summarized in Figure 8.21.

Sr. no.	Parameter	Value
1	Output bit rate	8.448 Mbps ± 30 ppm
2	Frame rate	9962.26415 frames/s
3	Frame duration	100.378788 μs
4	Maximum acceptable tributary bit rate	2.052 Mbps (approximate)
5	Minimum acceptable tributary bit rate	2.042 Mbps (approximate)

FIGURE 8.21
E2 bit stream parameters as per ITU-T G.742.

8.2.8 Bit Rate Adaptation

From the above discussion, we see that the minimum and maximum data rates permitted into an E2 multiplexer from the E1 tributaries are 2.042 to 2.052 Mbps.

We have seen that the E1 rate is 2.048 Mbps (64 kbps × 32 channels). Thus, the variation permitted is

On the positive side,

$$= +(2.052 - 2.048)$$
$$\text{or} = +0.004 \text{ Mbps}$$
$$\text{or} = +4000 \text{ bps}$$

Similarly, on the negative side, the permitted variation is

$$= -(2.048 - 2.042) \text{ Mbps}$$
$$\text{or} = -0.006 \text{ kbps}$$
$$\text{or} = -6000 \text{ bps}$$

Thus, the E1 rate variations that can be accepted are from +4000 to −6000 bps. This is sufficiently high to accommodate the variation due to clock inaccuracies (which is ±102 bps, as shown in Section 8.2.4) and other factors in transmission media.

8.2.9 Positive/Negative Justification

The discussion above shows that in the E2 frame (per G.742 of ITU-T) there "are" justification bits or there are "no" justification bits. Whenever there "are" justification bits, it is called positive justification, and when there are "no" justification bits, it is no justification. There is no negative justification. Thus, this E2 standard manages the positive and negative variations of incoming E1 tributaries with only positive justification. (There is another standard, G745 of ITU-T, which specifies the use of either positive, zero, or negative justification, but it is not widely adopted by manufacturers and will not be discussed here. The principles involved are, however, no different from those described above.)

8.2.10 Variable Bit Rates of E2

An important concept to understand here is that the bit rate at the output of an E2 multiplexer is constant at 8.448 Mbps; what varies is the ratio of stuff/justification bits to the tributaries data bits at the input of the multiplexer.

From what we have seen above, it can be easily concluded that if the "justification opportunity bits" in all the frames are declared as "justification bits" by the justification control bits (let us consider only one E1 tributary at a time), the tributary bit rate will be 2.042 Mbps; this will happen when the tributary bit rate is slower than the normal 2.048 Mbps rate. On the other hand, if the "justification opportunity bits" in all the frames are declared as "data bits," the tributary data rate will be 2.052 Mbps; this will happen when the tributary bit rate is faster than the normal 2.048 Mbps rate.

In practical situations, both these extremes are generally avoided because the margins available are very high. Depending on the tributary bit rate, the justification opportunity bits are declared as justification bits (to be removed at the time of demultiplexing) in some frames, and they are declared as data bits in some other frames (to be treated as proper data bits at the time of demultiplexing). Thus, the slow or fast tributary data rate is adapted into a constant bit rate, which is then multiplexed with another three similar tributaries to form a constant bit rate E2 output.

The function is achieved by means of a "bit rate adapter unit," which continuously monitors the bit rate of the incoming tributary and sets the values of justification control/service bits to declare the justification opportunity bits appropriately as data bits or justification bits as the case may be.

8.3 E3 Multiplexing

The PDH level 3 multiplexing is known as E3 (see Figure 8.9). An E3 multiplexer takes 4 numbers of E2 streams at its input at the rate of 8.448 Mbps and produces an output of 34.368 Mbps, which is commonly known as 34 Mbps. The principles involved are the same as those discussed in terms of E2 multiplexing. Again, a bit interleaving scheme is adopted for similar reasons. The accuracies desired for E2 and E3 bit streams are as follows per ITU-T standards:

$$E2 = 8.448 \text{ Mbps} \pm 30 \text{ ppm and}$$

$$E3 = 34.368 \text{ Mbps} \pm 20 \text{ ppm}$$

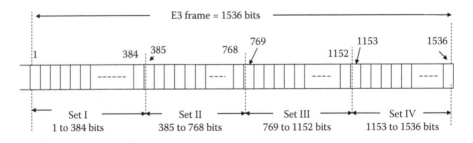

FIGURE 8.22
An E3 frame having 4 sets of 384 bits each.

8.3.1 E3 Frame Structure

The details of an E3 frame are displayed in Figure 8.22.

8.3.1.1 Set I

The arrangement of set I is the same as that of E2 set I (Figure 8.17), except that the number of data bits from tributaries are from bit 13 to bit 384, i.e., a total of 372 bits.

8.3.1.2 Sets II and III

These are also the same as those of E2 frames, except that the number of data bits from tributaries, which are from bit 5 to bit 384; hence, there are a total of 380 bits in the case of E3.

8.3.1.3 Set IV

Set IV is also the same as that of E2 frame, except that the tributary data bits are from bit 9 to bit 384, i.e., a total of 376 bits.

Thus, the summary of bit allocation in an E3 frame can be depicted as shown in Figure 8.23.

Other principles are the same as E2 multiplexing. Let us work out various parameters of E3 multiplexer.

(i) Number of frames per second = bit rate/number of bits per frame

$$= 34,368,000/1536$$

$$= 22,375$$

(ii) Duration of each time = 1s/22,375

$$= 44.69 \ \mu s$$

Set no.	FAS bits	Alarm bits	National use bits	Justification control/ service bits	Justification opportunity bits	Tributary data bits	Total bits
1	10	1	1	0	0	372	384
2	0	0	0	4	0	380	384
3	0	0	0	4	0	380	384
4	0	0	0	4	4	376	384
Total	10	1	1	12	4	1508	1536

FIGURE 8.23
Distribution of bits in an E3 frame.

(iii) Minimum tributary data rate per second (for E3 bit stream)

= minimum number of data bits per frame × number of frames/s

= 1508 × 22,375

= 33,741,500 bits/s

Therefore, the minimum bit rate per second for each incoming E2 tributary

= 33,741,500/4

= 8,435,375 bits/s

= 8435.4 kbps (approximate)

= 8.435 Mbps (approximate)

(iv) Maximum tributary data rate per second (for E3 bit stream)

= maximum number of data bits per frame × number of frames per second

= 1512 × 22,375

= 33,831,000 bits/s

Therefore the maximum bit rate for each E2 tributary

= 33,831,000/4

= 8,457,750 bits/s

= 8457.7 kbps (approximate)

= 8.457 Mbps (approximate)

To summarize, various parameters for an E3 bit stream are given in Figure 8.24.

Sr. no.	Parameter	Value
1	Out put data rate	34.368 Mbps ± 20 ppm
2	Frame rate	22375 frame/s
3	Frame duration	44.69 μs
4	Maximum acceptable tributary data rates	8.457 Mbps (approximate)
5	Minimum acceptable tributary data rates	8.435 Mbps (approximate)

FIGURE 8.24
E3 bit stream parameters as per ITU-T G.751.

8.3.2 Bit Rate Adaptation in E3

The minimum and maximum acceptable tributary data rates of E2 tributaries are 8.435 and 8.457 Mbps as brought out in Figure 8.24.

The E2 output bit rate is 8.448 Mbps ± 30 ppm as defined by ITU-T G.742.

Thus, we are able to accommodate the following bit rate variations of E2 tributaries:

(i) On the positive side $= 8.457 - 8.448$ Mbps

$$= 0.090 \text{ Mbps}$$

$$= 9 \text{ kbps}$$

(ii) On the negative side $= 8.448 - 8.435$ Mbps

$$= 0.013 \text{ Mbps}$$

$$= 13 \text{ kbps}$$

whereas the required variation is ±30 ppm, for clock inaccuracies.

Thus, the variation required is

$$\text{Number of bits per E2} = 8.448 \times 10^6 \times (\pm 30)/10^6$$

$$= \pm 253.44 \text{ bits/s.}$$

Hence, only 253.44 bits are required to be available for adjustment of bit rate on positive as well as on negative sides to cater for the clock inaccuracies. Some more bits will be required to cater for the impairments in the transmission media.

However, the number of bits available for adjustment is 9000 on the positive side and 13,000 on the negative side, which are much more than enough to cater for all such variations.

8.4 E4 Multiplexing

The PDH level 4 multiplexing is known as E4 (see Figure 8.9). An E4 multiplexer takes 4 E3 bit streams at its input, each at a rate of 34.368 Mbps and produces an output of 139.264 Mbps, which is commonly known as 140 Mbps. Alternatively, it may take at its input 16 tributaries of 8.448 Mbps. However, the output bit stream and structure have to be the same in both cases. The standards are defined by ITU-T G.751. The principles involved are the same as those explained earlier for E2 and E3 standards.

A bit interleaving scheme of multiplexing is deployed for similar reasons. The accuracy desired for E4 bit stream per ITU-T is

$$E4 = 139.264 \text{ Mbps} \pm 15 \text{ ppm}.$$

8.4.1 E4 Frame Structure

The details of an E4 frame are displayed in Figure 8.25. Unlike E2 and E3 standards, which have 4 sets in each frame, the E4 frame has six sets, each set consisting of 488 bits.

The assignment of functions to various sets is a little different from those of E2 and E3 frames.

8.4.1.1 Set I

The function of the 488 bits of set I are shown in Figure 8.26.

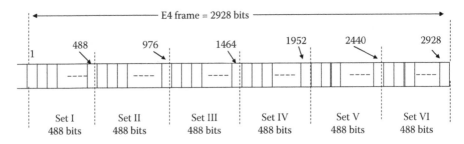

FIGURE 8.25
An E4 frame having 6 sets of 488 bits each.

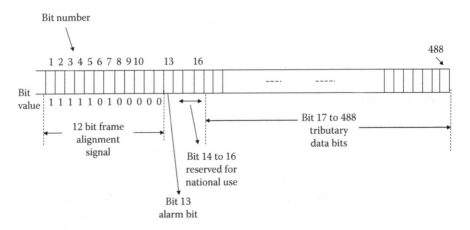

FIGURE 8.26
Set I of E4 frame as defined by ITU-T G.751.

Notice that the FAS word is 12 bits long as compared to a 10-bit FAS word of E2 and E3 frames. Since the bit rate of E4 is very high (4 times E3 or 16 times E2), it is considered that the probability of occurrence of the 10-bit long FAS pattern, i.e., 1111010000, is substantial in the data bit stream. To reduce this probability to acceptable levels, two more bits have been added to make the FAS word as 111110100000.

8.4.1.2 Bit 13

Bit 13 has been kept for remote alarm indication for "loss of signal," "power supply failure," etc.

8.4.1.3 Bits 14 to 16

These bits have been reserved for national use.

8.4.1.4 Bits 17 to 488

These 472 bits are used for tributaries data.

8.4.1.5 Set II to Set V

The next four sets, i.e., sets II–V, have identical details of bit allocations (see Figure 8.27).

The first four bits are used for justification control/service for the four E3 tributaries. The remaining 484 bits in each set are used for tributaries data.

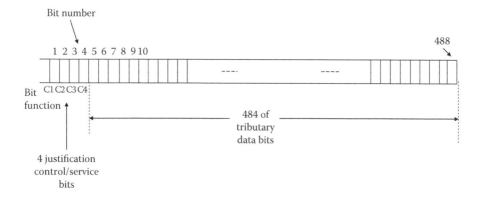

FIGURE 8.27
Output of sets II to set V of E4 per ITU-T G.751.

8.4.1.6 *Set VI*

The details are similar to the last set of E3 frames except the number of bits for tributaries data. This set is shown in Figure 8.28.

Thus, the summary of bit allocation in an E4 frame can be depicted as shown in Figure 8.29.

Note that although there are four justification opportunity bits for the four E3 tributaries, as in the previous cases, the number of justification control or service bits is 20. These 20 bits are distributed in 5 frames (from frames 2 to 6). Thus, a decision of a bit being a justification bit or a data bit in the case of E4 is taken by a set of 5 bits, one each in sets 2 to 6, in contrast with E2 and E3 frames, where it was decided by 3 bits. If the values of all these service bits are 1 (11111), the justification opportunity bit pertaining to that E3 is considered as a justification bit, and if the value is 0 for all the sets (00000), the bit is

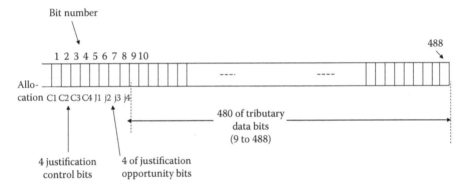

FIGURE 8.28
Details of set VI of E4 bit stream per ITU-T G.751.

Set no.	FAS bits	Alarm bits	National use bits	Justification control/ service bits	Justification opportunity bits	Tributary data bits	Total bits
1	12	1	3	0	0	472	488
2	0	0	0	4	0	484	488
3	0	0	0	4	0	484	488
4	0	0	0	4	0	484	488
5	0	0	0	4	0	484	488
6	0	0	0	4	4	480	488
Total	12	1	3	20	4	2888	2928

FIGURE 8.29
Distribution of bits in an E4 frame.

taken as a data bit. A majority decision is accepted in case the values differ from these.

Similar to the lower order bit streams such as E2 and E3, various parameters of E4 multiplexing can be calculated as under

(i) Number of frames per second = Data rate/number of bits per frame

$$= 139,264,000/2928$$

$$= 47,562.8415$$

(ii) Duration of each frame $= 1s/47,562.8415$

$$= 21.0248162 \ \mu s$$

(iii) Minimum bit rate per second (for E4 bit stream)

= minimum number of tributary data bits per frame

× number of frame per second

$$= 2888 \times 47,562.8415$$

$$= 137,361,486 \ \text{bits/s}$$

Therefore, the minimum data rate per second for each E3 tributary

$$= 1,373,361,486/4$$

$$= 34,340,371.5 \ \text{bps}$$

$$= 34.34 \ \text{Mbps (approximate)}$$

(iv) The maximum tributary bit rate per second (for E4 bit stream)

= maximum number of tributary data bit per frame × number of frames per second

$$= 2892 \times 47,562.8415$$

$$= 137,551,738 \ \text{bits/s}$$

Therefore, the maximum tributary data rate per tributary

$$= 137,551,738/4$$

$$= 34.38 \ \text{Mbps (approximate)}$$

The parameters are summarized in Figure 8.30.

8.4.2 Bit Rate Adaptation in E4

The minimum and maximum tributary data rates of the incoming E3 tributaries are 34.34 Mbps and 34.38, respectively, as worked out above.

As we have seen, the minimum "tributary data" rate occurs when we consider all the justification opportunity bits as justification bits and do not count them toward tributary data bits, whereas the maximum "tributary data" rate

Sr. no.	Parameter	Value
1	Output data rate	139.264 Mbps
2	Frame rate	47,562.8415 frames/s
3	Frame duration	21.248162 μs
4	Maximum acceptable tributary data rate	34.38 Mbps
5	Minimum acceptable tributary data rate	34.34 Mbps

FIGURE 8.30
E4 bit stream parameters as per ITU-T G.751.

occurs when we consider all the justification opportunity bits as tributary data bits and count all of them toward the tributary data.

The E3 output data rate is 34.368 Mbps ± 20 ppm as defined by ITU-T G.751.

We are able to accommodate the bit rate variation of E3 tributaries as follows:

(a) On the positive side = 34.380 − 34368 Mbps

$$= 0.012 \text{ Mbps}$$

$$= 12 \text{ kbps}$$

(b) On the negative side = 34.368 − 34.340 Mbps

$$= 28 \text{ kbps}$$

The required variation is −20 ppm for clock inaccuracies

= Number of bits required for adjustment per E3

$$= (34.368 \times 10^6) \times (\pm 20)/10^6$$

$$= \pm 687.36 \text{ bits/s}$$

Hence, only ±687.36 bits are required to be available for adjustment of bit rate, to cater for the clock inaccuracies. Some more bits will be required to cater for the impairments in the transmission media.

However, the number of bits available for adjustment, i.e., 12 and 28 kbps, respectively, on the positive and negative sides, are much more than enough to cater for all such variations.

Thus, the four E3 tributaries from different sources, having varying data rates are easily combined (into an E4 bit stream), transmitted, and properly received and separated at the other end.

8.5 Higher Bit Rates

Bit rates higher than E4 (i.e., higher than 139.264) Mbps have not been defined by ITU-T, and are not very popular. Whenever higher rates are adopted in PDH, the frame structures are proprietary of the manufacturers. However, the use of a higher-order system is no more required because of emergence of SDH standards, which have made these PDH standards obsolete, owing to numerous advantages that the SDH offers over PDH. SDH, of course, carries much higher bit rates too. Systems up to 10 Gpbs (STM-64) are deployed in abundance whereas systems up to 40 Gbps (STM-256) have begun to be installed. The recent trend is to go for more than one wavelength on the same fiber and deploy lower-order systems (up to STM-64), rather than going in for higher-order systems such as STM-256. Some of these terms may appear strange to you at this stage, but they will be amply clear when you go through the chapter on SDH.

8.6 Framing Stages, Tributary Data, Overhead, and Payload Bits

The preceding discussion brings out an important fact, that at each stage of multiplexing, i.e., E1, E2, E3, and E4, independent framing and other bits such as alarm, national use, justification opportunity, and justification control/service bits are added. These bits are called overhead bits. (The E1 frame reserves two TSs for framing, signaling, etc.) The E2 frame again adds similar bits to E1s, treating all the bits (2.048 Mbps) of E1 as "data bits." It is only when the E2 is demultiplexed into 4 E1 tributaries that the E1 overhead bits are visible as overhead bits. As long as the E1s are within an E2 frame, all the E1 tributary bits are "data bits" as far as the E2 multiplexer is concerned. A similar operation takes place while multiplexing 4 E2s into E3 and 4 E3s into E4. Thus, at every stage, the overhead bits keep on adding, thereby reducing the number of "user data" or "payload" bits. Figure 8.31 summarizes these rates in terms of basic 64-kbps user data channels.

Sr. no.	Multiplexing stage	No of 64-kbps channels carried	Total payload bits (kbps)	Total overhead bits	Final bit rate (kbps)	% of overhead bits
1	E1	30	1920	128	2048	6.25
2	E2	120	7680	988	8448	11.69
3	E3	480	30,720	3648	34,368	10.61
4	E4	1920	122,880	16,384	139,264	11.76

FIGURE 8.31
The ratio of payload and overhead bits in various PDH levels.

As can be seen, a substantial percentage of total throughput does not contribute to the channel capacity but is used as "overhead" for carrying the channels properly to the destination.

8.7 North American PDH Hierarchy

Although basic digitization of a voice channel was standardized at 64 kbps because of Nyquist's sampling theorem (see Section 3.2) and the principles of PCM and multiplexing are by and large the same all over the world, PDH hierarchical rates are substantially different in North America and Europe. While Europe and the rest of the world follows E1, E2, etc., levels, North America and Japan follow the T1, T2, etc., hierarchy.

The most fundamental difference in the multiplexing, signaling, etc. is at the first level of PDH. Although E1 has 30 voice channels of 64 kbps, T1 has only 24 such channels. E1 reserves two TSs (equal to two channels) for frame alignment and signaling, whereas T1 merely adds one bit to each frame to manage all the functions. Thus, as compared to an E1 frame of 32 × 8 = 256 bits, a T1 frame is only (24 × 8) + 1 = 192 + 1 = 193 bits. Obviously, frame alignment and signaling functions are managed in an entirely different way in either of the systems; nevertheless, the principles involved are the same as

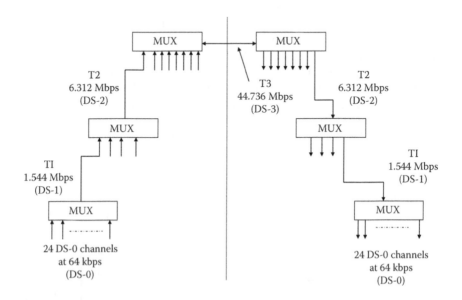

FIGURE 8.32
North American PDH hierarchy. The levels are also called digital signal (DS)-0, DS-1, DS-2, and DS-3 as shown.

those in E1s in general. Thus, if the reader understands the European hierarchy, i.e., E1, E2, etc., he will not have any problem in understanding T1, T2, etc. when he comes across them. But since the scope of this text is to cover the systems based on ITU-T standards, we will not discuss the American hierarchies in detail. The bit rates at various hierarchical levels, however, are as shown in Figure 8.32.

8.8 Types of Line Codes used in PDH

In Figures 8.1, 8.2, 8.3, etc., we have seen the bits or pulses of the E1 bit streams getting multiplexed to form higher-order bit streams such as E2. The bits have been shown as positive pulses with a gap in between only to make the illustration simple. In actual practice, the shapes of the transmitted pulses are a little different. The type of pulses is chosen depending upon the speed of transmission, transmission media, and standardization factors (see Chapter 4).

The type of pulses called "line code" used in PDH is HDB-3 (high density bipolar-3; see Section 4.2.4 for HDB-3 details) to obtain good timing content in the bit stream from E1 to E3 levels. However, at E4 level, a different line code known as coded mark inversion (CMI) is used (see Section 4.2.5). The reason is that HDB-3 is a return-to-zero (RZ) code, and when the bit rate is too high (like 140 Mbps of E4), the time for which the pulse remains at no potential or "zero" becomes increasingly less, and hence, it becomes more difficult for the equipment to judge a 0 bit. Therefore, line code CMI is generally chosen, in which a 0 bit is indicated by changing the polarity of the pulse from negative to positive midway between the pulse. Since CMI is a non-RZ code, the problem of 0 bit detection due to short pulse is eliminated (see Section 4.2.2).

8.9 Synchronization in PDH

In digital communication systems, nothing can work without synchronization. In other words, if there is no bit by bit correspondence in the transmitted and received signals/data/bit stream, the received signal cannot generate any useful information. Let us say if the bit 5 is received as bits 4 or 6, the whole sequence will be upset and what will be received is only garbage (see Chapter 7's introduction).

A PDH bit stream consists of framing bits, alarm bits, national use bits, justification opportunity bits, and justification control bits, besides data bits (see

Figure 8.29). All the bits occur at fixed positions or TSs in the frame. Thus, when the receiver receives an incoming E2 bit stream, it looks for the pattern of "FAS, i.e., 1111010000"; as soon as it locates this sequence of bits, it recognizes the following 838 (848-10) bits, per the predefined functions assigned to various bits per their numbers. The receiver counts the bits and recognizes bit 11 as the alarm bit, number 12 bit as the bit for national use, and so on. After bit 12, it comes across bit 13, which is a data bit from tributary 1, bit 14 from tributary 2, 15 from tributary 3, and 16 from tributary 4, and then the sequence repeats. Thus, various overhead and payload bits are distinctly recognized, and the E1 tributaries can be demultiplexed correctly. The synchronization is thus achieved. However, if the clock of the receiver does not match with that of the transmitter, the counting of bits by the receiver may not be accurate enough to maintain the one to one correspondence between the transmitted and received bits for long. This may cause loss of synchronization. Thus, to achieve and maintain synchronization, it is necessary that the receiver and the transmitter clock frequencies are the same. Exactly the same frequencies of transmitter and receiver are not realizable in practice, because of built-in inaccuracies of the clock-generating equipment and many environmental factors such as temperature, humidity, aging, etc. Thus, to achieve this one to one correspondence between the transmitted and received bits, one of the following two techniques are adopted in PDH systems:

(i) Using master–slave synchronization
(ii) Using high-accuracy clocks

The functions of various bits of an E2 frame are shown in Figure 8.33.

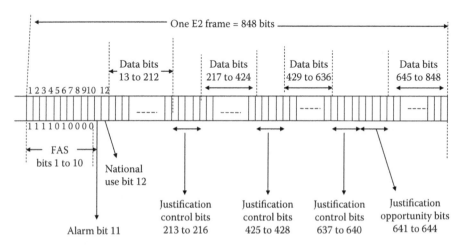

FIGURE 8.33
The details of bit allocation (or TS allocation) in an E2 frame. Data bit means the E1 tributaries bits. Set numbers are not shown to maintain clarity.

8.9.1 Using Master–Slave Synchronization

When two nodes are interconnected for digital communication, the synchronization between their bit streams can be easily achieved if one of them is made a master and the other one a slave for the purposes of operation of their clocks. This means that the slave clock will derive its timings from, and lock on to, the timings of the master clock. The arrangement is shown in Figure 8.34 (see Chapter 5 for a detailed discussion on clocks).

In this case, the node clock (node's local clock) of node A has been made as the "master clock" and that of node B has been made as the "slave clock." The master clock at node A drives its transmitter directly, whereas the slave clock locks itself on to the timing signals received at the node B from node A. Thus, the timings of occurrence of signal bits in the bit stream are the same at the transmitter of node A and the receiver of node B. Moreover, since the same timing signals come back to the node A receiver, there is a perfect matching again in the timings of bits transmitted at node B and received at node A. (Although there are no rules about which of the two clocks should be a master, generally, the clock with higher accuracy is made a master.)

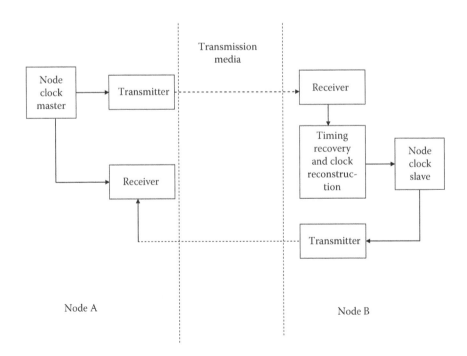

FIGURE 8.34
The master–slave operation of two nodes for synchronization.

8.9.2 Using High-Accuracy Clocks

The master–slave technique described above is a very good means of achieving synchronization; however, it is not possible to implement it in certain situations. For example, consider the situation depicted in Figure 8.35.

The node A has been configured as a master to nodes B and C, which have been configured as slaves to A for their clocks. This works fine for communications between A and B and between A and C. However, the problem comes when nodes B and C wish to communicate with each other. Since the clocks of both these nodes are slaves to node A, they cannot be made slave to any other node (including either of them), because one slave clock can lock on to only one master clock. Therefore, nodes B and C will have to work without synchronization with each other. (Another way they can communicate with each other is via node A, which is synchronized to both, but if geographically B and C are close by, it will be a waste of capacity to communicate via A.)

In this example, it is possible to define master–slave operation in a chain, say first make slave one of the nodes let us say B to node A and then slave the second node, i.e., C to node B, which will work. But this was an extremely simplified example, just for illustration. Generally, if the network is a linear chain, this arrangement may work, but the practical networks are never completely linear, as they have to cover an "area" with nodes spread over in all directions. The practical communication networks are much more complex, where there will not be any possibility of achieving a complete master–slave operation to make one of the nodes as master and the rest all as slaves.

One such network is shown in Figure 8.36.

Hence, in practical PDH networks, the master–slave operation is limited to a certain number of nodes where it is easily practicable.

The interconnections between the networks so formed are carried out without synchronization. The clocks used are of high accuracy, so that the systems can work without intolerable mismatches. This system of using

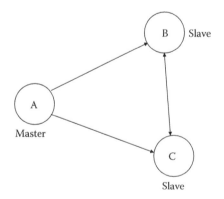

FIGURE 8.35
Three PDH nodes. The direction of the arrow shows the direction of timing supply.

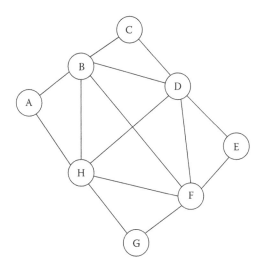

FIGURE 8.36
A more complex network where master–slave operation is not possible.

high accuracy is the one that generates the name "plesiochronous" (meaning nearly the same timings), and hence the name of the whole process of multiplexing and transmission was known as PDH.

Thus, in essence, the PDH systems work in partly synchronous (master–slave wherever possible) and partly asynchronous manner with plesiochronous clocks.

8.9.3 Clock Accuracy and Number of Slips in PDH Systems

We have talked about the use of highly accurate clocks for asynchronous operation in PDH systems. Let us see how accurate the clock should be and what are the consequences of poor accuracy.

Let us consider an E3 bit stream.

The bit rate of E3 bit stream is 34.368 Mbps ± 20 ppm (this accuracy translates to ±687.36 bits/s). Now if we use the clock of the same accuracy (i.e., ±20 ppm) in the transmitter as well as the receiver, and if both the clocks give their worst performance on opposite sides, then the total error will be (687.36) × 2 = ±1375 bits/s. This means that the receiver might receive 1375 bits more or less than 34.368 Mbps. These "missing or extra" bits per second are called slips. (For a detailed discussion on "slips," see Section 7.5.)

The slips are generally expressed in number of "frame slips." The effect of slips is minimized using a technique called "controlled slips" in asynchronous systems. The controlled slips application allows a certain number of frames to slip instead of allowing the whole system to go out of synchronization. If there is no synchronization in the transmitter and receiver of the E3 bit stream, there will be a large number of slips. Since the frame size in E3 is 1536 bits, the number of "frame slips" per second will be

= 1375/1536 (They will be on the positive or negative side depending upon the sign, either +1375 or –1375.)

= 0.89 frame slips/s

= 3223 frame slips/h (or simply slips/hour)

We can see that this is a very high number and is certainly unacceptable. The solution then is to increase the clock accuracy. The highest accuracy clocks are atomic "cesium beam" clocks with accuracies of better than ±1 part is 10^{11}.

This will translate to

$$\text{Number of bit errors per/second} = (1/10^{11}) \times 34.368 \times 10^6$$
$$= 0.0003468 \text{ bits/s}$$
$$= \pm 1.24848 \text{ bits/h}$$
$$= \pm 2.497 \text{ or say } 2.5 \text{ bits/h}$$
$$\text{(multiplied by 2, for worst side}$$
$$\text{performance of transmit and receive clocks)}$$

Therefore, the number of frame slips/hour = 2.5/1536 = 0.0016.

Hence the number of frame slips/day = $(2.5/1536) \times 24$
$$= 0.039,$$

which means that there will be one slip in 1/0.039 = 25.64 days.

When we use such highly accurate clocks, the asynchronous operation is called a plesiochronous operation.

This one slip is 25.64 days, which is much better than the minimum acceptable performance for voice as well as data applications. But this necessitates a very costly, cesium beam/rubidium clock to be used at every node, or at least at the boundaries of every network segment that has master–slave synchronization within itself (and thus not needing highly accurate clocks), and then connected to another similar network segment for the plesiochronous operation as stated above. For interconnecting two networks of different operators who will certainly have different clocks, leading to asynchronous operation or PDH, at least a few such clocks will be required with each operator.

8.9.4 Current Trends

The situation described above was practicable when the number and size of digital networks was small. As the number of nodes of each operator and the number of operators grew, they found that maintaining, upgrading (particularly), and adding additional nodes became very difficult in the asynchronous operation. They eventually started turning to establish synchronization for their complete networks. However, finally, SDH was developed in late 1980s, which provided a complete networkwide synchronization, besides eliminating many other problems of PDH systems. Thus, PDH is now obsolete and has given way to SDH. However, there is no other solution for providing synchronization between two operators; thus, the interoperator network connection continues to be on PDH, although each operator tries to build a complete synchronized nationwide SDH network of his own.

Nevertheless, the study of PDH is a necessary stepping stone to understand the SDH.

8.9.5 E1 Synchronization through Plesiochronous Network

The justification process allows the master–slave synchronization of two remotely located exchanges that are connected through an E1 in a PDH network, even though the PDH is an asynchronous system (see Figure 8.37). (It is important to understand that irrespective of the nomenclature, "PDH" is basically a definition of hierarchies such as E1, E2, etc., whereas "plesiochronous" is a synchronization methodology as explained in the preceding sections.)

Various E1s are received at different bit rates at the input of the E2 multiplexer. The bit rates are equalized through justification (see Section 8.2.3 for justification) and E1 tributaries are multiplexed to form E2 and then E3 levels. The bit rate of an E1 may be different for transmission through the network from its original bit rate due to the justification process, but on demultiplexing of higher PDH levels up to E1 level, the justification bits are removed, and the same average bit rate as that of original E1 at the sending end is obtained. The timings or the bit rate of the E1s are thus maintained at remotely located exchanges through the PDH network, leading to synchronization of the exchanges. This also implies that, if a master clock is used at one end of the E1, master–slave synchronization can be achieved, for the point to point E1 connectivity through the PDH asynchronous network.

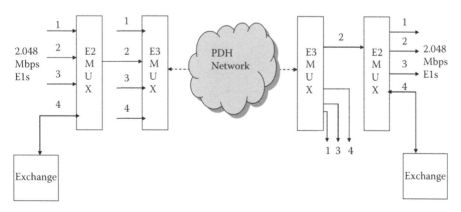

FIGURE 8.37
Interexchange synchronization through PDH network.

8.10 Asynchronous vs. Synchronous Multiplexing

This is one terminology that often creates confusion. The use of the term *synchronous/asynchronous* with *transmission* (or transport) and with *multiplexing* has different meanings. The PDH multiplexing of E1 to E2 to E3 and E4, which we have seen so far, are all examples of asynchronous multiplexing, and in Section 8.9, we have discussed asynchronous transmission and reception. Transmission and reception (one has to exist for the other to exist) together constitute a synchronous or asynchronous system. (We may call them synchronous or asynchronous operation or transport synonymously.)

As a reminder, we are discussing asynchronous and synchronous "multiplexing."

The E1 tributaries as we have seen reach the E2 multiplexer at varying data rates. Hence, they are asynchronous with each other and are called "asynchronous tributaries." Multiplexing these tributaries is called "asynchronous multiplexing." Even though their bit rates are equalized by means of "stuff" or "justification" bits in the E2 multiplexer, since the tributaries entering the multiplexer are at unequal rates, multiplexing them into an E2 stream is called "asynchronous multiplexing." As we have seen in Section 8.3.3, the bit rate adaptation unit equalizes the bit rates of all the E1s by bit stuffing, but their original bit sequence is lost in the process. Similarly, the other levels of PDH are also called asynchronous multiplexing (see Figure 8.38).

To make it clearer, let us see what synchronous multiplexing is.

If the bit rates of all four E1s are made equal before multiplexing them (of course, without the addition of stuff/justification bits), then multiplexing them will be called synchronous multiplexing. A schematic arrangement is shown in Figure 8.39.

The asynchronous E1 tributaries are received by an elastic store buffer unit, at the output of which the E1s emerge synchronized. This synchronization

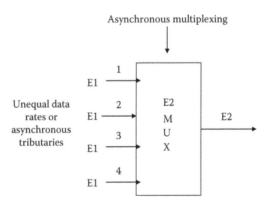

FIGURE 8.38
Asynchronous multiplexing of 4E1s to form an E2.

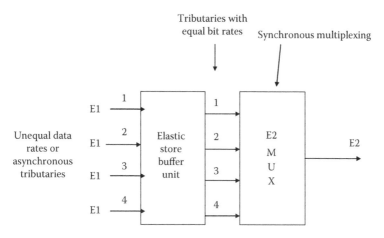

FIGURE 8.39
Schematic of a synchronous multiplexer.

is achieved with the help of controlled slips. These controlled slips (for a discussion on elastic store and controlled slips, see Sections 7.5 and 7.6) are allowed to occur at the input of the multiplexer to achieve the data rate equalization (or synchronization) of all the E1s. The synchronized E1s are then multiplexed to from the E2 stream. The example has been taken for illustration purposes only; actually, E2 stream is never formed like this. However, such a formation of synchronous E1s is done in few SDH designs, where the requirement is to derive the 64-kbps channels directly from STM bit stream. The derivation of 64 kbps or basic channels becomes possible due to correct identification of each bit in the STM bit stream because the stuff bits are not present. The presence of stuff bits makes the identification of channels difficult, as a stuff bit could actually be either a bit to be discarded or a part of the data, which will be known only after demultiplexing.

This multiplexing of synchronous tributaries is called synchronous multiplexing.

8.11 Skip Multiplexers

In PDH hierarchy, we have seen that 4 E1s get multiplexed to form an E2, 4 E2s get multiplexed to form an E3, and 4 E3s get multiplexed to form an E4. However, in some cases, it may not be desirable to adopt all the stages of multiplexing. In these cases, one or more stages of multiplexing are "skipped." Such multiplexers are called "skip multiplexers." In fact, in PDH, the skip multiplexer that multiplexes 16 E1s directly to form an E3 is more popular than the proper stage-wise multiplexer (see Figure 8.40).

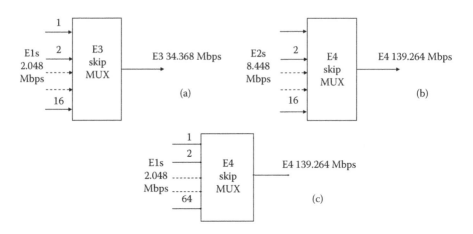

FIGURE 8.40
(a–c) Three types of skip multiplexers.

In all these cases, the final frame format and bit rates remain the same as the "stage-wise" multiplexer.

Review Questions

1. What is the necessity of multiplexing?
2. Why did the PCM stand for a 24-channel multiplexed system?
3. What is the relationship between a PCM and PDH? What do you understand about plesiochronous digital hierarchy?
4. What were the other names for the first level of PDH?
5. What are ITU-T and CCITT?
6. What are the differences between E1 and T1?
7. What do you understand about TS? How are the TSs distributed in an E1?
8. Explain the process of multiplexing through bit interleaving and byte interleaving.
9. What are the relative merits and demerits of bit interleaving and byte interleaving?
10. Explain the multiplexing structure of an E1 and work out its bit rate.
11. What is the role of various bits of TS 0 in an E1?

12. For what purpose is a multiframe used in an E1? What will happen if a multiframe is avoided and only normal single frames are used?

13. What is CRC and how is bit 1 of TS 0 used for it?

14. What is the need of FAS? How is it carried in an E1?

15. Which are the bits used for alarms and country-specific applications?

16. What is called signaling in the context of telecommunication?

17. How exactly is the TS 16 of an E1 used for signaling? What type of signaling is deployed?

18. Explain the terms "channel-associated signaling" and "common channel signaling."

19. What are the duties of the bits of the first frame of the multiframe of TS 16?

20. How and why was the PDH developed? What were the minimum and maximum bit rates? Explain the multiplexing structure with the help of a schematic diagram.

21. What do you understand about the terms "overhead and payload?"

22. Are the bit rates of 4E1s and one E2 the same? If not, why?

23. What problems can take place if the bit rates of all the E1s that are to be multiplexed into an E2 are not the same?

24. What are the reasons for different E1s to be at different bit rates?

25. What are "justification" and "bit stuffing" techniques? Where are they deployed?

26. Differentiate through an example the positive justification from negative justification.

27. How is bit dropping practically achieved?

28. Explain with illustration the bit rate adaptation process of the E2 multiplexer.

29. How many justification bits are required and how many of them are practically added and why?

30. Why do we need justification control bits? Where are they located? What is justification opportunity?

31. Explain the principle of majority voting in justification control.

32. Explain the E2 frame structure with the help of a diagram.

33. What is the significance of the concept of sets in E2 frame?

34. What are the functions of various bits of set I of an E2 frame?

35. What are the criteria for choosing between bit interleaving and byte interleaving?

36. What are the functions of sets II and III of an E2?

37. Summarize the distribution of functions of all the bits of an E2 frame.
38. What are the frame rate and frame duration of an E2 frame? Why is frame duration different from the standard 125 μsec?
39. Summarize various parameters of an E2 bit stream.
40. What is the bit rate variation expected in the E1 bit streams and how much is the accommodating capacity in an E2?
41. What are the bit rates and accuracy of E3 bit stream?
42. Explain E3 frame structure with the help of a diagram.
43. Summarize the bit distribution in an E3 frame.
44. Tabulate various parameters of an E3 bit stream.
45. Explain bit rate adaptation in an E3 frame.
46. What are the bit rate and accuracy of an E4 frame?
47. Explain E4 frame structure and bit assignments with a diagram.
48. Summarize bit distribution and various parameters of an E4 frame.
49. Bring out important differences between E3 and E4 frames.
50. How does bit rate adaptation in an E4 frame work?
51. Are higher rates than E4 possible or available in PDH? If not, why?
52. Summarize the payload and data rates for all levels of PDH hierarchy. Which levels have maximum and minimum overhead bits?
53. Draw a diagram showing various levels of the North American PDH hierarchy.
54. Which are the line codes used in PDH and why?
55. How is synchronization achieved in PDH? Explain master–slave synchronization with a schematic diagram.
56. How the synchronization is achieved by the use of high accuracy clocks?
57. Calculate the number of slips permitted by various clock accuracies in a PDH system.
58. What are the current trends in synchronization?
59. How is E1 synchronization achieved in a plesiochronous network? Explain through an illustration.
60. How do you differentiate between synchronous multiplexing and asynchronous multiplexing? What are their specific applications?
61. What do you understand about skip multiplexers? Where are they used and why?
62. Draw a diagram of a skip multiplexer other than what is shown in this chapter.

Critical Thinking Questions

1. What could have been alternative names suitable for plesiochronous hierarchy?

2. It is said that PDH is obsolete these days; however, what will be the repercussions on the telecommunication scenario of the world if PDH systems are totally removed from service?

3. If we use 20,000 frames/s instead of 8000 frames, what will be the repercussions on quality of speech, and channel capacity?

4. Could we provide signaling using only 3 bits of TS 16 instead of 4? If yes, how?

5. Find out from the Internet the details of "T1," like number of channels, type of signaling used, total number of bits in a frame, etc.

6. Which type of electronic devices or circuits can insert the justification bits in the appropriate positions?

7. What, according to you, should be the percentage variation in the bit rates of the tributaries attributed to different factors affecting it?

Bibliography

1. P. Moulton and J. Moulton, *The Telecommunications Survival Guide, Understanding and Applying Telecommunication Technologies to Save Money and Develop New Business*, Pearson Education, 2001.

2. S.V. Kartalopoulos, *SONET/SDH and ATM, Communications Networks for the Next Millennium*, IEEE Press, 2003.

3. P. Harikumar, *Teaching Notes on Synchronous Digital Hierarchy*, IRISET Secunderabad, 2002.

4. *In Service Course on SDH Systems*, Regional Telecom Training Centre, Hyderabad, 2000.

5. J.C. Bellamy, *Digital Telephony*, John Wiley and Sons, Singapore, 2003.

6. P.V. Shreekanth, *Digital Transmission Hierarchies and Networks*, University Press, 2010.

7. NIIT, *Advanced Digital Communication Systems*, Prentice-Hall, India, New Delhi, 2005.

8. B.P. Lathi, *Modern Digital and Analog Communication Systems*, Oxford University Press, Oxford, UK, 1998.

9. ITU-T Recommendation G.702, *Digital Hierarchy Bit Rates*, International Telecommunication Union, Geneva, Switzerland.

10. ITU-T Recommendation G.703, *Physical/Electrical Characteristics of Hierarchical Digital Interfaces*, International Telecommunication Union, Geneva, Switzerland.

11. ITU-T Recommendation G.704, *Synchronous Frame Structures Used at 1544, 6312, 2048, 8448 and 44 736 kbit/s Hierarchical Levels*, International Telecommunication Union, Geneva, Switzerland.
12. ITU-T Recommendation G.742, *Second Order Digital Multiplex Equipment Operating at 8448 kbit/s and Using Positive Justification*, International Telecommunication Union, Geneva, Switzerland.
13. ITU-T Recommendation G.745, *Second Order Digital Multiplex Equipment Operating at 8448 kbit/s and Using Positive/Zero/Negative Justification*, International Telecommunication Union, Geneva, Switzerland.
14. ITU-T Recommendation G.732, *Characteristics of Primary PCM Multiplex Equipment Operating at 2048 kbit/s*, International Telecommunication Union, Geneva, Switzerland.
15. ITU-T Recommendation G.751, *Digital Multiplex Equipments Operating at the Third Order Bit Rate of 34 368 kbit/s and the Fourth Order Bit Rate of 139 264 kbit/s and Using Positive Justification*, International Telecommunication Union, Geneva, Switzerland.

9

Plesiochronous Digital Hierarchy Maintenance Alarms

In any practical communication system, it is necessary to ensure that the signal or data that are being transmitted by a transmitting station is being received correctly by the receiving station. The timely knowledge of the problem is necessary to restore the services quickly in case of a disruption or incorrect reception.

There could be several impediments to the correct reception of the signal/ data. The following two situations can arise:

(i) No signal/data being received. (Digital signals are termed as *signal* or *data*, as these signals are always in the form of bits.)

(ii) The signal being received is not the exact replication of the transmitted signal.

The first situation can arise either due to a fault in the equipment, its associated power plant, etc., at the transmitting or receiving nodes or due to a break in the transmission media such as optical fiber or copper cable.

On the other hand, the incorrect reception of data could be due to various impairing factors in the transmission media (mostly, it is the media causing the problem; very rarely do we come across an incorrect reception problem due to equipment failure), as we have seen in detail in Chapter 6.

The plesiochronous digital hierarchy (PDH) standards cater for detection and reporting of both the above types of faults in the system. While going through the previous chapter on PDH, we have come across the description of the roles of various bits of the E1, E2, E3, and E4 frames. Aside from the bulk of the bits that carry user data, there are some bits dedicated to alarms, indication of national/ international origin, and cyclic redundancy check (CRC) applications. These bits, in addition to some more pattern recognition mechanisms built into the equipment, help us to locate the fault and determine its nature.

Let us have a look at these features of PDH systems.

9.1 Types of Alarms

The following alarms are provided in the PDH systems:

(i) *Loss of signal (LOS):* This is a local alarm that gets activated when the signal is "not received" from a transmitter due to any reason.

(ii) *Loss of frame (LOF):* This is a local alarm at a node that gets activated when the framing sequence (frame alignment signal [FAS] word) is not properly received.

(iii) *Alarm indication signal (AIS):* This is an alarm signal sent to the next station downstream when no signal is being received from an upstream node, or if there is an LOF.

(iv) *Remote defect indication (RDI):* This is an alarm indication sent by a node that receives AIS from the previous node. This indication is sent back to the node that had sent AIS to this node. RDI is also sent to the upstream node by a node having a LOS or LOF alarm.

(v) *Loss of multiframe alignment (LOMF):* This signal is generated locally at a node where the multiframe alignment is lost due to equipment failure.

(vi) *Multiframe alignment AIS (MFAIS):* This alarm indication is sent to the next node downstream when the multiframe alignment is lost.

(vii) *Multiframe alignment RDI (MFRDI):* This is an alarm indication sent by a node that receives an MFAIS from the previous node. This indication is sent back to the node that had sent the MFAIS.

(viii) *Remote end block error (REBE):* This alarm indication is generated when the CRC check-sum mismatch takes place, indicating bit errors in the received data. This indication is sent back to the transmitter node.

All the above alarms and indications are from the standards defined by the International Telecommunication Union, Telecom Standards group (ITU-T) in their recommendation G.704, which guide all the manufacturers of the systems. The individual manufacturers, however, usually offer some more indications by using the spare bits available in the standard formats.

Now we will have a look at each of these alarms in more detail.

9.2 Loss of Signal

A local alarm is generated at a receiving node when it stops receiving any data. The LOS may be due to a disruption in the transmitting media such as copper cable or optical fiber cable (OFC) or even due to a high amount of

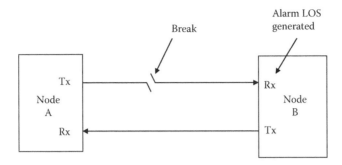

FIGURE 9.1
LOS alarm.

fading in terrestrial microwave or satellite communication. The LOS may also take place due to an equipment failure at the transmitting node (Figure 9.1).

But when and how does the equipment at node B decide to set the alarm? Or how long does it have to wait on a situation of "no received signal" before activating the alarm? Fortunately, there is no confusion on this account, as the criterion has been clearly defined by ITU-T in their recommendation number G.775. The criterion goes like this:

> If there are "no transitions" in the received signal for more than 10 and less than 255 "pulse intervals," an LOS is declared. A "no transition" is a signal level that is less than or equal to 35 db below nominal signal level (for all PDH levels from 2.048 to 139.264 Mbps).

The equipment also needs to know a criterion when it can "clear" or stop the alarm. This clearance criterion is "to have 'transitions' in the received signal for more than 10 and less than 255 'pulse intervals.'" A transition is considered to be a signal level that is greater than or equal to 9 db below the nominal signal level for the 2.048-Mbps signal (for 34.368 and 139.264 Mbps, it is 15 db).

9.3 Loss of Framing

Referring to Figure 9.1, if the signal is being received by node B (there is no LOS), but if the framing sequence (or the FAS word) is not being received properly, it will not be able to use the received data, leading to the same consequences as LOS. This type of error is quite common when a temporary or permanent fault is present in the transmission media or the transmitting equipment or due to any other impairment to the signal.

The criterion for equipment at node B to decide and generate the LOF alarm is to "not receive the framing sequence properly for 'three consecutive frames'" or "receive the bit 2 of the time slot 0 (TS-0) of odd frames incorrectly

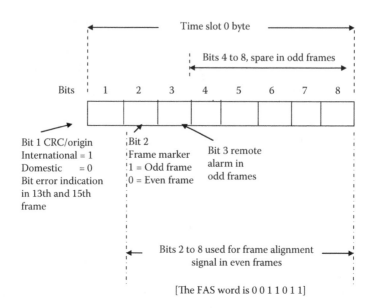

FIGURE 9.2
Assignment of various bits of TS-0 in the E1 frame.

for three consecutive frames." To understand it clearly, let us redraw the bit allocation diagram of TS-0 in a 2048-kbps bit stream of an E1.

As can be seen in Figure 9.2, the FAS word is defined in bits 2 to 8 of the TS-0 of even-numbered frames. The FAS word is "0011011." If this sequence is not received correctly for three consecutive frames, the LOF alarm is generated by the receiving equipment. Bit 2, which is a frame marker bit, is always set to 1 in the odd frames and to 0 in the even frames. The odd frames are not supposed to carry the FAS word; thus, if odd frames have a value of bit 2 set to 0, it is a signature of erroneous framing information. Hence, three consecutive wrong values of bit 2 in odd frames will also entitle the equipment at the receiving end to generate an LOF alarm. Figure 9.3 illustrates the principle.

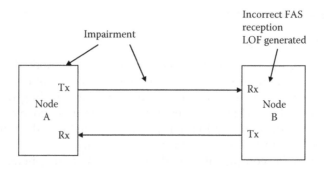

FIGURE 9.3
LOF alarm.

(There were other criteria being used by various vendors for different alarms before the ITU-T recommendation G.775 was made. They are not explained here because our objective is to impart an understanding of the concept rather than providing a detailed reference. The PDH by itself is obsolete today. We are discussing the details in this text only to develop an understanding of the prevailing synchronous digital hierarchy [SDH] technology, for which the understanding of PDH is necessary. In fact, the alarms and management features of SDH are a further development of PDH features only.)

9.4 Alarm Indication Signal

When an LOS or LOF is detected at a node in accordance with Sections 9.2 and 9.3, the node generates an AIS and sends it to the next node downstream. Figure 9.4 shows the AIS generation and transmission.

The format of the AIS signal sent to the downstream node (node C in this case) is "all 1s" in an "unframed" format. The continuous "transitions" in the 1s bit stream help in maintaining the timings in node B.

The detection criterion for the alarm at node C is the detection of "less than or equal to two 0s" in the bit stream in each of the two consecutive double frame periods (i.e., 512 bits). Naturally, when all 1s are being transmitted by node B, no 0s will be detected at node C, and thus, an AIS alarm will be activated. The defect is cleared and the alarm is stopped when node B starts sending the normal signal containing data bits. The criteria for clearance adopted by node C are reception of three or more 0s in each of the two consecutive double frame periods (which is very normal for any real data) or on the detection of proper FAS.

When node C receives an AIS signal from upstream, it does not have any real data to send to the downstream node. It can pass on the same AIS signal to the next node. However, carrying the alarm to more nodes is neither useful nor necessary, thus node C sends a bit stream of all 1s to the downstream nodes, so that the timing pulses at the downstream node can be sustained in the absence of a

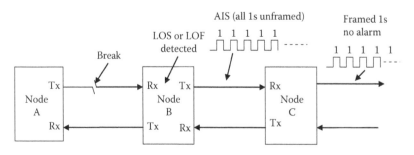

FIGURE 9.4
AIS generation and transmission.

PDH level	Bit rate (Mbps)	Number of 0s in two consecutive frames, for detection	Number of bits in each frame	Number of 0s, in two consecutive frames, for clearance
E1	2.048	2	512 (double frame)	3
E2	8.448	4	848	5
E3	34.368	4	1536	5
E4	139.264	5	2928	6

FIGURE 9.5
The detection and clearance criteria for various PDH levels.

real data signal, but this time, this "all 1s" signal is "framed" to make it different from the AIS signal. The framed 1s will have a proper FAS word, thereby avoiding activation of an AIS alarm, per the criteria mentioned in the earlier paragraph.

The detection criteria for higher PDH levels like E2, E3, and E4 are similar, except the number of 0s to be detected in the bit stream for activation or clearance in two consecutive frames. The details are shown in Figure 9.5.

9.5 Remote Defect Indication

When AIS is detected at a node, it generates a local AIS alarm and sends an alarm signal to the node from where it received the AIS signal. Let us consider Figure 9.6 for the discussion.

Node C receives AIS (sequence of all unframed 1s) from node B and activates an AIS alarm locally. Generation of the AIS alarm sets the position of "remote indication alarm bit" to 0, which is otherwise set to 1. Let us recall Figure 9.2, redrawn in Figure 9.7 for better appreciation.

Bit 3 of odd frames of TS-0 is reserved for the "remote indication alarm" or RDI. Thus, all the frames being transmitted from node C to node B, after the generation of AIS alarm at node C, will contain 1 in this bit. Node B will notice this change and will activate the RDI alarm.

The criterion for the detection of the RDI alarm (in this case at node B) per ITU-T G.775 is the detection of a 1 in this alarm bit position for two to five consecutive double frames, in case of an E1 bit stream. The defect will be cleared when a 0 is detected in two to five consecutive double frames. In case of E2, E3, and E4, it is three to four consecutive frames. The alarm bit in case of E2 and E3 is bit 11 of set I, and in case of E4, it is bit 13 of set I (refer to the previous chapter for details of these PDH levels).

The generation of LOS or LOF at node B, in addition to generating a signal for transmission to node C, also generates the RDI indication for transmission to node A. Thus, RDI is received by both node B and node A.

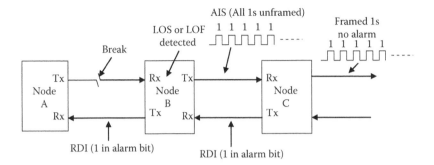

FIGURE 9.6
Generation and transmission of RDI alarm.

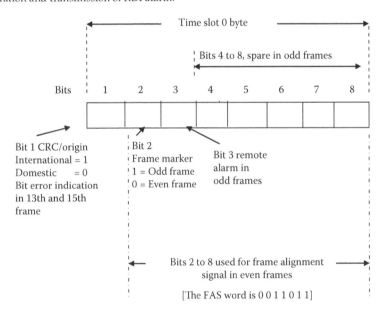

FIGURE 9.7
The alarm bit in TS-0 of odd frames of an E1 (2.048-Mbps) frame.

9.6 Determination of Fault Location

From the position of alarms at various nodes, we can diagnose the faulty section (refer to Figure 9.6).

In the depicted situation, node A is receiving an RDI alarm from nodes downstream to node A, i.e., node B onward. Node C is receiving an AIS signal from an upstream node and also sending a RDI alarm, meaning that the fault is beyond the upstream node. This establishes that the faulty section is between node A and node B.

9.7 Loss of Multiframe Alignment

At times, at a particular node, multiframe alignment may be lost even if the frames are received properly. It means that the multiplexer equipment does not know the frame number. This type of defect may come up due to some problem in the node equipment. Let us recall Section 7.1.3 of Chapter 7, where we have seen the details of multiframing. Figure 9.8 is redrawn from this section for easy reference.

TS-16 carries the information about multiframe alignment, and signaling information about all the 30 voice channels. The first of the 16 frames of a multiframe carries the multi-FAS as its first four bits ("0000"). If there is a corruption of any or all of these four bits or there is a problem in recognition, the multiframe alignment will be lost, meaning that the frame numbers will not be identifiable. This will result in the signaling information being rendered useless.

The node then generates an LOMF alarm locally.

The criterion for detecting the LOMF as defined in ITU-T G.706 is that the multiframe alignment should be considered to be lost if it does not accomplish within 8 ms (for E1 bit stream).

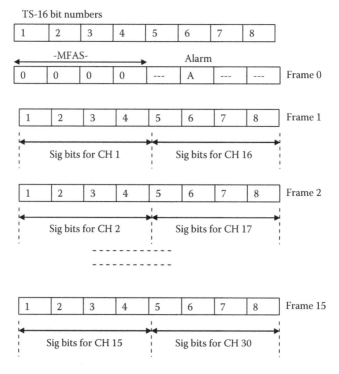

FIGURE 9.8
TS-16 bit assignment of an E1.

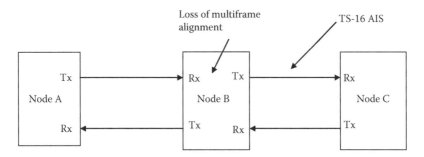

FIGURE 9.9
Detection of LOMF and transmission of AIS.

9.8 Multiframe Alignment AIS

When a local alarm on the loss of multiframe is generated at a node, as explained in Section 9.7, an AIS signal is sent down the stream to the next node. This AIS signal is sent using the complete TS-16. The signaling information contained in the TS-16 is rendered useless as soon as the multiframe alignment is lost. Thus, TS-16 becomes free from its normal duties and is used to send AIS, also called TS-16 AIS, to the next node. The format of this TS-16 AIS signal is "setting all the bits of TS-16 to 1." Figure 9.9 illustrates the situation.

9.9 Multiframe Alignment

Upon receipt of the MFAIS in TS-16, node C sets the value of the alarm bit of TS-16 to 1 in the signal being transmitted to node B (refer to Figure 9.9). We have already learned in Section 8.1.5 of Chapter 8 (and also seen in Figure 9.8) about the bit assignment of TS-16 in an E1 bit stream. Figure 9.10 shows the bit assignment of TS-16 of frame 1 of a multiframe of 16 frames.

Bit 6 used for the alarm signal is shown in Figure 9.10. The bit is normally a 0 and is set to 1 in case of alarm, i.e., when node C receives a TS-16 AIS from node B. The alarm is again called RDI; let us call it MFRDI to differentiate it from the RDI defined earlier in this chapter. The MFRDI is thus sent to the upstream node B (Figure 9.11).

As stated earlier in the beginning of this chapter, this is an alarm signal that is sent back to the transmitter, when there is a mismatch in the CRC checksum at the receiver, indicating bit errors in the received data.

CRC is a very powerful tool to detect errored reception of data without taking the system out of service (CRC has been explained in detail in Section

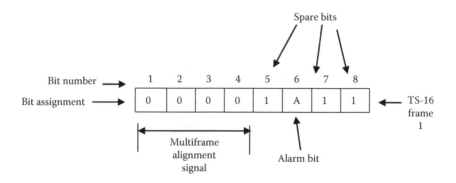

FIGURE 9.10
The bit assignment of TS-16 of frame 1 of a multiframe of E1.

6.7). Let us see how CRC is implemented in PDH level 1, i.e., E1 level, and how the REBE bit is accommodated and transmitted.

The diagram showing the principle of operation of CRC is redrawn here for the ease of reference as Figure 9.12.

The frame check sequence (FCS) or as it is normally called, the *CRC word*, is generated by performing a complex logical manipulation (usually through a polynomial implementation) on eight frames of the E1 bit stream at a time. These eight frames are actually half of a multiframe of 16 frames. For this purpose, the multiframe is divided into two sub-multiframes (SMFs) called SMF I and SMF II. This is depicted in Figure 9.13.

The CRC manipulations are performed on all the bits of an SMF and a 4-bit CRC word (FCS) is generated. (Use of a 4-bit word for CRC gives it

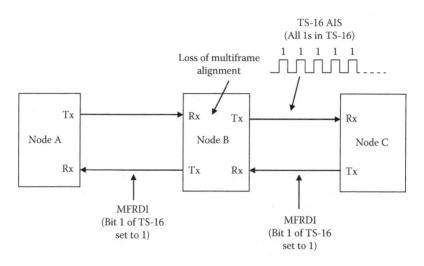

FIGURE 9.11
Generation and transmission of MFRDI.

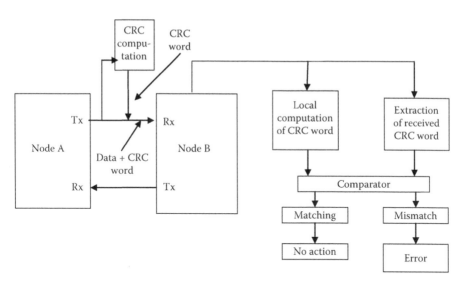

FIGURE 9.12
Principles of operation of CRC.

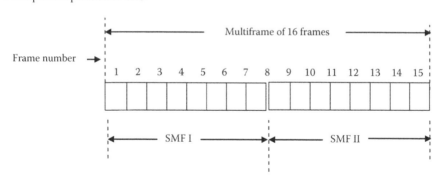

FIGURE 9.13
The multiframe of E1 is divided into two sub-multiframes SMF I and SMF II.

the name CRC-4, for use in E1 transmission. For different transmission schemes, the CRC word length may be different, i.e., 5, 6, 7, 8, etc., and accordingly the CRC scheme will be called CRC-5, CRC-6, CRC-7, CRC-8, etc., respectively.)

This CRC word is transmitted in the next frame. At the receiver, the same manipulations are performed again on respective SMFs and a local CRC word is generated. The received and locally generated CRC words are compared with each other. If they are found matching, then the received data has no errors, else the data is interpreted as "errored" by the receiver.

Let us see exactly how the CRC word is carried from the transmitter to the receiver.

Since the CRC manipulations take place simultaneously with the transmission of the data, the CRC word cannot be carried in the same SMF on which the CRC manipulations are being done. The natural choice thus for carrying the CRC word is the next sub-multiframe. Hence, the CRC word of SMF I of the current frame is carried in the SMF II of the current frame, and the CRC word of the SMF II of the current frame is carried in the SMF I of the next frame, and so on. Figure 9.14 illustrates the concept.

Thus, SMF is correct, but which are the bits which actually carry the CRC word?

We have discussed earlier that bit 1 of TS-0 (refer to Figure 9.6) is used to indicate the origin of the E1. If E1 is of international origin, the bit is set to 1; if it is of domestic origin, the bit is set to 0. The same bit is used for carrying the CRC word in even-numbered frames of a multiframe, when CRC is deployed. In this situation, the identification of (international or otherwise) the origin is lost, and it is considered to be an acceptable penalty for giving way to the implementation of CRC. (CRC may not be deployed always by all vendors, as it is considered to be an add-on feature for enhancing system availability through better diagnostics, without which also the basic functioning of the system may continue.)

There are four numbers of even-numbered frames in an SMF, and thus 4 bits are available for accommodating the 4-bit CRC word of the previous SMF. Each even-numbered frame of a multiframe thus carries 1 bit each of CRC words. Figure 9.15 shows the details.

As stated earlier in this section, the received and locally generated CRC words are compared with each other. If they match properly, no action is taken, but in case of a mismatch, an error signal is sent back to the transmitter by the receiver. This signal is called REBE. The situation is depicted in Figure 9.16.

The REBE signal is sent back to the transmitter indicating that there are errors in the received data. Although it is not clear as to how many

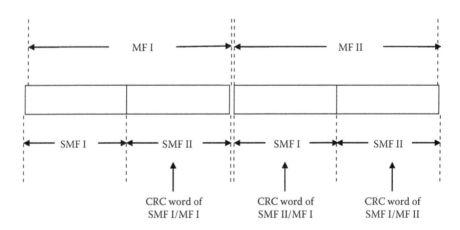

FIGURE 9.14
Carriage of CRC word by SMFs.

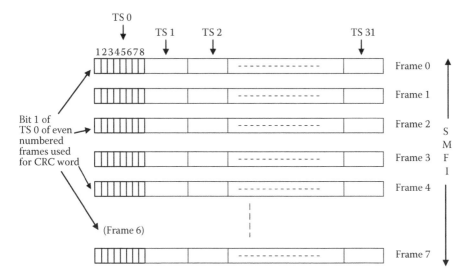

FIGURE 9.15
Bit 1 of TS-0 of even-numbered frames carries the CRC word in each SMF.

errored bits are received and where they are located, it is still considered very useful because its operation is "in circuit," i.e., it does not call for the system to be taken out of service. The maintenance engineers are alerted on receipt of the REBE alarm and take necessary remedial action to correct the bit errors.

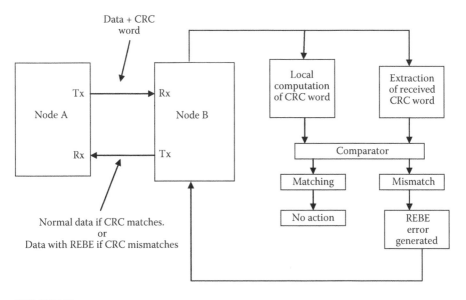

FIGURE 9.16
Generation and transmission of REBE in case of CRC word mismatch.

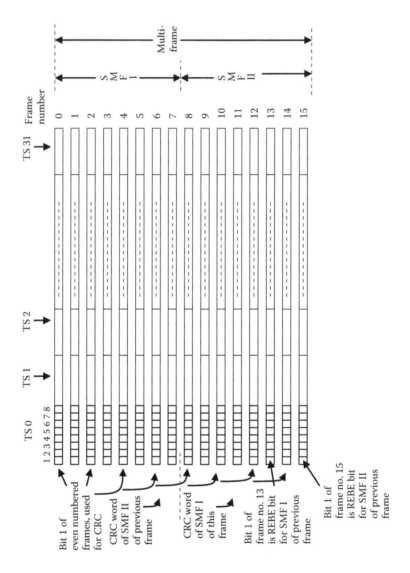

FIGURE 9.17
Use of bit 1 of TS-0 of multiframe for CRC word and REBE.

The REBE signal is conveyed by setting bit 1 of TS-0 of frame number 13 of the multiframe of 16 frames to a value of 0, which is normally set to 1 otherwise. Let us have a look at the bit assignment of TS-0 of frame number 13, which is shown in Figure 9.17.

On receipt of bit 0 in this position, the transmitter interprets it as a REBE signal and generates a local alarm. This REBE indication is for SMF I of the previous frame. The REBE indication for SMF II of the previous frame is carried similarly in frame 15 of a multiframe through the same bit 1 of TS-0.

You may notice from Figure 9.17 that bit 1, which was reserved for marking the (international or otherwise) origin of the E1, has been used in all the even-numbered frames for CRC word and in odd-numbered frame 13 and 15 for CRC error indication. The bit in remaining odd frames is also not used for "origin identification" when CRC is deployed, rather it is used for a 6 bit CRC sequence, which is used for interworking with the nodes that do not deploy the CRC procedure.

Review Questions

1. Why are alarms needed in a communication system?
2. Which are the main contributors to the incorrect reception of digital signal?
3. How is the detection and reporting of errors done in PDH?
4. List various alarms and their functions in brief.
5. What do you understand by LOS? How does the PDH system generates this alarm?
6. What do you understand about LOF and what is its significance?
7. Explain with the help of a schematic the generation and clearance of LOF.
8. How does a receiver know about the AIS? What is the criterion for clearance of AIS?
9. Explain the generation and clearance of an AIS alarm through an illustration.
10. What will happen if a framed E1 is used for AIS instead of unframed E1 and if the number of 0s are increased or reduced? (Critical)
11. Are the detection and clearance criteria for AIS of all the levels of PDH the same? If not, what are the differences?
12. What is RDI and why is it required?
13. Bring out the relationship among AIS, LOS, LOF, and RDI clearly.
14. How is the location of a fault determined in PDH?

15. LOMF indicates the loss of many frames or the loss of multiframe alignment. Explain the generation of this alarm.

16. What is the relation between MFAIS and TS-16 AIS? How is MFAIS generated?

17. Differentiate between RDI and MFRDI alarms.

18. Which time slot is used for MFRDI and how?

19. What do you understand about REBE?

20. Explain the principles of CRC with special reference to the generation of REBE alarm.

21. Describe, with the help of diagrams, how the CRC word is carried on a multiframe and how the bits are inserted in TS-0.

22. Draw a diagram showing the use of TS-0 of a multiframe for transmission of CRC word and REBE.

23. What is the sacrifice one has to make to deploy the CRC in a PDH system?

Critical Thinking Questions

1. Compare the merits and demerits of "in-circuit" detection of faults with that of "out of circuit."

2. What alarms more than the ones discussed in this chapter would you like to have if you are a maintenance engineer and how can you accommodate in the existing PDH structure?

Bibliography

1. ITU-T Recommendation G.775, *Loss of Signal (LOS), Alarm Indication Signal (AIS) and Remote Defect Indication (RDI) Defect Detection and Clearance Criteria for PDH Signals*, International Telecommunication Union, Geneva, Switzerland.

2. Trend Communications, *Trend's E1/T1 Guide, Pocket Guide for E1 and PDH Testing*, Trend Communications, Buckinghamshire, UK.

3. ITU-T Recommendation G.704, *Synchronous Frame Structures Used at 1544, 6312, 2048, 8448, and 44 736 kbit/s Hierarchical Levels*, International Telecommunication Union, Geneva, Switzerland.

4. ITU-T Recommendation G.706, *Frame Alignment and Cyclic Redundancy Check (CRC) Procedures Relating to Basic Frame Structures Defined in Recommendation G.704*, International Telecommunication Union, Geneva, Switzerland.

10

Synchronous Digital Hierarchy

When the PDH hierarchy was developed, it was considered to have enough capacity, which was then good enough for the requirement of voice and data traffic. However, as time passed and the demand for the services grew, the need for higher rates of transmission emerged very fast, largely because of the boom in data traffic and a steep growth in the consumer base of mobile telephones. Designing still higher rates in PDH was highly complex and uneconomical—uneconomical because more stages of multiplexing would have to be added to achieve a higher data rate, and consequently, a similar number of stages of demultiplexing were to be deployed wherever even one tributary was required to be dropped or inserted (see Chapter 8 for more details on PDH). In fact, the International Telecommunication Union (ITU) did not work out the standards for PDH systems beyond 140 Mbps, and whatever systems existed for higher data rates were proprietary of the vendors. After the emergence of optical fiber technology in the 1980s, which had enormous channel capacity, the limited capacity systems of PDH standards became a real bottleneck in transport of larger number of channels on the optical media.

10.1 Evolution of Synchronous Digital Hierarchy

The digital networks began with small segments, replacing the analog links. The number of segments grew as the demand for traffic increased. These segments, even though possessing master–slave synchronizations within themselves, operated on PDH philosophy at the interconnections with each other. Even all the network segments of one operator could not be integrated on a master–slave synchronization philosophy, leading to too many interconnections on the plesiochronous level. Each such interconnection gave rise to slips in the network, which were highly undesirable. Eventually, the number of slips in the network became intolerable (see Section 7.9 for a detailed discussion on this issue). The telecommunication operators then started establishing synchronization for their complete networks during the 1970s. The cost of high-precision atomic clocks and subsidiary timing distribution accessories was also gradually coming down, making network-wide synchronization affordable.

Also, the integration of the networks pertaining to different vendors was not possible at the optical level because of the nonstandard optical interfaces associated with PDH. This became a big bottleneck because the growing voice and data traffic worldwide demanded more and more interconnections between various networks irrespective of the vendor. Another problem for the international traffic was the different standards of electrical levels of digital signals in Europe, Japan, and America (E1 in Europe and the rest of the world and T-1 in North America and Japan).

These factors necessitated a new standard that could overcome all these bottlenecks and that could last for a reasonable period in the future. In developing the new standards, many more factors were considered, which offered a number of advantages over the PDH standards (these will be discussed in subsequent sections).

The network-wide synchronization in the proposed new standard offered the feasibility of bit by bit identification of digital signals in the network. This feature had a phenomenal advantage—we could pinpoint and identify any time slot in the bit stream. Thus, any channel (pertaining to any time slot) could be directly dropped or inserted. Although the possibility exists to directly drop even the 64-kbps channels, commonly used systems deploy up to E1 level drops and inserts. But this could be achieved only if we did direct multiplexing of tributaries instead of hierarchical multiplexing and resort to a common (only one set of) overhead bits. These requirements forced the hierarchy to be changed to a different type of hierarchy. Figure 10.1 illustrates the concept.

The E2 bit streams (also called tributaries) arrive at the E3 multiplexer at their own time, which means that the bit numbers or time slot numbers of each of them will be different at the time of arrival at E3 multiplexer. (This is not the same as different bit rate, which we have seen earlier in Chapter 8,

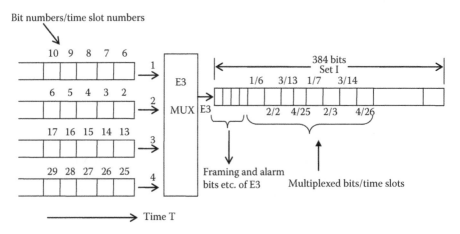

FIGURE 10.1
An E3 multiplexer multiplexing the E2s arriving at the input at varying times.

for which stuffing or justifications are used.) After the necessary rate equal-izations by justification and others, the bits are multiplexed as they arrive. The outgoing E3 bit stream thus contains the sequence of first arrived bits of each E2, then second arrived bits of each E2, and so on. In the example in Figure 10.1, at a given instant in time, the following bit numbers are available for multiplexing: 6th of first E2, 2nd of second E2, 13th of third E2, and 25th of fourth E2. They get multiplexed in the same sequence. These multiplexed bits are placed in the bit positions marked for tributary data bits, set I of E3 stream after framing, alarm bits, etc. (see Section 8.3 for E3 details).

Now, by looking at this E3 bit stream, we can see that it is easily possible to separate the 4 E2s because they are all treated as "tributary data" in fixed positions in the E3 bit streams. However, we will be at a loss to know about the function of individual bits of E2, i.e., whether any of these bits is a fram-ing bit of E2, justifications bit, tributary data bit, or something else. We will be able to know these details if we use an E2 demultiplexer, which will rec-ognize the frame structure of E2, and after observing the framing bits for a few frames, will ascertain itself that it is receiving a proper E2 stream, and only then can it identify all the bits of E2 bit stream from their positions and demultiplex them into E1 tributaries.

On the other hand, if we use direct multiplexing of E1 tributaries, for example, with a common set of framing bits, the individual channels can be easily separated through the knowledge of their positions in the bit stream (see Figure 10.2).

The direct multiplexer shown in Figure 10.2 is a hypothetical model and is only for illustration of concept. Let us say 63 E1s are multiplexed in a com-mon frame (63 is chosen just to match the number with that used in STM-1). Thus, through the knowledge of bit positions, the tributaries can be identified

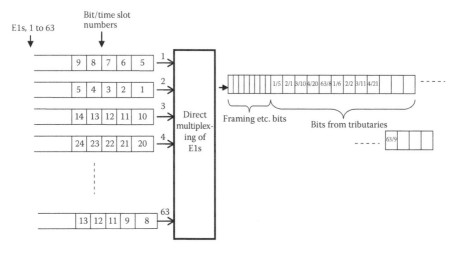

FIGURE 10.2
A hypothetical direct multiplexer for E1s.

and separated. Multiplexing could be at bit level or at byte level (one byte at a time is called "byte multiplexing").

You can see that if you have synchronous network throughout, you can identify individual bits without mistake, and with a direct multiplexing structure, you can drop or insert any tributary without having to demultiplex the whole bit stream.

This concept was applied to design a completely new standard of digital multiplexing hierarchy called synchronous digital hierarchy (SDH). Initially, in February 1985, Bell-Core Technologies USA proposed a new standard named synchronous optical network (SONET), which is in use in North America and Japan. The principles were later adopted by ITU-T and they brought out the SDH standard. The first recommendations of SDH, i.e., G.707, G.708, and G.709 were approved in February 1988, and their improved versions were created in 1990. While G.707 defined the standardized levels of SDH, G.708 described the principles and basic frames. The multiplexing structure of basic frames with each other and within themselves was defined in G.709. Subsequently, G.708 was used for defining the sub-STM-0 networks, G.709 was used for defining a new transport standard called optical transport network (OTN), and SDH was redefined in recommendation G.707/Y1322 in December 2003. Amendments continued further, and the latest version is G.707/Y1322 (2007) Amendment I released on July 29, 2007.

SDH finally evolved, ensuring that it not only meets current needs but reasonable future needs as well. The standard was made to ensure that it is fully compatible with all the digital hierarchies prevailing in the world at the time of its introduction, whether European, American, or Japanese. It was also made compatible with its counterpart, the SONET, and many other new technologies such as ATM, IP, etc.

Finally, SDH has a lot of advantages over PDH, which can be summarized as follows.

10.2 Advantages of SDH

10.2.1 Data Rates

Much higher data rates (or bit rates; "data rate" or "bit rate" are used synonymously in the discussion of SDH, and it should not be confused with data communication) are possible. PDH had defined data rates of up to 140 Mbps only. Beyond this, the systems were not defined by ITU-T, but some proprietary systems still existed. However, in SDH, the data rates up to STM-256, i.e., up to 40 Gbps, have been conceived. STM stands for "synchronous transport module"; the data transport units in SDH are called transport modules, the smallest being STM-1 at a data rate of 155 Mbps. Presently, systems up to

10 Gbps (STM-64) only have gained popularity, although STM-256 (40 Gbps) have also begun to be deployed.

10.2.2 Direct Drop/Insert of Tributaries

Due to the synchronous nature of multiplexing, as explained in the previous section, it is possible to drop or insert the tributaries (a tributary in a digital transmission system is a stream of data that could be an E1 = 2Mbps, T-1 = 1.544 Mbps, or any other higher-order bit stream), without having to go through the multiple stages of multiplexing/demultiplexing. This has considerably brought down the cost of the equipment per unit capacity, and hence, the total cost of the systems has come down drastically as well.

This, in fact, is one of the most important advantages of SDH. This, along with the need of higher data rates with compatibility worldwide, has been the prime force responsible for the birth of SDH.

10.2.3 Automatic Protection Switching/Self-Healing

The SDH network is generally constructed in the form of rings wherever possible (we will see this in more detail in Chapter 12). The design of SDH provides for the so-called automatic protection switching or self-healing feature, which means that if there is a breakdown of services, to a node or to a number of nodes, due to a fiber cut or any other problem, the traffic will be routed to the affected node(s) via an alternative route automatically. This provides almost instantaneous restoration of traffic, resulting in a very high reliability and a cool repairing of the fault by the maintenance teams, who can do a neat and proper job of repairing without the pressure of traffic restoration. This is why it is called the "self-healing" feature of the SDH network.

10.2.4 Compatibility with the Prevailing Standards

Just because a new standard that is better than the existing ones has come up does not mean that the existing networks could be abandoned overnight. Establishing those networks has involved huge costs, and many of them might not have rendered sufficient service life yet. Thus, for many years to come, they will continue to find deployment somewhere. Thus, the SDH has been designed to be able to accommodate almost all the existing standards, allowing the connectivity to all European, Japanese, and American standards. Hence, it is able to penetrate all the international boundaries and seamlessly carry traffic. The owners of the legacy networks had thus no cause to worry on account of emergence of SDH.

Various PDH hierarchies in use worldwide are shown in Figure 10.3. All these hierarchies are easily accommodated in SDH. SDH is also fully compatible with its American counterpart SONET.

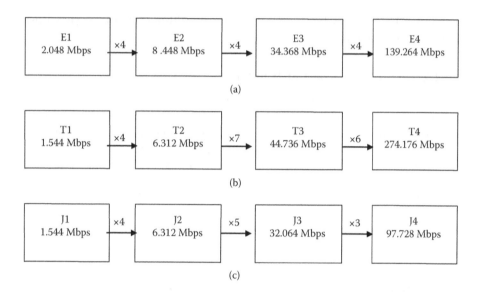

FIGURE 10.3

Various PDH hierarchies in use worldwide. (a) European PDH hierarchy; (b) North American PDH hierarchy; (c) Japanese PDH hierarchy.

Although higher rates than those shown in Figure 10.3 were also sparingly used, they had been mostly proprietary designs of the manufacturers and were not standardized by ITU-T.

10.2.5 Vendor Compatibility

The optical interface has been standardized in SDH. Before SDH, they were not standardized, thus every vendor was using a different interface, leading to incompatibility with the other. With the optical compatibility between the equipment of various vendors, the equipment pertaining to different vendors can be connected together without any special interface device. This has proven to be a revolution in speedy and economical interconnectivity worldwide.

10.2.6 Compatibility with Data Communication Protocols

Data communication has been growing faster than the voice traffic in recent years, but there are no transport networks available exclusively for data traffic. The data traffic backbone for long distances has therefore to be built on the available infrastructure. SDH provides for highly efficient mapping of almost all the data communication protocols such as HDLC, Ethernet, PPP, IP, ATM, LAPS, etc.

10.2.7 Excellent Operations, Administration, and Management

PDH has a very limited set of O&M functionality, with a few bits indicating the failure or errors as maintenance aids. There are very limited management features in PDH. In contrast, the frame structure of SDH includes a large number of bytes as overhead bytes for O&M. Although these overhead bytes slightly reduce the efficiency of transmission, they greatly enhance the operations, administration, maintenance, and management capabilities of the networks. A large number of alarms are generated for quick and precise locations of faults. In addition, a large number of bytes are reserved for data communication to facilitate better management of the entire network. This includes provisioning (configuring and providing the tributaries of the desired level between any two or many points), monitoring, disconnections, cross connections, drops and inserts, billing, etc.

10.3 Price to Be Paid for the Advantages

Of course, all the advantages as discussed in Section 10.2 have not come for free. There is a price to be paid, in terms of (i) establishing timing distributions throughout the network, (ii) added jitter due to pointer adjustment, (iii) poorer framing efficiency, and (iv) virus threat.

10.3.1 Establishing Timing Distribution throughout the Network

Since the entire network needs to be synchronous, i.e., the locations of bits pertaining to individual tributaries have to be known precisely, a highly accurate timing distribution has to be established throughout the network. Highly accurate atomic clocks have to be used at strategic locations in the network, which have to be backed by highly accurate synchronization supply units (SSUs) (see Chapter 5 for a detailed discussion on clocks). This is a costly affair; however, the costs have come down drastically in the recent past, to make it more affordable.

10.3.2 Added Jitter Due to Pointer Adjustment

The exact locations of tributaries are indicated in SDH by means of something known as "pointer" (we will subsequently discuss this in detail in this chapter). The pointer needs to be adjusted frequently to adjust the data rates to accommodate variations due to jitter and wander, due to the equipment instabilities, and due to problems in the transmission media and also to keep track of the locations of tributaries. Every time a pointer adjustment takes place, additional jitter is introduced in the bit stream. This jitter causes difficulties in the demultiplexing of the low rate tributary signals, leading to more complex circuitry and filters deployment.

10.3.3 Poorer Framing Efficiency

Due to a large number of bits used in SDH for the purposes of enhanced OAM (the O&M of PDH becomes OAM, i.e., operations administration and maintenance/management in SDH) features, the actual payloads carried by SDH systems are less as compared with PDH systems as a ratio of total capacity. For example, a PDH, E4 signal (or bit stream) carries 64 E1s as payload, when the E4 bit rate is 139.264 Mbps, whereas STM-1 (lowest configuration of SDH) carries 63 E1s as payload, when the bit rate is 155.520 Mbps.

However, with the easy transportability of higher rate signals on "Optical Fiber Links," these additional bits in the "Overhead" are easily affordable, particularly in view of the numerous advantages of extensive OAM facilities.

10.3.4 Virus Threat

Since the SDH systems are highly software dependent for the OAM functions, they are vulnerable to the outbreak of viruses, particularly if the OAM functions are operated through common facilities of computers, which are used for other general purposes. The security of the SDH systems has thus to be of a very high standard.

Nevertheless, again, the enormous features offered by the software management of SDH easily outweigh the threats.

In the overall scenario, the advantages heavily outweigh the disadvantages, clearing the stage for the exit of PHD and the entry of SDH systems.

10.4 Synchronous Transport Module

Transport in the context of telecommunications has the same meaning as it has otherwise. It means sending and receiving the commodities across distances. When the commodity is sent from one end and received at some other end, it is supposed to have been transported. So is in the case of telecommunication—the transmission and receptions of the communication signals or bit streams is called transportation.

In SDH, the bit stream is synchronous and is transported in frames, which are called modules. Let us consider the concept of direct multiplexing explained in Section 10.1 (see Figure 10.4).

As shown, the direct multiplexing (instead of multistage multiplexing such as PDH) of 63 E1s produces the smallest hierarchical module of SDH

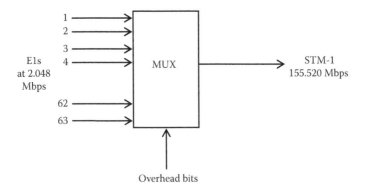

FIGURE 10.4
Direct multiplexing of 63 E1s with overhead bits to form an STM-1.

called STM-1. Note that the bit rate is much higher than what is required for
63 × 2.048 Mbps.

$$63 \times 2.048 = 129.024 \text{ Mbps}$$

$$\text{whereas STM-1} = 155.520 \text{ Mbps}$$

(The smallest module in SDH is actually STM-0, which carries 21 E1s, and
has a bit rate of 51.840 Mbps. STM-0 is equivalent to synchronous transport
signal [STS-1] of SONET standards, which are in use in North America and
Japan. STM-0 can accommodate 21 E1s [which are at a nominal bit rate of
2 Mbps] or 28 T-1s [which are at a nominal bit rate of 1.5 Mbps]. Other levels
of SDH and SONET are also matched for compatibility, for example, STS-3
is equivalent to STM-1 at a bit rate of 155.520 Mbps, STS-12 is equivalent to
STM-4 at a bit rate of 622.080 Mbps, and so on. STM-0 is, however, not popular with the manufacturers, and the smallest configuration generally available in the market is STM-1.)

The difference goes for overheads such as justification (as we have seen for
PDH in Section 8.3), frame alignment, and various other activities, which we
will subsequently see in this chapter.

The higher data rates in SDH are defined by higher rate STMs, such as
STM-4, STM-16, etc. Although they are in exact multiples of STM-1, they
are not multiplexed in a hierarchy such as PDH, but they use the concept of
direct multiplexing shown in Figure 10.4 at each stage. Various SDH rates
defined by ITU-T are as shown in Figure 10.5.

Figure 10.5 shows all the modules defined by ITU-T for SDH hierarchy,
their exact data (bit) rates, and their commonly known data rates. STM-1 is
the basic module, and the rest are simply multiples of STM-1, as for as the
data rate or bit rate goes.

STM module	SONET equivalent	Multiplying factor for basic module STM-1	Bit rate (Mbps)	Commonly known as (bit rate)
STM-1	OC-3	1	155.520	155 Mbps
STM-4	OC-12	4	622.08	622 Mbps
STM-16	OC-48	16	2488.32	2.5 Gbps
STM-64	OC-192	64	9953.28	10 Gbps
STM-256	OC-768	256	39,813.12	40 Gbps

FIGURE 10.5
Various STM modules defined by ITU-T G.707/Y1322 for SDH hierarchy.

10.5 Formation of STM-1

Now let us see how STM-1 is formed from 63 E1 tributaries and how the bit rate of 129.024 Mbps (2.048 × 63) becomes 155.520 Mbps. Where are all these extra (or overhead) bits used, how they are used, and why?

Let us reconsider the direct multiplexing concept of Section 10.1 (Figure 10.2), where a hypothetical case of direct multiplexing of 63 E1s was taken. The figure is redrawn here as Figure 10.6.

This is exactly what is done, in principle, to obtain an STM-1 frame from 63 E1s. Unfortunately, however, things are not that simple. There are problems that need to be overcome to form practicable systems. Let us have a look at them one by one.

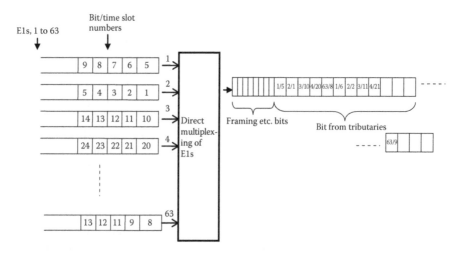

FIGURE 10.6
Direct multiplexing of 63 E1s for a synchronous system.

10.5.1 Justification of E1 Tributaries

As we have seen during the discussions on PDH multiplexing (Section 8.2.1), the bit rates of all the E1 tributaries are not exactly the same, probably because of different sources of their originations and different media over which they arrive at STM-1 multiplexer. Thus, a justifications process similar to that described in Section 8.2.3 has to be adopted.

A sufficiently high bit rate is therefore chosen, such that the E1 with the highest bit rate is easily accommodated. This high rate automatically accommodates the lower bit rate tributaries by declaring the required number of justification opportunity bits as either justification bits or data bits. However, before we start exploring the details of the justification process in STM-1, we need to know the "frame rate" of STM-1.

10.5.2 Frame Rate of STM-1

Recall the basic sampling theorem of Nyquist (explained in Section 3.2), which states that any analog signal can be faithfully reconstructed through its samples if the sampling is done at more than twice the maximum frequency content of the signal. Accordingly, the speech is sampled at 8 kHz (more than twice the maximum frequency, i.e., 3.4 kHz). Thus, 8000 samples are produced every second, which translate to a time of 125 μs between each sample. Thus, this 125 μs is adopted as a standard framing period in PCM or E1 (PDH level 1, see Section 8.1) multiplexing. Although in higher levels of PDH multiplexing such as E2, E3, etc., the frame rates were quite different, while designing SDH hierarchy, it has been found to be most convenient to use 125 μs as the standard frame rate not only for STM-1 but for all STM levels. As a matter of fact, the frame rate has got nothing to do with the sampling rate. The data transmission has to take place at the predefined bit rate such as 2.048 Mbps for E1 and 8.448 Mbps for E2, etc., which neither changes due to nor depends upon the framing pattern. On the contrary, the frame rate is so chosen as to transmit the data at the given rate error-free, as far as possible. Accordingly, a convenient frame duration of 125 μs was chosen for PCM or E1 because it facilitated easy understanding owing to the fact that the 8000-Hz sampling rate provided one sample for each of the 30 multiplexed channels during 125 μs. The number of bits that can be sent in a frame can neither be too low nor too high. Fewer numbers of bits per frame leads to poor framing efficiency because the number of "framing bits" (which are overhead bits) that are transmitted in every frame in addition to data bits (data or information or payload bits) is constant. A very high number of bits in a frame may lead to higher numbers of bit errors due to transmission media problems or due to poor accuracy of the receiver clock. Thus, a reasonable balance is normally provided while designing the frame rate.

In case of STM-1, the number of bits per frame is 155,520,000/8000, whereas in STM-4, they are 4 × 155,520,000/8000 per frame. STM-1 rate gets multiplied by 16 and 64 in case of STM-16 and STM-64. Thus, the number of bits per frame becomes very high in the case of higher-order STMs because the frame duration is kept constant at 125 µs. However, these high bit rates do not cause much of a problem in correct transmission and reception of the bit stream due to very high accuracy of the clocks used and efficient error detection and correction mechanisms.

Now let us see the justification process. Per ITU-T recommendations, the clock accuracy permitted for E1 nodes is ±50 ppm (or ±50 × 10⁻⁶). Therefore, the total number of bits that could be more or less

$$\frac{\pm 50 \times 2.048 \times 10^6}{10^6} = \pm 102 \text{ bits/s.}$$

Thus, if we keep a margin for stuffing 1 bit per frame for E1 tributaries in STM-1, there will be

$$\text{Number of extra bits per second} = 1 \times 8000$$
$$= 8000$$

However, the required extra bits are only ±102 per second (plus some more on account of transmission impairments, but the variations will not be much). Thus, 1 bit per frame appears to be too much. Thus, to economize on the number of stuffing/justification bits (and consequently improving framing efficiency), 1 bit is inserted in every fourth frame, which gives us a margin of 8000/4 = 2000 bits/s, which too is much more than our requirement of ±102 bits/s.

To accommodate the variations of both positive and negative sides, two bits are, in fact, provided in every fourth frame. One bit is used for the so-called positive justification and the other for negative justification. When the bit rate of incoming E1 tributary is less than 2.048 Mbps, a justification bit is added (in fact, the justification opportunity bit called "S1" provided for this purpose is declared as a justification bit rather than actually adding any extra bit), this is called "positive justification." When the bit rate of incoming E1 tributary is higher than 2.048 Mbps, a justification bit is deleted (in fact the justification opportunity bit called "S2" for this purpose is used as data or information bit rather than actually deleting any bit), this is called "negative justification."

The multiframe structure combining four E1 frames so produced is shown in Figure 10.7.

Note that the "S1" bit is the last bit of byte number 103 and "S2" bit is the first bit of the 104th byte. Also notice the presence of "justification control bits" "C1 and C2" in three of the bytes. These are the bits that indicate whether the bits S1 and S2 are data bits or stuff/justification bits. When all the three C1 bits are 0, bit S1 is a data bit (i.e., C1C1C1 = 000) and when all the

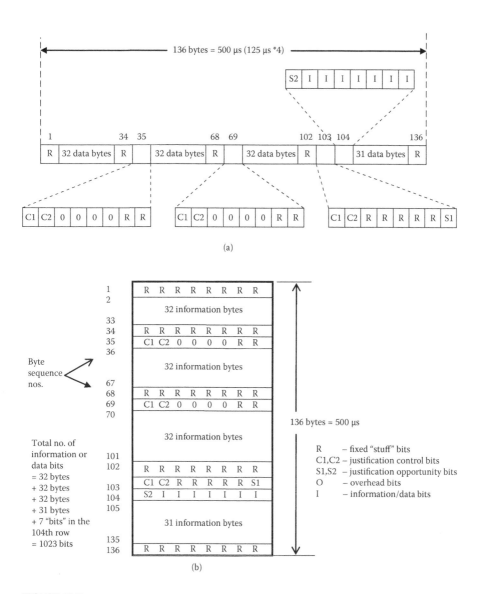

FIGURE 10.7

The 500-μs multiframe structure combines four E1 tributaries. The bits are transmitted from left to right and top to bottom. (a and b) Two types of representations of the multiframes.

three C1 bits are 1 (i.e., C1C1C1 = 111), bit S1 is a justification bit. Similar indications are given by the three C2 bit about S2 bit (see Section 8.2.3 for more details of the justification process).

You may be wondering about the rest of the details of the frame structure. We will see about them very soon. Let us first see the details of the justification rate calculations.

The total number of data bits is 1023 in each frame, and there are 2000 such frames every second. Thus, the number of bits per second = 1023×2000.

Or the data rate is = 2,046,000 bits/s

Now if bit S1 is also a data bit, then total number of data bits will be $1023 + 1 = 1024$, and therefore,

The data rate = 1024×2000
= 2048 kbits/s, which is the normal E1 rate

If S2 bit is also a data bit, then total number of data bits will be

$1023 + 2 = 1025$, and therefore,
The data rate = 1025×2000
= 2050 kbits/s

The actual data rate will be somewhere in between the limits of 2046 and 2050 kbps, giving a total justification range of up to 4 kbps against the requirement of $\pm 102 = 204$ bps or so.

The other bits in the frame structure are

- *R bits:* These are fixed stuff (justification) bits in this frame structure. The purpose is to make the frame of a particular size by adding some redundant bits, so that it gets accommodated easily along with other types of frames in the STM-1 multiplexing scheme. For example, STM-1 frame may include various tributaries such as E1, T-1, E3, E4, etc. (this will be subsequently discussed).
- *O bits:* These are "overhead" channel bits used for various maintenance and alarm features. The details of these bits are dealt with in the next chapter on OAM features of SDH.
- *S1, S2 bits:* These were already discussed.
- *C1, C2 bits:* These were already discussed.
- *I (or D) bits:* These were already discussed.

The utilizations of various types of bits to form the frame structure we have just seen is called "mapping," and the frame structure itself is called a "container." These are two of the very fundamental concepts in SDH. Let us see them in a little more detail.

10.6 Container

A container is one of the fundamental building blocks of SDH. The 136-byte frame structure we have seen in Figure 10.7 is actually a defined SDH container called "C."

The container concept can be very well understood from what it means literally. Consider a case where a truck has to be loaded with different commodities, varying in sizes. If they are loaded in the truck in their original sizes, they may not "pack" properly in the truck, leading to damages during transportation. Thus, we need to design containers, of adequate size, which will fulfill dual objectives of, first, accommodating the commodities of different sizes while keeping them properly packed inside the containers with the help of stuffing material which can be varied in quantity and shape depending upon the size of the commodity, and second, fitting the dimensions of the truck properly when they are stacked side by side and on top of one another in a certain number of rows and columns. Incidentally, for SDH application (and of course, many other applications of telecommunication systems), the transportation of telecommunication traffic is perfectly analogous to the systems of transportation of physical commodities.

Container C of Figure 10.7 is designed so that it can carry an E1 tributary of varying data speeds (varying data speed implies that E1 may contain a different number of bits in any given unit of time, say, 1 s, analogous to the commodities of different sizes). The justification bits S1 and S2 act as stuffing material. The fixed stuff bits R are stuffing material to achieve the container size of 136 bytes, which will fit well in the truck of STM-1 size to allow 63 such containers to be carried. Some readers might have a doubt here as to why not reduce the number of R bytes and reduce the container's size, so as to allow more containers to be carried in the truck of a given size. The answers lies in the built-in flexibility of the STM-1 truck, which has to accommodate not only the container of this size, but the containers of various other sizes for accommodating various existing transmission standards such as T-1 (North American PCM standard of 1.544 Mbps, consisting of 24 channels of 64 kbps), E3 (PDH level 3 at 34.366 Mbps), T-2 (North American PDH level 2 standard at 6.312 Mbps), E4 (PDH level 4 at 139.264 Mbps), etc. Naturally, if a truck has to be able to properly pack the containers of these various sizes (one type at a time), then the container sizes for various commodities have to be decided in such a way that the truck is able to pack at any given time the required type of containers properly, and, therefore, some containers might carry too much stuffing material, while others may carry too little or no stuffing material.

Following are the types of containers used in the SDH for various tributary rates (see Figure 10.8).

It can be seen in Figure 10.8 that while E1, E3, and E4 are European PDH standards, T-1, T-2, and T-3 (or DS1, DS2, and DS3) are North American PDH standards. This signifies the flexibility of SDH to accommodate all the existing standards prevailing in the world. It is also possible to adapt data communication standards such as ATM, HDLC, PPP, Ethernet, etc., by mapping them into some of these containers.

You can also see that the containers are assigned various levels in increasing order, as the data rate of the tributaries goes up. The container level 1 is

Container type	Used for tributary	Tributary data rate (kbps)	Number of containers in STM-1
C11	T-1/DS1	1544	84
C12	E1	2048	63
C2	T-2/DS2	6312	21
C-3	E3	34,366	3
	T-3/DS3	44,736	3
C4	E4	139,264	1

FIGURE 10.8
Different containers defined in SDH for various standards of tributary data rates.

meant for T-1 or E1, level 2 is for T-2, level 3 is for T-3/E3, and level 4 is for E4. However, level 1 container is divided in two parts: level 1 type 1, i.e., C11 for carrying a T-1, and level 1 type 2, i.e., C12 for carrying an E1. For level 3, only one container size has been designed for T-3 or E3 (separate containers would have made sense if we were able to carry 4 E3s, but that is not possible in the given STM-1 rate of 155.520 Mbps, after keeping a margin for sufficient number of overhead bits, thus a little extra stuffing is used for E3, while maintaining a common container size for E3 and T-3).

10.7 Mapping

Mapping is another most fundamental concept in SDH. Let us again consider the "container" of Figure 10.7.

We have seen in the previous section (Section 10.6) above, that the containers are so designed that they are able to pack the commodities (let us start calling these commodities "payloads" as they are normally called in telecommunication parlance) of various sizes inside them, by readjusting the quantity of stuffing material. What is adjusted in this process is the "fixed stuff bytes R," the quantity of which is varied to accommodate the different sizes of the payloads. However, to be able to recognize: the payload bits, stuff bits, and other overhead bits individually, for the purpose of separating them at the receiver, the location of each of these bits has to be fixed. This adjustment of fixed stuff bits and fixing their locations in the frame structure for various types of payloads is called mapping.

For example, the container of Figure 10.7 (redrawn as Figure 10.9a) shows the mapping of "E1" tributary into the container "C12." The same container C12, when mapped for ATM cells, looks like what is shown in Figure 10.9b.

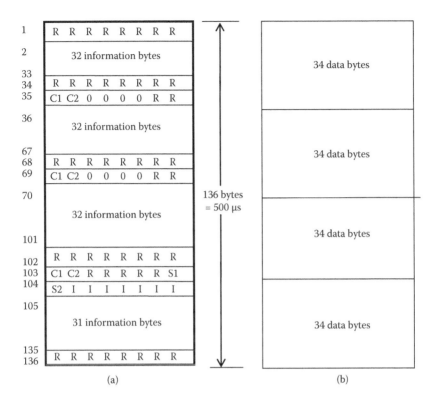

FIGURE 10.9
Mapping of (a) E1 and (b) ATM cells into container C12.

You can see the difference in the number of locations of R (fixed stuff) bytes in the mapping of "E1" as compared with the mapping of "ATM cells" into the container "C-12." While the E1 mapping contains 49 R bits, the ATM mapping contains no R bits. (This is due to the nature of ATM transmission, which is a data transmission system, not requiring fixed bit positions for achieving synchronizations.) The ATM cells that are 53 bytes each can begin and end anywhere in the container. In other words, the ATM cells can cross the boundary of the container. One ATM cell can be partly in one container and partly in another container. Thus, the ATM payload carried by the container need not be a fixed structure; it can be of any length and kind of a continuous payload. Hence, no stuffing or justification bytes are required. The representation is shown for the purpose of illustration only in the actual mapping the allocation of bytes is slightly different owing to the creation of "virtual containers," which are created from containers. Let us have a look at these virtual containers.

10.8 Virtual Containers

The container that is to be loaded on a truck, as described in Section 10.6, should have a label about the contents of the container, so that the receiving party can appropriately separate the containers carrying different materials. Also imagine a situation where there is no telephone available on which the receiving party can convey to the sending party a message about proper receipt of the material. In such a situation, the container that is being dispatched to the destination in the opposite direction just after the receipt of a container from the same destination may be requested to carry this information about the proper (or otherwise) receipt of the contents. This is what is exactly done in the case of the SDH containers.

Container C12 (or any other container for that matter) is loaded with some additional information about the content (type of payload, empty, etc.) and the transmission performance (proper delivery of the payload). The container so loaded with this information is called a "virtual container." A total of 4 bytes are added to the container to make it a virtual container. The "payload" remains the same because all these 4 bytes are "overhead" bytes. This virtual container is called VC-12 and is shown in Figure 10.10. The virtual container is the actual entity that is transported from end to end (user to user) in the SDH network, while the container becomes visible only during the stages of multiplexing and demultiplexing. (Quite interesting that the "virtual" container is the "actual" container.)

We will see the purpose and application of these four "overhead" bytes a little later. Let us first know what these bytes are called as a group and why.

All these four bytes (V5, J2, N2, and K4) together as a group are called "path overhead (POH)" bytes, or to be precise, "lower-order path overhead bytes" (LO-POH). They are overhead bytes because they are not the user data or the payload and are used for proper transmission and monitoring of the payload. They are called lower order as VC-12 (or C12) is a lower-order container in the STM-1 multiplexing scheme. Figure 10.8 lists various types of containers used in STM-1. VC-11, VC-12 are called lower-order containers whereas VC-4 is called a higher-order container. Other containers can be classified into either higher- or lower-order depending upon the multiplexing structure chosen. (This will be clear in subsequent sections where we see the complete multiplexing structure of SDH.) Thus, the overhead bytes used to carry the VC-12 properly to the destination are called LO-POH.

Now what is this "path"? Here we come across another very important concept used in SDH to define the various stages of the network systematically. We will see about "path" soon.

10.8.1 Data Rates

One important point to be noted here are the changes taking place in the data rates as we add the overhead bytes. Note that the frame duration of container

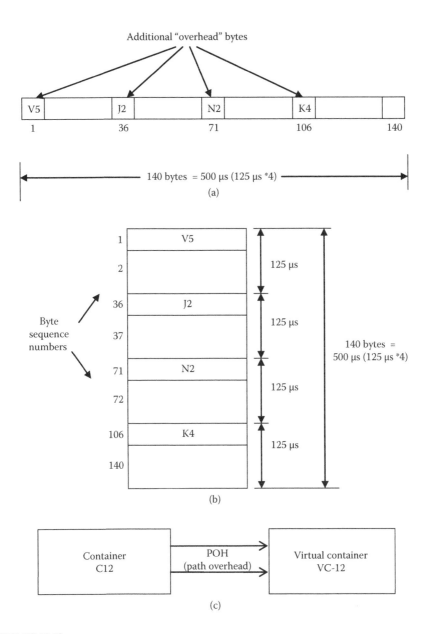

FIGURE 10.10
Four bytes are added to the container of Figure 10.7 to make it a virtual container VC-12. (a–c) Three type of representation of the same VC-12. One overhead byte each is inserted at the beginning of each subframe (125 µs) of the multiframe.

"C" and that of the virtual container is the same (125 µs for the individual frame and 500 µs for the multiframes). Thus, when we add additional bytes into the frame and wish to maintain the same frame duration, the data rate has to go up. Let us calculate the data rates of C12 and VC-12:

$$\text{Data rate of C12} = \text{No. of bits in a frame/duration of the frame}$$
$$= 136 \times 8/500 \times 10^{-6}$$
$$= 2.176 \text{ Mbps}$$

and

$$\text{Data rate of VC-12} = 140 \times 8/500 \times 10^{-6}$$
$$= 1120/500 \times 10^{6}$$
$$= 2.240 \text{ Mbps}$$

10.9 Path and Section

The "path" in SDH is defined as the end-to-end connection between two users (Figure 10.11).

As can be seen in Figure 10.11, the path is basically a transmitter–receiver concept of normal telecommunication terminologies. The SDH/MUX used here have to be either add/drop multiplexer (ADMs) type or terminal MUX (also called PTE or path terminating equipment) where the user tributaries can be added or dropped.

But if it is so simple and obvious, then what is the need to define it?

Consider a distance between the users that is too large (say, more than 200 km) to manage by the driving capability of a single transmitter. The

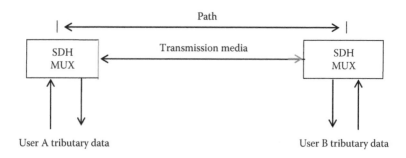

FIGURE 10.11
A "path" in SDH is basically the end-to-end customer link.

distance that a transmitter's signals can reach varies depending upon the system used. While in radio systems the distance is limited to 40–50 km owing to the line-of-sight transmission, in optical fiber cables (OFC), the distances of 30 to 100 km are feasible. Hence, if the distance is 200 km, we will have to use a "repeater" for the SDH signal (called regenerator in SDH terminology) to amplify the signal, so that when it reaches the receiver, its strength or amplitude is above the receiver's threshold level. The number of regenerators will depend upon: the transmitted power, receiver threshold, and the total distance to be covered. Figure 10.12 shows one such arrangement.

The portion of the SDH network between the path terminating nodes (or ADMs where user tributary is inserted/dropped) and the regenerator node is called "regenerator section."

There could, however, be situations when we need to drop/insert some other tributary, originating somewhere else, at a station, which is somewhere in between the "path" defined in Figure 10.12. At such a location, we will need to have an ADM instead of a mere regenerator; the multiplexer in this case will, however, perform the job of the regenerator (i.e., strengthening the signal) also. Such a situation is depicted in Figure 10.13. The section between two multiplexers is called a "multiplex section."

Any practical communication network, whether PDH, SDH, or anything else, will actually have all these three sections, but they are so properly defined and distinguished only in SDH. The purpose of clearly and distinctly defining these sections separately is to properly and systematically organize the management capabilities in the network. The management information is separately added into the transmitted bit stream for each of these three types of sections, thereby facilitating distinctive OAM (operations administration and maintenance/management) features at each level separately. For

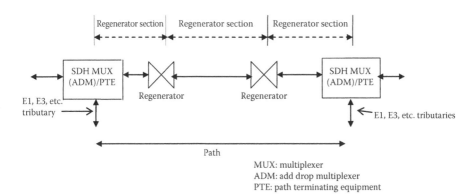

FIGURE 10.12
A portion of the SDH network showing the "path" and "regenerator sections."

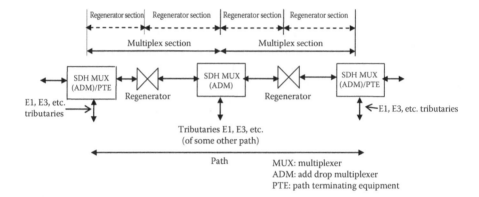

FIGURE 10.13
A portion of SDH network, showing path, multiplex sections, and regenerator sections.

example, in case of a failure, its location is easily identified with the help of the overhead bytes belonging to each of these sections. There is a clear indication available as to whether the failure is in the regenerator section, multiplex section, or at the path level. The overheads are accordingly identified with their sections. Let us see in brief about these overheads; the details will be discussed in the chapter on operations and maintenance in SDH systems.

10.9.1 Regenerator Section Overhead

This is a specific number of bytes with a defined position in the STM-1 frame. It is used for the operations, administration, and maintenance/management (OAM) of the regenerator section.

10.9.2 Multiplex Section Overhead

This is again another set of a specific number of bytes with defined position in the STM-1 frame. It is used for the operations, administration, and maintenance/management (OAM) of the multiplex section.

10.9.3 Section Overhead

The RSOH and MSOH together are called SOH.

10.9.4 Path Overhead

This again is a specific number of bytes, but rather than being defined in overall STM-1 frame, they are defined in the virtual container. They are carried transparently through any number of regenerator or multiplexing nodes but are taken out only when a tributary is dropped.

10.10 SDH Layers

The SDH network diagram we have seen in Figure 10.13 is actually a representation of what are called "SDH layers." The overheads (RSOH, MSOH, and POH) in SDH are so defined that they interact only with their counterparts in various nodes. The RSOH of one node will interact with the RSOH of the next node; similarly, the MSOH of one multiplex node will interact with the MSOH of the next multiplex node and the POH will interact with the POH of the end-to-end nodes (at one end where the tributary is inserted and the other end where it is dropped). This concept makes the interactions between various SDH nodes to happen in a kind of layers, with each layer of one node interacting with the similar layer of the other node. Figure 10.13 is redrawn here as Figure 10.14 illustrating the layers.

The concept of layers facilitates ease of understanding and management of the interaction between the nodes. The path layer can actually be broken into a higher-order path (VC-4 and others) layer and lower-order path layer (VC-12 and others), and a physical interface layer can also be defined.

Having seen what is called path and section in SDH, let us get back to see how our virtual container finally gets multiplexed into STM-1 and gets transported.

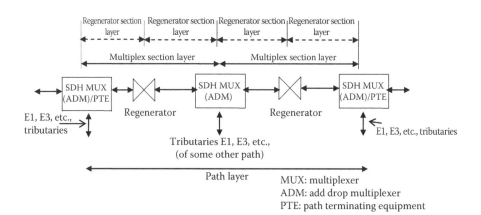

FIGURE 10.14
SDH layers.

10.11 Tributary Unit

We have seen in Section 10.8 that the virtual container consists of 136 bytes of data from four E1 tributaries and 4 bytes of POH to make a multiframe of 140 bytes. This multiframe in the form of virtual container can be multiplexed with many other similar VCs to form an STM-1 frame. However, since it is a multiframe carrying common information about 4 tributaries, the beginning of the VC has to be identified before demultiplexing the tributaries at the receiver, for these two reasons:

(i) To separate the overhead bytes from the data bytes and thus to locate the tributary data, and

(ii) To know the exact location of justification bits and justification control bits, so that on demultiplexing, the tributaries can restore their original data rates

(Recall, in Section 10.8, that there is one positive justification opportunity bit and one negative justification opportunity bit for each container. Also there are three justification control bits for each justification bit, i.e., a total of six justification control bits. Unless the exact locations of all these bits are known, it will not be possible to identify them because all bits look alike to the receiver.)

Thus, to identify the beginning of the VC^S a pointer is added which pinpoints the location of the VCs within STM-1 bit streams. VC added with a pointer is called a "tributary unit" (see Figure 10.15).

The tributary unit (for VC-12, it is TU-12) is thus self signifying. It is a unit that belongs to tributaries, i.e., whenever you wish to do anything with a tributary (dropping or inserting from or to STM-1 bit stream), you have to deal with its unit called a tributary unit. So what is actually stacked into STM-1 (or any higher-order STM for that matter) is a set of tributary units.

This pointer (called TU pointer or TU-PTR) consists of 4 bytes named V1, V2, V3, and V4. (The detailed explanation of them is deferred for the time

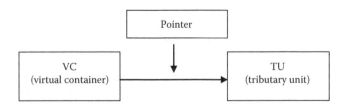

FIGURE 10.15
A tributary unit is formed by adding a pointer to the virtual container.

being and will be seen later in this chapter.) Thus, the tributary unit consists of a total of 144 bytes. These additional four bytes are also accommodated in the same multiframe of 500 μs by suitably augmenting the bit rate. The bit stream of the TU-12 is shown in Figure 10.16.

Now, what exactly is this pointer? It is a set of bytes containing the address of the beginning of the VC-12. Since the beginning of the VC-12 is always with V5 byte, the pointer always contains the address of the V5 byte. Why use a pointer and why not? We will see how it functions in this chapter, but before that, let us see what happens to the tributary unit in the STM-1 bit stream.

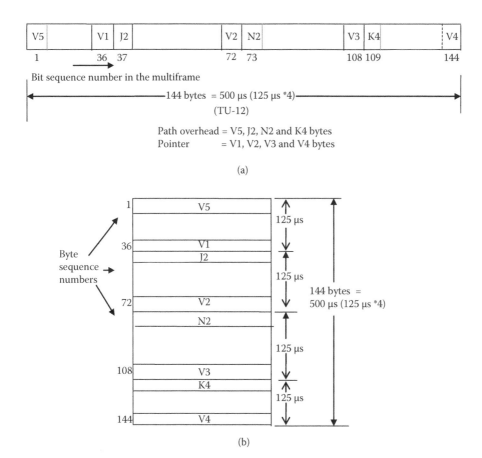

(a)

(b)

FIGURE 10.16
A TU-12 containing 144 bytes in 500-μs multiframe. The details of remaining bytes are omitted here to maintain clarity. Two types of representations are shown in (a) and (b).

10.12 Further Multiplexing

So the tributary unit is the entity that is actually packed with other similar tributary units into the transport module for actual transportation from the transmitter to the receiver. In the case of E1s, 63 tributary units are multiplexed together to form an STM-1 transport module (see Figure 10.17).

In the whole process of multiplexing, it is important to note the changes in the bit rates, i.e., number of bits per second, as we progress to form the container, virtual container, and tributary unit, from E1 tributary. Let us have a look at them.

10.12.1 Bit Rates

> E1 = 2.048 Mbps (32 bytes = 256 bits in 125 µs)
> C12 = 2.176 Mbps (see Figure 10.6; 136 bytes in 500 µs or 34 bytes in 125 µs)
> VC-12 = 2.240 Mbps (see Figure 10.10; 140 bytes in 500 µs or 35 bytes ion 125 µs)
> TU-12 = 2.304 Mbps (see Figure 10.16; 144 bytes in 500 µs or 36 bytes in 125 µs)

Note that the frame rate is kept constant at 125 µs for individual frames and 500 µs for this multiframe; thus, the bit rate keeps on increasing due to the additional overhead bits. In other words, we stuff more and more bits in a fixed period, say, every second, leading to smaller periods available for every bit, making the pulses thinner and thinner. This fact is significant because at higher levels of multiplexing, i.e., STM-4, STM-16, or STM-64, the frame width is also maintained at 125 µs only, which makes the pulses extremely thin, with very small pulse widths. The smaller the pulse width, the more

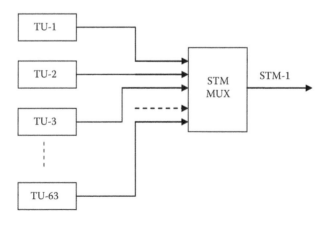

FIGURE 10.17
Multiplexing of tributary units to form STM-1.

difficult it is to process it accurately, which adds to cost. This actually is the cause of limitations for still higher levels of multiplexing, and it is due to this reason that STM–256 systems, although being offered by some vendors, are still not very popular.

10.12.2 Tributary Unit Groups

Tributary units, however, are not multiplexed exactly as shown in Figure 10.17, but there is a two-stage multiplexing process adopted as shown in Figure 10.18. The first three TUs are multiplexed to form a "tributary unit group" that is called TUG-2 or tributary unit group level 2, and then seven TUG-2s are multiplexed to form a "tributary unit group level 3," or TUG-3 (see Figure 10.18).

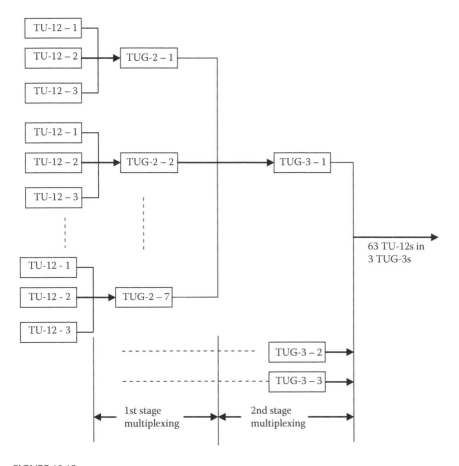

FIGURE 10.18
The two-stage multiplexing of tributary units to form tributary unit groups of levels 2 and 3.

The three TU-12s are byte multiplexed to form a TUG-2, seven TUG-2s are byte multiplexed to form a TUG-3, and again three TUG-3s are byte multiplexed to form a higher-order group.

10.12.3 KLM Numbering

The tributaries in STM-1 are numbered in terms of K, L, and M for quick and unambiguous identification. K stands for the serial number of TUG-3, and hence it has 3 values from 1 to 3. L stands for the serial number of TUG-2, and hence it has 7 values from 1 to 7. M stands for the TU serial number and thus has 3 values from 1 to 3 in the case of the above scheme. A tributary number 123 will thus be the third TU-12 (TU-12-3) of second TUG-2 (TUG2-2) of the first TUG-3 (TUG3-1), referring to Figure 10.18.

An alternative representation of Figure 10.18 is shown in Figure 10.19.

We can see that the TU-12 gets multiplied by three to form TUG-2, which gets multiplied by seven to form TUG-3, which finally gets multiplied by three to form a higher-order group. Thus, the total number of multiplexed TU-12s is

$$\text{Number of TU-12s} = 3 \times 7 \times 3$$
$$\text{(and hence, E1s)} = 63$$

Thus, this higher-order group contains 63 tributary units, which contain 63 E1 tributaries in turn.

This higher-order group is then added with a higher-order POH (HO-POH) to form the entity called a higher-order virtual container (VC-4). The reasons for adding a POH are the same as we have seen in Section 10.8, as those for adding a lower-order POH (LO-POH) to the C12 to form a VC-12. (In fact, the VC-4 is formed from two routes, one is by adding HO-POH to the higher-order group as we have defined above and another is by adding HO-POH to container C4, which is used for multiplexing an E4 bit stream. That is why the entity is called VC-4. This will be clearer as we proceed.) Thus, our higher-order group of 3 TUG3s takes the shape as shown in Figure 10.20.

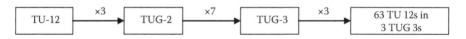

FIGURE 10.19
At alternative representation of tributary unit multiplexing.

FIGURE 10.20
Formation of VC-4.

This HO-POH is of "9" bytes. Details of functions of these bytes will be seen in the next chapter.

Let us now have a look at the further processing that takes place before reaching the STM-1 level.

10.12.4 Administrative Unit

The VC-4 is added with a pointer to facilitate its identification in the STM-1 frame. The pointer is called an AU pointer (administrative unit pointer), and the unit so formed is called the administrative unit 4, or AU-4. Figure 10.21 shows this.

Finally, the "section overheads" are added to the AU-4 to make it what is called STM-1. (The POHs at the lower level, i.e., VC-12, and at the higher level, i.e., VC-4, have already been added.) The section is divided into "regenerator sections" and "multiplex sections," and so are the overheads, called regenerator section overhead (RSOH) and multiplex section overhead (MSOH). Figure 10.22 shows the STM-1 formed.

Thus, the entire chain of multiplexing becomes as shown in Figure 10.23.

The "AUG" called administrative unit group is not a separate entity; it is just a stage of aggregating the AUs. It appears redundant for AU-4 but will become relevant when we see AU-3s subsequently in this section, which are to be multiplied to form AUG from a different route.

Let us now see the reasons for the multiplexing scheme shown above (we could have multiplexed all 63 TU-12s in one go, but we have not done it).

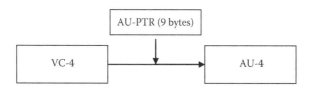

FIGURE 10.21
AU pointer is added to the higher level virtual container to form AU-4.

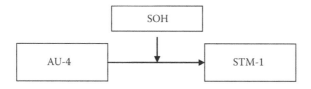

FIGURE 10.22
The section overheads are added to get the STM-1 bit streams from AU-4.

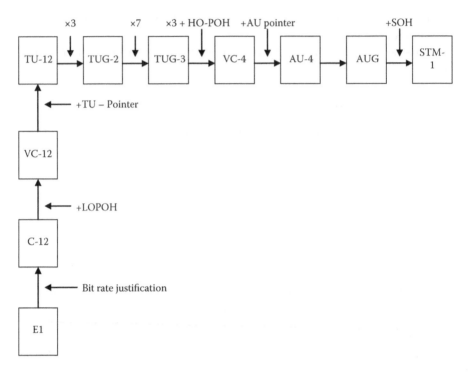

FIGURE 10.23
The complete multiplexing chain for multiplexing 63 E1s into the STM-1 frame.

Recall that in the beginning of this chapter (Section 10.2) we saw that one of the best strengths of SDH is its ability to adopt and accept the tributaries from all the existing PDH standards, be they European, American, Japanese, or whatever. Through the multiplexing stages and bit rates (by adding the necessary number of stuff bits) that it adapts by the time we reach the multiplexing stage of STM-1, it is possible to accommodate all such types of tributaries, which carry any type of payload. Let us see how our other rate tributaries (other than E1s for which it has been shown) are multiplexed into STM-1.

We have seen earlier in this section that the bit rate of E1 is 2.048 Mbps, and when it is loaded into the container for justification of bit rate to make a container C-12 out of it, POH is added to form VC-12 and further added with TU pointer to make a TU-12, with its bit rate becoming 2.304 Mbps.

Similarly, when we deal with a tributary "T-1" of North American PDH hierarchy, its basic rate is 1.544 Mbps, and after containerization, addition of POH, and addition of pointer, its bit rate becomes 1.728 Mbps. ("T-1" stands for "telecommunications standards entity number 1" in the North American

PDH hierarchy. It is the equivalent or counterpart of E1 or European hierarchy based on ITU-T standards. Similar to E1, a T-1 also consists of 64-kbps channels, but only 24 in number [as against 32 of an E1]. It is further multiplexed [4T-1s] to form a T-2, and T-2s are further multiplexed to form a T-3 [7 T-2s]. The bit rate of a T-1 is 1.544 Mbps, that of a T-2 is 6.312 Mbps, and that of a T-3 is 44.736 Mbps. T-1, T-2, and T-3 are also synonymously called DS1, DS2, and DS3 for digital signal levels 1, 2, and 3, respectively.)

Its container is called C11, virtual container is called VC-11, and the tributary unit is called TU-11. The process of forming C11, VC-11, and TU-11 from T-1 is exactly the same as that of E1 for forming VC-12, TU-12, etc. The only difference is the bit rate, because the number and positions of R bits differ.

Now if we multiply TU-11 bit rate, i.e., 1.728 Mbps by 4, we get a bit rate of 6.912 Mbps. If we multiply the TU-12 bit rate, i.e., 2.304 Mbps by 3, then also we get a bit rate of 6.912 Mbps. Thus, 4 numbers of TU-11s can be multiplexed to get the same bit rate that is required for multiplexing 3 numbers of TU-12s.

What a coincidence.

No, it is not a coincidence, but it is by design. This is the reason why we see so many "R" bits or "fixed stuff bits" in various containers. The overall sizes of containers have been designed so that they accommodate various tributaries with their justification bits, and by adjusting the number of "R" bits, they are able to produce bit rates that are common at higher levels, permitting the multiplexing of all such types of tributaries into a final bit rate of STM-1. Figure 10.24 shows the multiplexing scheme of T-1 and T-2 with that of E1s.

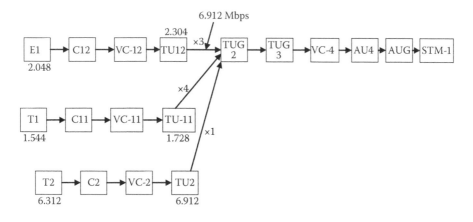

FIGURE 10.24
Multiplexing of T-1 and T-2 with E1, getting into the chain at TUG-2 level. Note that the TU-2 bit rate is the same as TUG-2 at 6.912 Mbps. Other details have been omitted for the sake of clarity.

10.13 Multiplexing of Higher-Order PDH Tributaries

We have seen the multiplexing structure up to T-2 PDH rates in the previous sections. However, the prevalent PDH rates are up to 139.264 Mbps (E4). The lowest basic rate of STM, i.e., STM-1 has been designed at a bit rate of 155.520 Mbps so as to accommodate up to this highest prevalent PDH rate. Let us see how the other PDH rates such as E3, i.e., 34.368 Mpbs, D-S3 (or T-3) i.e., 44.736 Mbps, E4, etc. are accommodated in STM-1.

Similar to the process of forming the containers, VCs, etc. for E1, the higher PDH rates are also processed. The container for E3 is called C-3 and is formed after bit rate justification, similar to what we have seen for E1 (Figure 10.25).

The container for E3, i.e., C-3, is deliberately kept large enough so that it can accommodate the American PDH level 3 tributary, i.e., T-3 (or DS3) also, which is at a bit rate of 44.736 Mbps. A container C-3 of size 48.384 Mbps is not only able to accommodate both these rates easily, but it also provides a common bit rate for both the tributaries to facilitate their multiplexing into STM-1. VC-3 and TU-3 are formed by adding the POH and the pointer as explained for earlier cases. Thus, the complete path of multiplexing followed by E3 or T-3 is as shown in Figure 10.26.

However, container C-3 is allowed to follow another path (called route) to STM-1, as shown in Figure 10.27.

Note that in Figure 10.24, VC-3 is a lower-order container and hence assigned a LO-POH (LO-POH), whereas in Figure 10.25, VC-3 is a higher-order container and has been accordingly assigned a HO-POH. Thus, VC-3 can be either a lower-order virtual container or a higher-order virtual container, depending upon the multiplexing structure and the path/route followed.

10.13.1 Multiplexing Structure of E4

The multiplexing structure of E4 is shown in Figure 10.28 on similar lines.

There is no lower-order tributary unit in case of E4. VC-4 is basically designed to accommodate E4 tributary, but it has to act as a higher level container in case of multiplexing of smaller PDH tributaries, to offer a common rate of multiplexing to STM-1

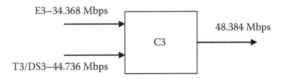

FIGURE 10.25
The PDH level 3 tributaries of European and American standards are framed into a common rate container C-3.

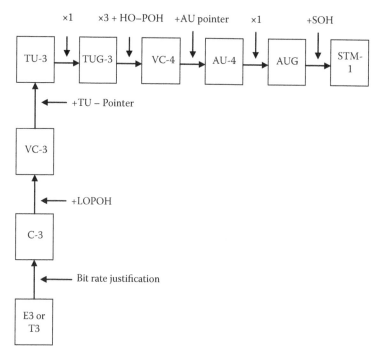

FIGURE 10.26
The complete multiplexing path of E3 or T-3 up to STM-1.

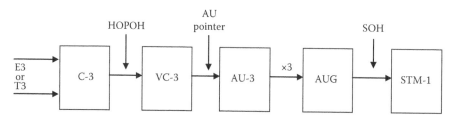

FIGURE 10.27
Alternative path or route followed by E3/T-3 to STM-1 during multiplexing stages.

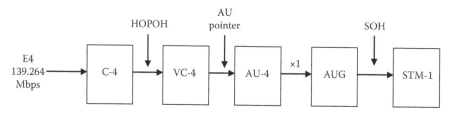

FIGURE 10.28
Multiplexing structure of E4 tributary.

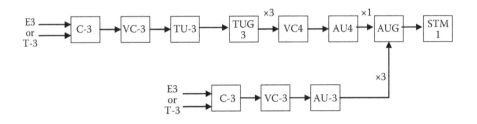

FIGURE 10.29
The multiplexing structure for E3/T-3 into STM-1 showing two different paths/routes.

10.13.2 Multiplexing of E2 Rate (8.448 Mbps)

Note that we have not discussed the multiplexing structure of E2 or PDH level 2 into SDH. Yes, E2 rate multiplexing is not provided in SDH, owing to its poor popularity. You will find that E1 and E3 rates are more popular in PDH European hierarchy. You might have come across the term Skip-MUX. It is a popular name for the PDH systems, which multiplexes E1s directly into E3, skipping level E2 (see Section 8.11 for skip MUX). Skip MUX's popularity shows the nonpopularity of E2 MUX.

A look at Figures 10.26 and 10.27 reveals that the path beyond and including AUG is common. The figures can therefore be combined and drawn as Figure 10.29.

10.13.3 Multiplexing Routes

As shown in Figure 10.28, the multiplexing of E3 or T-3 into STM-1 can follow two different paths. These paths are called routes. The various multiplexing routes possible have been defined by ITU-T recommendation G.707 for all types of data rates. However, the routes to be followed by a country have to be unique and predefined. The choice may depend upon the type of tributaries in use. For example, in case of a European hierarchy, if we have a mix of tributaries of lower level (say, E1s) and higher level (say, E3s) then the first route is preferable, as it allows the mixing of streams at the TUG-3 level (we will see in the next section), whereas if only E3 levels are present in tributaries, then the second route is preferable, as it will involve fewer stages of multiplexing.

Figure 10.29 shows a combined multiplexing structure for two routes followed for multiplexing E3/T-3, however a look at Figure 10.24 reveals that it too has a stage of multiplexing as TUG-3. The data rates of TUG-3 of Figure 10.24 and that of Figure 10.28 are the same and therefore they can also be represented by a combined diagram. Also, AUG, when multiplied by "N," gives us STM-N. Thus, STM-N is nothing but a simple combination of STM-1s by suitable (byte) multiplexing.

10.13.4 Significance of AUG

The AUG appeared redundant when we derived it from AU-4 (Figure 10.23). However, the relevance is clear from Figure 10.28, where the AUG is really a group of 3 AU-3s, from the second multiplexing route of C-3. AUG is no separate entity, but it is simply a grouping of either 1 AU-4 or 3 AU-3s.

10.14 Complete SDH Multiplexing Structure

If we combine Figures 10.28, 10.29, and 10.24 at the TUG-3 level, we get the complete multiplexing structure or complete multiplexing routes of the SDH as shown in Figure 10.30. The bit rate at the TUG-3 level is the same for both cases. In fact, the bit rates for all the levels and stages defined in any other previous diagrams or sections are unique, and as explained earlier, they have been designed as such to allow the multiplexing of various types of tributaries into a common transport entity, called STM-1. Figure 10.30 shows these details.

However, to make the multiplexing structure really complete, we will have to include the lower- and higher-order STMs in the diagram. The lower-order

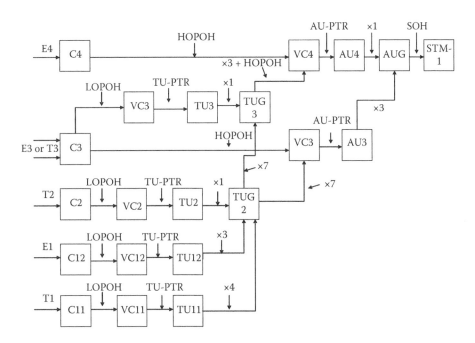

FIGURE 10.30
The multiplexing routes followed by various types of tributaries in SDH multiplexing scheme.

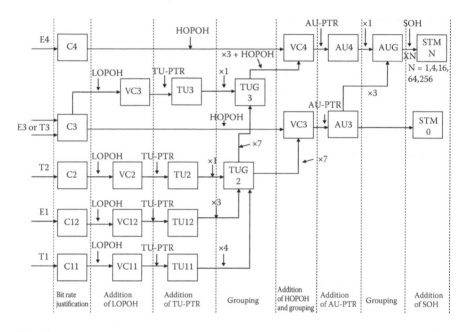

FIGURE 10.31
The complete SDH multiplexing structure defined by ITU-T-707/Y 1322 showing various stages and operations thereupon.

STM is "STM-0" and the higher-order ones are STM-4, 16, 64, and 256. STM-0 is obtained directly through AU-3 (instead of multiplexing it by 3 to get AUG), and higher-order STMs are obtained by simply multiplexing the AUG by the order of STM, before adding the SOH. The details are shown in Figure 10.31 along with the operations performed at various stages of multiplexing.

10.14.1 STM-0

As can be seen from Figure 10.31, STM-0 consists of a single administrative unit of level 3, i.e., AU-3. Although STM-0 is defined in the ITU-T multiplexing hierarchy, it is rarely used by the manufacturers owing to its poor popularity. The basic SDH is called STM-1, the bit rate of STM-0 is 51.84 Mbps. As can be seen, STM-0 can accommodate one number of E3 or T-3, or 7 numbers of T-2s, or 21 numbers of E1s, or 28 numbers of T-1s.

As we have discussed earlier, at each stage of SDH multiplexing, some bytes are added, but the frame duration is kept at a constant 125 µs, which matches with the standard sampling rate for voice signals, i.e., 8000 frames per second. Thus, as we reach the final stage of STM-N, the bit rate considerably increases, as compared with the bit rate of tributary (or tributaries). Figure 10.32 shows the multiplexing structure again, but this time with the number of bytes per 125 µs at each stage and the bit rate so generated. The

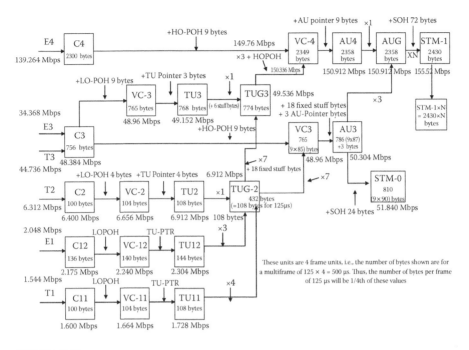

FIGURE 10.32
Number of bytes and bit rates at various stages of SDH multiplexing.

matching bit rates produced by different types of tributaries at the merging points is one of the most important features of SDH, allowing a common transport module to emerge at the optical level irrespective of the type of tributaries at the input.

Since the mapping of E1 and T-1 tributaries are done at the level of a multiframe of $125 \times 4 = 500$ μs, the number of bytes shown in Figure 10.32 for these tributaries has been taken for the complete multiframe of 500 μs for convenience. The bit rates, however, will remain the same, irrespective of the pattern of representation. The number of bytes per frame will be one fourth of the number of bytes per multiframe, keeping the number of bits per second as the same, which is the bit rate.

You will also notice that many of the figures (such as addition of fixed stuff bits, size of the AU pointer, etc.) are multiples of 9. The reason for this is the presentation pattern of the SDH frame structure (we will see the details in the next section). The frame is represented in a pattern of 9 rows and 270 columns (for STM-1; for STM-N, the columns are multiplied by N, whereas the rows are kept constant at 9). This representation makes things easier to understand if they are in multiples of 9, as it results in the addition or deletion of complete columns. Surprisingly, the implementations are driven by the convenience of presentation, and rightly so, as there are no other compelling parameters because there is enough margin available in the number of overhead bytes. The enough, or we can even call it more than enough, margin in the number of overhead bytes is a boon to the manager of the network, which has been made possible because of the huge payload capacity of the optical fiber systems. Thus, even after keeping more than sufficient numbers of bytes for overhead, enormous amount of payload is carried by the systems.

10.15 Frame Structure of STM-1

Having seen the complete multiplexing scheme of SDH per ITU-T standards, it is time to have a look at the frame structure of SDH. The STM-1 frame is called the "basic frame" (although STM-0 is also defined in the SDH scheme, the basic frame is STM-1; STM-0 is treated as a special case owing to its poor popularity) in the SDH, which means that all other frames such as STM-4, STM-16, STM-64, etc. are derived frames (multiple of) from STM-1.

The bit rate of STM-1 is 155.520 Mbps and the frame duration is 125 µs. To draw the diagram representing STM-1 for these many bits will be difficult, so let us calculate the number of bytes in each frame.

$$\text{The number of bytes in each frame} = (155.520 \times 10^6)/(8 \times 8000)$$
$$= 2430 \text{ bytes}$$

Since the multiplexing in SDH is performed using byte interleaving, the representation in bytes poses no problem. But even this number, i.e., 2430 bytes, is too big to allow its presentation on the paper of the size of this text. Thus, let us try to draw it in a way that we are able to visualize all the details of the frame structure and yet we are able to fit it within the space available on this paper. Figure 10.33 shows this representation.

The bits and bytes in the frame can be verified as follows:

$$\text{Total number of bytes} = 9 \times 270$$
$$= 2430$$

$$\text{Total number of bits} = 2430 \times 8 = 19{,}440 \text{ bits per frame}$$

$$\text{Total number of bits per second} = 19{,}440 \times 8000$$
$$\{8000 \text{ frames of } 125 \text{ µs in 1 s}\}$$
$$= 155.520 \text{ Mbits}$$

$$\text{Total number of payload bytes} = 260 \times 9 = 2340$$
$$\text{Total number of overhead bytes} = 9 \times 9 = 81$$

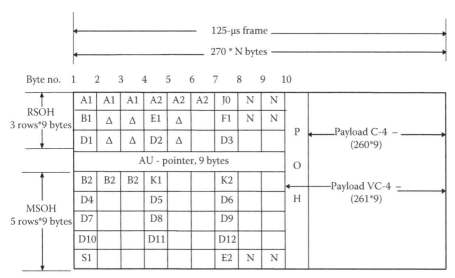

Legend: N bytes denote the bytes reserved for national use

Δ bytes are media dependent bytes

Blanks are the bytes reserved for future use

FIGURE 10.33

Basic SDH frame (STM-1) structure. (From the International Telecommunication Union, Fig. 9.3 STM-1 SOH, p. 66, ITU-T rec G.707/Y.1322, January 2007. With permission.)

Various components of the frame, i.e., RSOH, MSOH, POH, and payload, have been explained in detail in the next chapter. The AU pointer (Figure 10.33) will be explained in detail in subsequent sections of this chapter.

The beauty of this frame structure is that although it contains a large number (2430) of bytes (or 19,440 bits), we are able to represent it with a convenient diagram like that in Figure 10.33. Note that the RSOH is organized in the frame so that its first 9 bytes are the first 9 bytes of the frame itself, but its 10th to 18th bytes are the 271st to 279th bytes of the frame. If we draw it linearly, it will appear as shown in Figure 10.34.

We can see that the bytes pertaining to RSOH have been so arranged that they fall in the first nine columns of the diagram (Figure 10.33) in the first three rows. Similarly, the bytes pertaining to the AU pointer fall in the first nine columns in the 4th row, and the MSOH falls in the first nine columns in the 5th to 9th rows. The POH falls in the 10th column. Thus, the convenience of representation and understanding had been the driving criteria for deciding the functions of the different bytes of the STM-1 stream, rather than any operational requirement, which had been rightly so in the absence of any other more compelling reason for arranging the bytes otherwise.

Before looking into the further details of multiplexing, mapping, and other aspects, let us have a look at the "pointer," the term that we have been coming across since we started our discussion on multiplexing.

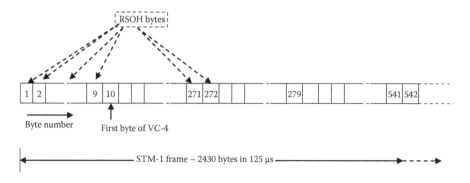

FIGURE 10.34
STM-1 frame represented in a linear diagram, showing the locations of RSOH bytes.

10.16 Pointer

The pointer is perhaps the most important and unique concept in SDH, which differentiates it from the other multiplexing hierarchies.

The pointer in SDH is a "set of bytes" containing the address of the beginning (the address of the first byte) of the payload in an STM frame.

Two levels of pointers are in use in SDH, the higher-level pointer called AU pointer and the lower level pointer called TU pointer. As the names signify, the AU pointer locates the AU payload and the TU pointer locates the TU payload.

10.16.1 AU Pointer

The AU pointer consists of 9 bytes. The 9 bytes are located in the 4th row of the first nine columns (overhead columns) of the STM-1 structure (see Figure 10.33), which means that it occupies bytes 811 to 819 [(270 × 3) + 1 to (270 × 3) + 9] in the STM-1 frame of 2430 bytes. The job of the pointer is to pinpoint the location of the payload, i.e., the first byte of the payload (the byte number 10). Note that the byte number 10 is the beginning of the VC-4. Although VC-4 also contains one column of POH, the administrative unit considers the whole VC-4 as payload. (This is generally true for the entire structure of the multiplexing, that even though the overheads are added at several lower and higher stages of multiplexing, the higher stages always consider the lower stages as payloads.)

The administrative unit that includes the AU pointer is shown in Figure 10.35.

The formation of AU-4 with VC-4, in principle, is shown in Figure 10.35a, while Figure 10.35b shows the location of the AU pointer bytes in the AU-4, along with the structure of AU-4 in an STM-1 frame.

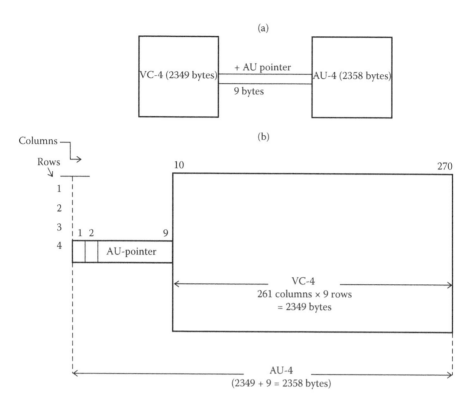

FIGURE 10.35
The AU-4 organization in a 9 × 270 STM-1 frame. Two types of representations are shown in (a) and (b).

However, before getting into the details of functions of the nine bytes of the AU pointer, let us first see the need of the pointer and the purpose it serves.

As already stated, the pointer is used for pinpointing the location of the payload within the frame. But what is the need? As we have seen in Figure 10.33, the VC-4 payload begins at byte number 10 in the STM-1 frame structure. Thus, if we generate the VC-4 in our STM-1 multiplexer and then add the pointer to it and add the other overheads such as SOH at the appropriate positions as shown in Figures 10.33 and 10.35, then the location of the beginning of the VC-4 is known (i.e., the 10th byte in sequence). But when we wish to demultiplex the signal at the receiver, how will the receiver know about the 10th byte, because the STM-1 bit stream will be just a stream of bits (or bytes), and it is very much unlikely in any practical situation that the receiver is able to catch the bit stream, right from first bit. It is very simple to understand and quite obvious that the practical transmitters and receivers will be switched on and off at random instants in time due to various reasons such as new installations, failures, maintenance shutdowns, upgrades, etc. This causes the bit streams of various tributaries and final frame structures to be available to the multiplexer and demultiplexers at varying instants of time, which (times) need not be the same as the beginning of the frames. In fact, it is practically impossible to do these switching on and off operations to synchronize with the frame beginnings. In other words, the phases of all the tributaries (in case of multiplexing) will be different; therefore, the design of the demultiplexers has to be able to know the precise beginning of the frames in the bit streams to extract meaningful data from the bit stream. This is achieved through the process of "framing" (see Section 7.1.3 for a discussion on framing).

Thus, if we add framing bits (or bytes) to the transmitted frame, we should be able to know the beginning of the frame and thus we would be able to easily identify the 10th byte. The framing bytes called A1 and A2, which are a total of 6 bytes (3 A1 and 3 A2), are the first six bytes of the STM-1 frame. Hence, our demultiplexer in the receiver will be able to interpret these bytes due to their predefined structure and will count the 10th byte to know the beginning of the VC-4. Then what is the need of the pointer? To know, let us imagine a situation where we have multiplexed various tributaries to form a VC-4, and after adding the overheads including the framing bytes, transmitted it, ensuring that the VC-4 bytes start at the 10th byte of the overall frame. Now when this frame reaches an intermediate node, from where it has to take another route, after demultiplexing, it will be "read" (demultiplexed) into the intermediate node using the clock derived from its own bits, thus the clock speed will remain the same.

However, it may be transmitted (after add/drop operation) using a different clock—why? Let us see a typical SDH network shown in Figure 10.36.

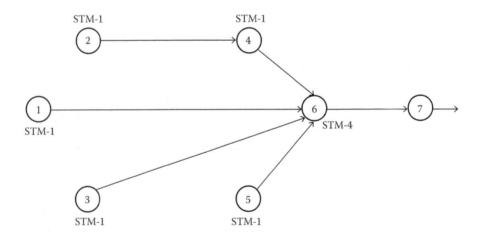

FIGURE 10.36
A part of a typical SDH network where the STM-1s generated at nodes 1, 3, 4, are 5 are to be combined at node 6 into an STM-4.

STM-1 generated at node (1) has to be carried to node (7) and beyond via node (6), which is an STM-4 node. Similarly, node (6) has to carry STM-1 received from nodes (3), (5), and (4) also. Each of these STM-1 streams may arrive at slightly different speeds at the STM-4 nodes due to jitter and wander caused by the clock inaccuracies and impairments in the transmission media (a detailed discussion on this can be seen in Section 6.5).

The varying bit rate STM-1 tributaries have to be combined to form a common bit rate STM-4 stream. In synchronous networks, the node clock is synchronized to the clock derived from the incoming bit stream. Thus, it gets synchronized to the master clock. But the question is, of the 4 incoming bit streams (STM-1), which one will it synchronize with? Each one of the bit streams may come at a different clock rate, and the node will get synchronized to the best clock (in terms of stability and accuracy) out of the four or it may get synchronized to one of the four clocks per a predefined priority. Thus, one of the STM-1s bit rate will be exactly equal to one fourth of that of the outgoing STM-4, and the rest of the three STM-1s could be having a difference of the bit rate with respect to the outgoing STM-4. (One possible way to accommodate the difference in speed could be the bit/byte justification as explained in detail in Section 8.2.3, like in PDH networks, but if we attempt to do it here, our network will also become PDH network, and we will lose all the advantages of an SDH networks. Justification is done in SDH networks also, but at the initial stage, for accommodating the varying rates of incoming PDH tributaries, but after the justification process, the bit rates of all tributaries (e.g., TU-12) become the same, and further multiplexing results in a synchronous bit stream of STM-1, from which each and every bit can be identified without stage wise demultiplexing. Thus, there has to be another way of adjusting the varying data rates of STM-1 incoming tributaries at the STM-4 node, which will not affect the synchronous nature of the outgoing bit stream.)

The pointer provides us with the means to accommodate such difference of speeds. Let us see how.

When we use the pointer, we need not begin our VC-4 payload at a fixed location, i.e., byte number 10 of the STM-1 frame. In fact, we can start the payload (VC-4) anywhere in the frame and pinpoint its location with the help of the address of the 1st byte of VC-4. Figure 10.37 shows this aspect.

With the help of the pointer, the VC-4 can begin at any byte within the STM-1 frame (excepting the section overhead and pointer bytes whose locations are fixed). In Figure 10.37b, the VC-4 begins at the 18th byte, whereas in Figure 10.37c, it begins at the 855th byte. What happens then to those bytes of VC-4 that could not fit into this frame (this frame will accommodate 854 bytes less than the length of VC-4)? This means that each of the STM-1 frames will contain parts of two VC-4s. Thus, the VC-4 payload is not fixed in the frame, but it actually floats across two frames, i.e., it may start anywhere and occupy parts of two frames. Thus, the VC-4 payload is accordingly called "floating payload." Figure 10.37c is redrawn as Figure 10.38 to highlight this aspect of "floating payload."

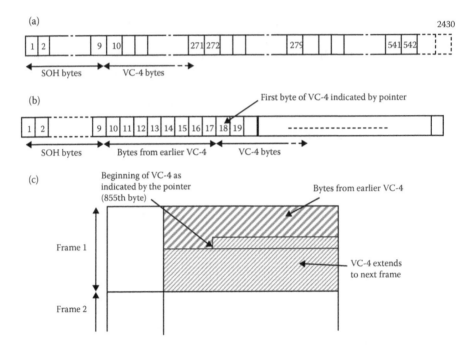

FIGURE 10.37
Location of first byte of VC-4 (a) in a STM-1 frame without pointer, (b) with a pointer indicating the beginning of VC-4 at 18th byte of the frame, and (c) the complete frame showing the beginning of VC-4 at 855th byte.

10.16.1.1 Floating Frames

As explained above and in Figure 10.38, the VC-4s within an STM-1 do not occupy a fixed position with respect to the framing bytes (or overhead bytes in general), but the VC-4 can begin at any location in the frame. Thus, the position of the 1st byte, which is the start of the VC-4, "floats" with respect to the STM-1 frame. This is called "floating payload." However, synonymously, it is also called "floating frame." This provides us with a means of flexible and dynamic alignment of the payload inside the STM-1 frame—flexible, as it can begin anywhere in the frame, and dynamic, as it can change its position anytime depending on the need. This essentially means that the virtual container VC-4 will be partly in one frame and partly in the next frame. Although it looks a little tricky, it is actually not, as the delimitation of each and every bit in the STM (1 and 4 and so on) is defined.

Hence, the pointer provides us with a means of starting the VC-4 payload anywhere in the STM-1 frame and knowing the address of the starting byte. Thus, if we are able to accommodate a few extra bits for slower STM-1 tributaries and drop a few bits from the faster STM-1 tributaries and increment/decrement the pointer accordingly, we can maintain a uniform bit rate for

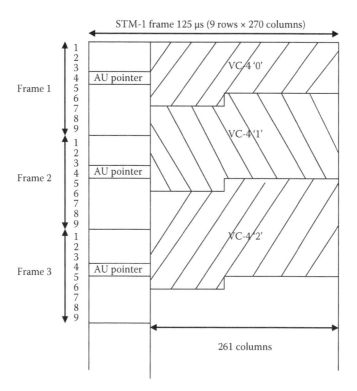

FIGURE 10.38
The VC-4 payload occupying portions of frame 1 and frame 2, called floating frames.

the outgoing STM-4, without losing the framing of the VC-4 payload. This is what is exactly done in SDH, and let us see how it works.

The AU-4 pointer consists of 9 bytes, i.e., 3H1 (H1, Y, Y), 3H2 (H2, F, F), and 3H3 (H3, H3, H3) bytes in sequence. The bytes are shown in Figure 10.39.

The H1 and H2 bytes contain the address of the VC-4's first byte in the STM-1 frame, whereas H3 bytes are used as justification bytes for adjusting the bit rates of incoming STM-1 to that of outgoing STM-4 (referring to Figure 10.36).

If the incoming bit rate is slower than the outgoing bit rate, we need to add additional bits to the outgoing bit stream to make the rate equal. This is known as positive justification (or stuffing or filling). If the incoming bit rate is faster than the outgoing bit rate, then we need to drop some of the bits to make the rate equal—this is known as negative justification. Let us see how it works.

The process of pointer interpretation and regeneration is shown in Figure 10.40.

As shown in Figure 10.40, VC-4 is recovered from the incoming STM-1 by removing the SOH and pointer. These operations are performed using the clock recovered from the incoming bit stream. The pointer value is

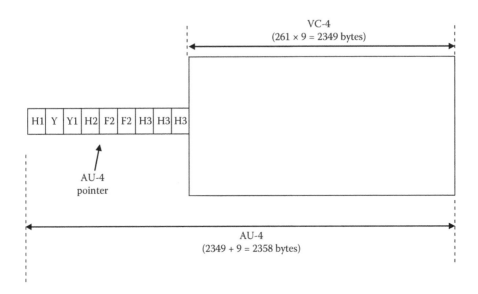

FIGURE 10.39
The AU-4 and the AU-4 pointer.

interpreted and stored. The average frequency of the incoming clock is compared with that of the local clock. If both the clocks have the same average frequency, no change in the pointer value is carried out. However, if there is a difference in the clock frequencies of the recovered clock with respect to the local clock, the pointer value is changed. The new pointer is then inserted into the VC-4, SOH is added, and the STM-1 is retransmitted, all using local clock. The pointer adjustment procedure and the role of H1, H2, and H3 bytes are as follows.

The H3 bytes generally contain a pseudorandom bit sequence. In case the incoming bits of STM-1 arrive faster than the outgoing STM-1 rate, then the 3 H3 bytes are dropped by decrementing the pointer by one unit, and the

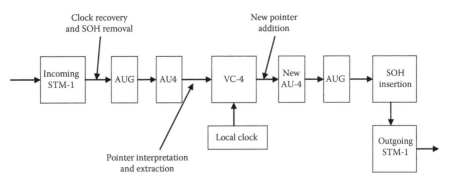

FIGURE 10.40
The process of pointer interpretation and regeneration at an intermediate node.

positions occupied by them are filled with the VC-4 bits. By dropping the H3 bytes, there is no loss of data, as H3 bytes contained only a pseudorandom bit sequence. The knowledge that H3 bytes have been dropped facilitates correct recovery of payload bytes at the receiving node. This knowledge is conveyed by H1 and H2 bytes (the details of which will be given shortly). This process is called "negative justification" because a few justification bytes are actually "dropped."

A similar case, but in reverse, is a slower incoming STM-1 bit stream, as compared with the outgoing STM-4 bit stream. In this case, however, three extra stuff or justification bytes are added, next to the H3 bytes, by incrementing the pointer by one unit. Adding justification or stuff bytes next to the H3 bytes means that these 24 bits, i.e., 3 bytes are filled with pseudorandom bit sequence. The receiver at the next node is conveyed this knowledge of stuff bytes through H1 and H2 bytes, so that it can ignore them, while taking out the payload. The process is called positive justification, because three justification bytes are "added."

When the bytes are added or dropped as brought out above, the location of the VC-4 within the AU-4 payload changes, the pointer value is incremented or decremented accordingly by three bytes (three bytes make one pointer unit) or one unit, so that it continues to pinpoint the payload and synchronization is maintained. The details of "pointer units," "incrementing," etc. will be discussed in this chapter.

To conclude, we can say that the use of pointer provides a means of adjusting the bit rates of SDH transport modules at intermediate nodes, to maintain the synchronization, and it provides an easy means of locating the payload within the transport modules (STM-1, STM-4, etc.).

We will discuss shortly the bit assignment (duties or functions of the bits) of each bit of the H1, H2, and H3 bytes of the AU pointer, but before that, let us have a look at the TU pointer.

10.16.2 TU Pointer

Recall our discussion in Section 10.11 where we defined a TU. A TU consists of a lower-level virtual container and a pointer. Figure 10.15 is redrawn here for convenience as Figure 10.41.

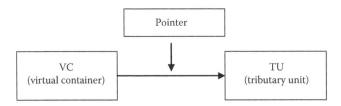

FIGURE 10.41
A tributary unit is formed when a pointer is added to a virtual container.

Bit sequence number in the multiframe

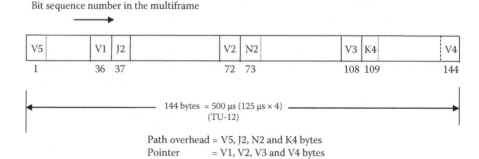

Path overhead = V5, J2, N2 and K4 bytes
Pointer   = V1, V2, V3 and V4 bytes

FIGURE 10.42
Location of the TU-12 pointer bytes in the TU-12 multiframe structure.

The VC could be of any type or size permitted by ITU.T G-707, i.e., it could be VC-11, VC-12, VC-2, or VC-3, combining them with pointer will form the TU-11, TU-12, TU-2, and TU-3, respectively. However, to explain the working of TU, we will take the case of VC-12, which is the most prevalent standard in Europe and the rest of the world (except America and Japan).

The TU-12 pointer consists of 4 bytes, namely V1, V2, V3, and V4. As we have seen earlier in this chapter, the VC-12, and hence the TU-12, is a multiframe structure. The locations of V1 to V4 bytes in the TU-12 frame structure are shown in Figure 10.42.

10.16.2.1 Functioning and Purpose

The purpose and functioning principles of the TU pointer are the same as that of the AU pointer. It provides a means of flexible and dynamic alignment of the payload (i.e., lower-order VC in this case) within the TU-12 multiframe and provides for the bit rate adjustment of tributaries at intermediate nodes (see the discussion in Section 10.16.1 of AU pointer). However, there is one difference.

When the STM-1 or STM-4 are multiplexed into higher bit rates, i.e., STM-16, STM-64, etc., the individual SOH is removed at the ADM or cross-connect nodes, AUGs are recovered, and a common SOH is again inserted for the outgoing higher-order bit stream. This involves framing of incoming tributaries to know the exact locations of the SOH beginning. The frames of tributaries coming from different sources and different media do not all arrive at the node at the same time. Thus, to recover the AUs from the STMs (by removing SOH), the node has to wait for 125 µs (one complete frame) before every incoming tributary has been framed (as the A1, A2 bytes of some tributaries may come at a time distance of maximum of 125 µs, from some other tributaries). In practice, this 125 µs is actually multiplied by 3 to 4 because the framing of a tributary is considered to have been confirmed after 3 or 4

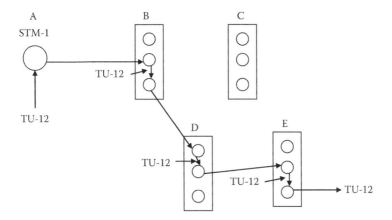

FIGURE 10.43
A TU-12 inserted at an STM-1 node A is taken out from the STM-1 at nodes B and D and reinserted to reach the node E, where it is again diverted on another route.

incoming frames coincide with each other. Thus, a time delay of about 500 μs is introduced at each intermediate node.

However, this is not the same for TUs. A tributary unit undergoes no change at the intermediate nodes. It simply remains embedded in the VC-4 and gets carried to the next node. The bit rate adjustments and others are done at the VC-4 level by the AU pointer. Even if the TU (let us say, TU-12) has to be dropped and reinserted because it has to take another route, it gets multiplexed with the TU-12 of other tributaries in the outgoing STM. The location of the VC-12 continues to be pinpointed during all this process, hence no framing of VC-12 is required before inserting this TU-12 along with other tributaries for another direction at the intermediate node.

In the absence of the pointer, the VC-12 framing would have been necessary, resulting in a time delay of 500 μs at each intermediate node (and, of course, the requirement of large buffer memories and a fair amount of additional processing at each node). This factor plays an important role in the transmission process, as the time delays in the network should be avoided as far as possible. The situation is depicted in Figure 10.43.

Thus, the use of the TU pointer has the additional advantage of saving a large amount of network delays to the associated VCs.

10.16.3 Summary of Pointer Advantages

The overall advantages of the use of pointer in SDH can thus be summarized as follows:

(i) The synchronization process becomes much simpler because the bit rate adjustment is possible at intermediate nodes to cater to the temporary variations in the bit rates of various tributaries due to jitter and wander (as explained in detail in Section 10.16.1). The AU pointer takes care of the variations of STM-1 level tributaries and TU pointer takes care of the variations of the TU-12 (or TU-2, TU-3, etc.) level tributaries.

(ii) The use of the TU pointer avoids the otherwise necessary delays due to framing of VCs at each intermediate node, thus minimizing the overall network delay. The requirement of buffer memories or elastic stores is also minimized, as no framing is performed, and hence, no large storage is required (as explained in Section 10.16.2).

(iii) In any SDH node, if the VC-4 does not use the same clock as that of STM-1, then also the synchronizations can be maintained by dynamic adjustments of the AU pointer (as explained in Section 10.16.1).

(iv) Synchronization at PDH boundaries becomes very simple. PDH boundary is a situation where two SDH networks meet. This happens very frequently, as each country has several network operators, thus an international call has to pass through the networks of many operators. All the operators have their own master clocks. Thus, the SDH bit streams encounter a different clock as they enter the network of another operator every time. The pointer adjustment process explained in Section 10.16.1 takes care of the differences in the clock frequencies and maintains the synchronization. The only other alternative for tackling the problem of clock mismatch at PDH boundaries is to allow controlled slips per ITU-T recommendations G-823. In this process, even though the slips are controlled, they are there to cause poor quality of reception. If the ITU-T G-823 is followed strictly, the number of slips will be 1 frame slip in 70 days for each PDH boundary. It is interesting to note, however, that the performance achieved by the operators is generally better than this owing to the fact that the clock accuracies of the commercially available atomic clocks are better than what is specified.

10.16.4 Disadvantages of Pointer

The only disadvantage of the use of pointer is that, every time a pointer adjustment is made, it introduces jitter in the transmitted bit stream.

However, in a synchronous network, generally, the STM clock and VC-4 clock are the same, leading to no pointer adjustment. If the transmission media is OFC, which introduces much less jitter and wander, it leads to very infrequent pointer adjustments. Thus, in practice, the pointer adjustment is not very frequent. Hence, the practical SDH systems do not experience too much of a pointer-generated jitter.

Having seen the need, utility, and functioning of AU and TU pointers, it is time to have a look at how the bits and bytes of the pointer actually work. We will first see the details of AU pointer and then TU pointer.

10.16.5 AU-4 Pointer Details

The AU pointer is located in the 4th row of the SDH frame structure. It is 9 bytes long, as shown in Figure 10.44.

The H1 and H2 bytes contain the address of the location of the 1st byte of the VC-4 payload, i.e., J1 byte; three H3 bytes are for justifications as explained earlier. The Y bytes contain a fixed value, i.e., 1001SS11 (SS are not defined), and the F bytes contain all 1s. (The bytes Y and F are defined for AU-4 pointer. In case of AU-3 pointer, these are replaced by H1 and H2 bytes, respectively, leading to three numbers each of H1, H2, and H3 bytes, and each set of H1, H2, and H3 bytes makes up for the pointer of each of the three AU-3s. We will discuss the AU-3 pointer a little later in this chapter.)

The details of the H1 and H2 bytes of the AU-4 pointer can now be seen a little more elaborately. Figure 10.45 shows this.

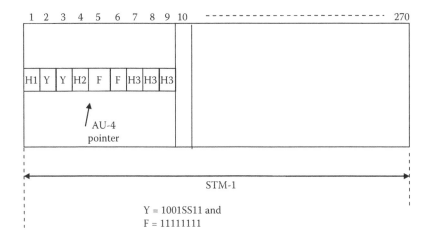

FIGURE 10.44
The AU-4 Pointer of 9 bytes.

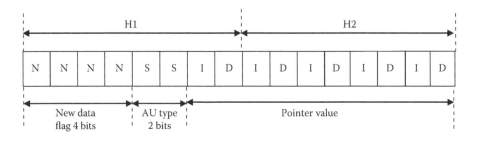

FIGURE 10.45
The bit assignment of the H1 and H2 bytes of the AU pointer.

The last two bits of the H1 byte and all the eight bits of the H2 byte together, i.e., a set of 10 bits, contain the address of the first byte "J1" of the VC-4. For example, for a few pointer values, the starting byte of the VC-4 will be located as follows:

(i) For the pointer value 0000000101, VC-4 will start at a distance of 5 "pointer units" from the last H3 byte

(ii) For the pointer value 0001010111, VC-4 will start at a distance of 87 "pointer units" from the last H3 bytes

Note two important aspects in the above expressions:

(i) The pointer count starts from the first byte after the last H3 byte

(ii) The pointer value indicates the distance from the last H3 byte, not in number of bytes, but it indicates it in number of "pointer units"

The pointer unit is a 3-byte unit, and this is the pointer resolution. The pointer cannot move by less than 3 bytes at a time. Thus, if the 10-bit count of the H1 and H2 bytes increases by 1, the pointer increments by 1, but the VC-4 moves by 3 bytes in the frame.

Our pointer value in the H1 and H2 bytes is 10 bits long, which means it can pinpoint up to 1024 locations. Then, how does it suffice for our complete VC-4? Let us see.

$$\text{Number of bytes in VC-4 payload} = 261 \times 9 = 2349$$

Therefore, the number of pointer locations required = 2349/3 = 783, as each pointer unit is of 3 bytes.

Thus, a 10-bit pointer is sufficient to locate all the 783 locations of the VC-4. However, the design counts these units not from 1 to 783 but from 0 to 782. Figure 10.46 depicts the situations.

The first pointer location called 0 location is just next to the last H3 byte of the concerned frame. Counting down in the VC-4 payload we can easily find that the first three bytes of the next frame are pointer location 522, and the last pointer location, i.e., 782, is just before the beginning of the pointer of the next frame, i.e., H1 byte of the next frame.

A question that may arise in the reader's mind is why the pointer is 3 bytes. To know, let us consider a transreceive scenario of SDH.

When the pointer value is changed, i.e., it is either incremented or decremented, or it gets an altogether new value, the receiver does not recognize the new pointer value as a valid pointer location, unless it is received in three consecutive frames (as is normally done for all framing processes). Thus, the pointer value cannot be allowed to be changed at the transmitter in less than

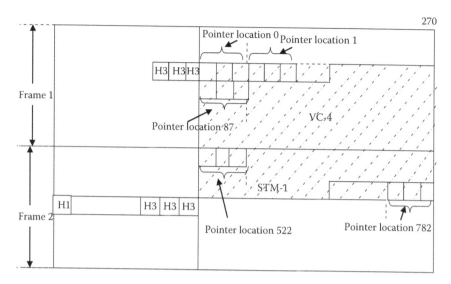

FIGURE 10.46
The locations of various pointer units in the STM-1 frame structure.

three frames, and hence there is no harm in keeping the pointer unit 3 bytes long (although the actual requirement for bit rate adjustment may be much less; we will see these calculations shortly). In fact, the pointer adjustment is not allowed in the next three frames, the frame in which the pointer value is either incremented or decremented is regarded as the first frame. The next three frames cannot carry a new pointer value. Thus, only the fifth frame can again carry a new pointer value.

We have seen the mechanism of bit rate adjustment through the use of pointer movement in detail. Let us now see which bits do what and how to achieve the results.

10.16.5.1 Justifications and Pointer Adjustment

In the case of an AU-4 administrative unit carrying a payload of VC-4, the bit rate adjustment is done through the use of three numbers of H3 bytes located in the AU pointer. The process is illustrated in Figure 10.47.

When the VC-4 payload's bit rate is faster than the normal bit rate, the 3 H3 bytes are stuffed with data bytes (the contents of H3 bytes are dropped and payload data stuffed instead). This is called negative justification because the "justification bytes" are actually "removed" from the outgoing bit streams.

On the other hand, when the VC-4 bit rate is slower than the normal bit rate, 3 stuff bytes filled with pseudorandom bits are inserted into the VC-4

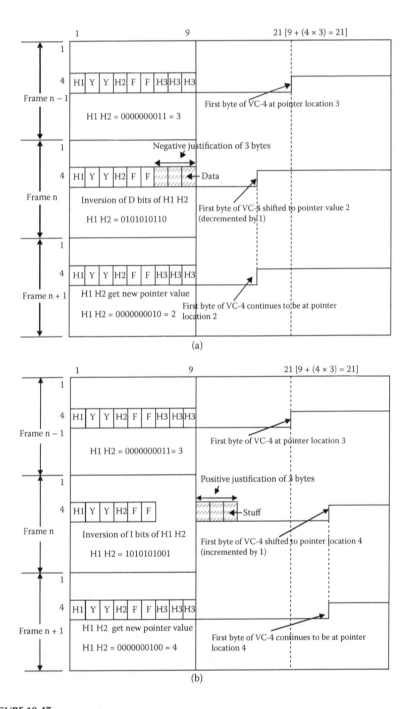

FIGURE 10.47
(a) The pointer operation in case of negative justification. (b) The pointer operation in case of positive justification.

area, just next to the last H3 bytes of the frame (actually, these bytes will get inserted into the previous VC-4, but it is immaterial because the adjustment is made in the overall bit rate, and the action for insertion or dropping of stuff bytes is taken once after many frames, depending upon the difference in the bit rate). This is called positive justification.

The three bytes next to the last H3 bytes belonging to payload area, where the stuff bytes are inserted, are also known as justifications opportunity bytes. This is unlike justification opportunity bytes of PDH (see Section 8.2.3), where the bytes are always present but declared as either data bytes or stuff bytes. Here only the location is specified (to be next to last H3 byte), and additional bytes (stuff bytes) are inserted at that location and the payload shifted accordingly by using the pointer.

The pointer action is initiated by temporarily inverting the "I" or "D" bits of the H1H2 bytes. Figure 10.45 is redrawn here as Figure 10.48 for convenience. The "I" bits, also called increment bits, are inverted when the pointer has to be incremented for positive justification in case of slower VC-4.

The "D" bits, called decrement bits, are inverted when the pointer has to be decremented for negative justifications in case of a faster VC-4. The process is already illustrated in Figure 10.47a and b.

Bit numbers 5 and 6 of H1 byte indicate the type of AU or TU. Presently, the value SS = 10 is defined for indicating an AU-4, AU3, or a TU-3.

In both cases, positive justification and negative justification, the inversion of "I" bits or "D" bits, as applicable as stated above, is recognized by the receiver node on majority voting. That means, if 3 or more of the 5 "I" or "D" bits are found inverted, when compared with the previous frame, the justification operation is taken as confirmed by the receiver. In case fewer than 3 bits are found inverted, no action is taken. This is to take care of any probable bit errors in these bytes during the process of transmission and reception in the transmission media. It is not expected that more than 2 bits will get inverted due to a transmission error.

10.16.5.2 New Data Flag

Refer to Figure 10.48 again. The first four bits of H1 byte, i.e., NNNN are called new data flag. True to its name, the new data flag indicates to the system that new data has arrived. In case of normal continuous flow of VC-4, the NDF will remain constant at 0110, but when the flow of VC-4 stops for some reason like switching off, maintenance, equipment replacement, etc., the NDF inverts its bits to indicate this situation. The NDF becomes 1001. This indicates to the system that new data has arrived, which means that the VC-4 could be anywhere with respect to the STM-1 frame and the previous value of the pointer is no longer valid. Thus, at this stage, the first byte of the VC-4 is located by framing and the value of the pointer is set equal to the new location of the VC-4.

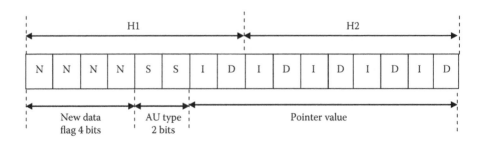

FIGURE 10.48
The bit assignment of H1 and H2 bytes.

Again, a majority voting is perceived by the receiver as the arrival of the NDF, in the sense that the reversal of 3 or 4 bits is taken as NDF received; fewer than 3-bit reversal is ignored.

10.16.6 Bit Rate Adjustment Range of Pointer

We have seen in the preceding section that the pointer adjustment is not allowed for the three subsequent frames, after the pointer is adjusted once. It means that we can adjust the pointer once in every four frames at the most. That works out to an addition or deletion of 3 bytes (3H3 bytes or 3 adjustments bytes) to the number of bytes being carried by the payload in 4 frames.

The number of bytes in a VC-4 payload are 2349 (refer to Figure 10.33).

Therefore, the number of bytes in 4 VC-4 frames = 2349 × 4

= 9396 bytes

Now, 3 bytes can be either added or deleted from these 9396 bytes by pointer action; thus,

Minimum possible number of bytes in 4 VC-4 frames
= 9396 − 3 = 9393 bytes, and

Maximum possible number of bytes in 4 VC-4 frames
= 9396 + 3 = 9399 bytes.

Hence,

Minimum possible bit rate = 9393 × 8 × 8000/4   [1 byte = 8 bits
= 150.288 Mbps   1 s = 8000 frames
Bit rate = bits/s]

Maximum possible bit rate = 9399 × 8 × 8000/4
= 150.384 Mbps

Thus, the possible adjustment range is

$$= 0.096 \text{ Mbps}$$
$$= 150.394 - 150.288 \text{ Mbps}$$
$$= 96 \text{ kbps} = \pm 48 \text{ kbps}$$

If we calculate the bit rate adjustment tolerance in parts per million, it will be

$$\text{Bit rate tolerance} = \pm 48{,}000/150.336 \qquad [\text{Normal rate} = 9396 \times 8 \times 8000/8$$
$$= \pm 319.28 \text{ ppm} \qquad\qquad = 150.336 \text{ Mbps}]$$

Thus, a tolerance of ±319.28 ppm is available for SDH nodes. Per ITU-T recommendations, the SDH equipment clock should have an accuracy of ±4.6 ppm (9.2 ppm, in worst case) or better. Thus, the available tolerance, i.e., ±319.28 ppm, is much more than enough to adjust not only the variations due to clock inaccuracies but also the variations if any are due to transmission media jitter and wander. In actual practice, the SDH node clocks are of much better accuracies, and they are, most of the time, synchronized to the master clock. Thus, the nodes hardly ever need to make such large adjustments in the bit rates, and consequently, the pointer is very rarely incremented or decremented. Minimum requirement of pointer adjustment is a desirable feature in any network as every time the pointer is adjusted, it introduces jitter.

The convenience of implementation rather than the required tolerance has in fact been the major factor in deciding the number of bytes used by the pointer for bit rate adjustment. In SDH, since byte multiplexing is used, it is preferred to add or delete bytes rather than bits. In the AU-3 pointer, a minimum of 1 byte is used for adjustment, and in the AU-4 pointer, a set of 3 bytes is used for the same purpose. (There is only one H3 byte is the AU-3 pointer. We will see about AU-3 pointer in the next section. SONET, the American counterpart, also uses only 1 byte for its payload, which is equal to AU-3. Since AU-4 is three times the size of AU-3, the number of H3 bytes in AU-4 pointer is also 3.)

10.16.7 AU-3 Pointer

Recall from Section 10.14 that an STM-1 (in fact, AUG) can consist of either 1 number of AU-4 or 3 numbers of AU3s. When it consists of AU3s, we need to have three pointers because each AU-N is supposed to consist of a VC-N and its pointer.

Accordingly, the 9 bytes of AU-4 pointer are divided into three pointers of three bytes each. The arrangement is shown in Figure 10.49.

The only difference from the AU-4 pointer is that instead of "Y" bytes in 2nd and 3rd column of the 4th row, there are H1 bytes, and in place of "F" bytes in the 5th and 6th columns, there are H2 bytes. The pointer for the first AU3 consists of the first H1 byte clubbed with the first H2 byte (i.e., 4th pointer byte) and first H3 byte (i.e., 7th pointer byte). Similarly, the pointers

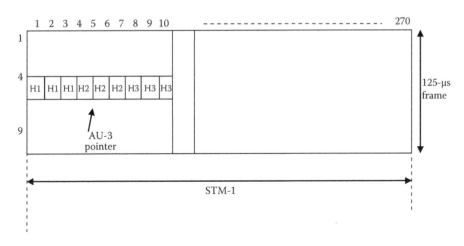

FIGURE 10.49
The AU-3 pointer details.

for the 2nd and 3rd AU-3s are formed by clubbing the 2nd and 3rd H1 bytes to the 5th and 6th H2 bytes and 8th and 9th H3 bytes, respectively. The justification for each AU-3 is of one H3 byte.

The rest of the bit allocations and functioning are the same as that of the AU-4 pointer.

10.16.8 Details of TU Pointer

We have seen the requirement, functioning, and general shape of the TU pointer in Section 10.16.2. Let us have a closer look at the other details of the TU pointer.

Similar to that of the AU pointer, the address contained in the TU pointer pinpoints the location of the payload, i.e., VC-12, VC-11, etc., within the STM-1 frame. Since the TU-12 is a multiframe unit (let us discuss the TU pointer with respect to TU-12, and because we have already seen the TU-12 structure, other TU-Ns will be simpler to understand in any case), the pointer also applies to the same multiframe.

The TU-12 pointer consists of 4 bytes called V1, V2, V3, and V4. The location of these bytes is shown in Figure 10.50. Bytes V1 and V2 contain the address of the first byte of the VC-12 payload, i.e., that of byte V5, in exactly the same way as the AU pointer's address is carried by H1 and H2 bytes.

In fact, the complete functioning including the bit allocation for NDF, type of TU, information and data bits (I&D bits), etc., is the same as that of the H1 and H2 bytes of the AU pointer. Figure 10.51 depicts the situation.

Byte V3 is used as a negative justification byte, exactly in the same manner as that of H3 bytes of the AU-4 pointer. The only difference is that, here, only one V3 byte is available for justifications of the bit rate. (This is actually

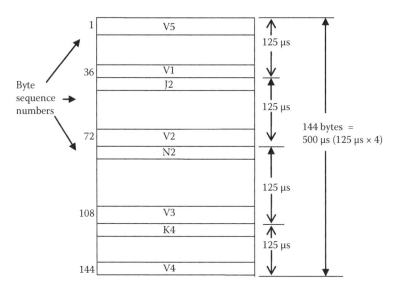

FIGURE 10.50
The locations of the TU-12 pointer bytes in the multiframe.

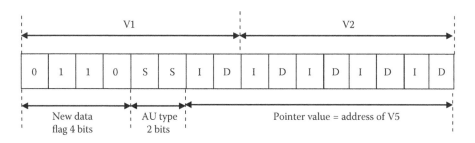

FIGURE 10.51
The TU-12 pointer V1 and V2 bytes bit allocation.

much more than enough, considering the low bit rate at the TU-12 level, as compared with that at the AU-4 level; see Section 10.16.6 for calculations of required and available margins.)

Byte V4 is not used at present.

Again similar to the AU-4 pointer, the byte just next to the V3 byte in the VC-12 payload area is treated as a positive justifications opportunity, although no byte is actually reserved for it. The byte is used if pointer movement is required; otherwise, it carries the payload as usual.

The counting of TU-12 pointer starts from the V2 byte, i.e., the first byte after the V2 byte, is designated as the 0 location (see Figure 10.52).

Thus, if there is no pointer movement, the 0 byte just follows V2, and thus the location of V5 byte (first byte of VC-12 payload) is 70.

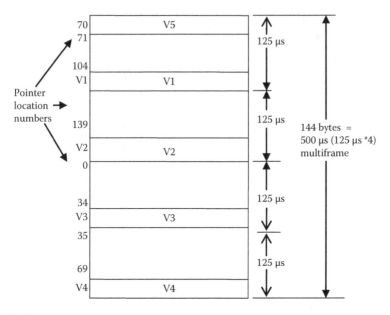

FIGURE 10.52
The counting of bytes for TU-12 pointer in a TU-12 multiframe. Since there are a total of 140 bytes in the VC-12, the pointer values will be from 0 to 139 (the "pointer unit" is "1 byte" against that of 3 bytes in case of AU-4 pointer).

Other rules for pointer application are similar to those explained earlier for AU-4 pointer.

The pointers for other TU-Ns such as TU-2, TU-3, etc., function in a similar manner. The details are omitted here, as the objective here is to understand the concept. The interested reader is referred to the ITU-T recommendation G.707 for all the details.

10.17 Formation of Higher-Order STMs

The higher-order STMs (STM-4, -16, -64, and -256) are formed by multiplexing the required number of STM-1s. The multiplexing can be done in two ways: AUG level multiplexing and STM-1 level multiplexing.

10.17.1 AUG Level Multiplexing

The process is depicted in Figure 10.53. A pointer is added to each of the VC-4s that are to be multiplexed together to form the respective AU-4s. The AU-4s as AUGs are multiplexed by byte interleaving to form a common byte stream, to which RSOH and MSOH are added to finally form the STM-N.

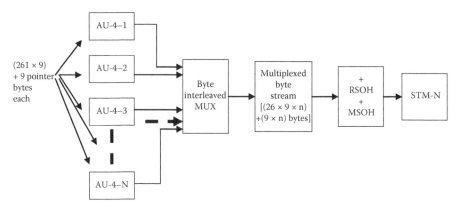

FIGURE 10.53
Formation of STM-N from AU-4s.

10.17.2 STM-1 Level Multiplexing

Another way to form a higher level STM is to byte multiplex the STM-1s after they are completely formed as STM-1s, i.e., the overheads are added to the AUGs. In this case, the overhead bytes have to be defined freshly, as the STM-N so formed has to work on a single set of overhead bytes (see the STM-4 overhead in the next section). Figure 10.54 illustrates the principle.

This is the type of multiplexing that is normally deployed. The frame structure of the higher-order STMs is nothing but a multiple of the STM-1's frame structure. However, there are small differences that are explained in the next section.

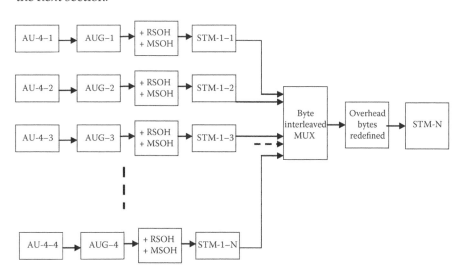

FIGURE 10.54
Formation of STM-N from STM-1s.

10.18 Frame Structure of Higher-Order STMs

As stated earlier, the frame structure of higher-order STMs is nothing but a multiple of the STM-1's frame structure. However, there are small differences. The framing bytes A1 and A2 of each of the STM-1 are retained in the multiplexed higher-order STM because framing at the STM-1 level will be required at each regenerator and multiplex section. The B2 bytes are also retained, as they convey the information about the errors in each individual STM-1 (B1 bytes detect the errors of the entire STM-N frame and hence are required only once). The other overhead bytes, on the other hand, are required only in one set, as the regenerator sections and multiplex sections become common to all STM-1s of the STM-N, requiring common management and hence a common set of management bytes. The overhead bytes of the rest of the STM-1s are left unused in the STM-N. The functions of all the overhead bytes are explained in detail in the next chapter on operations and maintenance of SDH.

The structure of the STM-4 overhead is shown in Figure 10.55. The payload follows the overhead in a way similar to what is shown in Figure 10.33.

As can be seen in Figure 10.55, many of the bytes that are marked for specific overhead functions in individual STM-1s such as B1, E1 and E2, F1, D1 to D12, K1 and K2, J0, S1, and M1 appear only once in the STM-4 frame structure, and in their place, the bytes belonging to other STM-1s become spare. Similar is the case with other higher-order STM-s such as STM-16, STM-64,

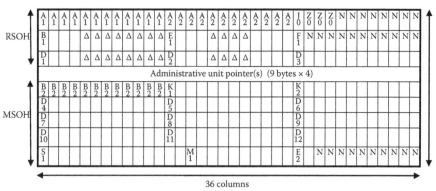

36 columns

where:

A1, A2, B1, B2 and other marked bytes are for RSOH and MSOH.

Δ : Bytes are for media dependent use.

N : Bytes are reserved for national use.

Unmarked bytes and Z0 bytes are reserved for future standardization; however some of them have been used in STM-16 and above; for FEC (Forward Error Correction).

Functions of all the overhead bytes are discussed in detail in the next chapter.

FIGURE 10.55

The structure of STM-4 overhead. (From the International Telecommunication Union, Fig. 9.4 STM-4 SOH, p. 66 of ITU-T rec G.707/Y.1322, January 2007. With permission.)

etc. Some of the spare bytes created by nonutilization of overheads, as stated above, are utilized for FEC (which is discussed in Section 6.8) in higher-order STMs.

The details of frame structures of higher-order STM-s are omitted here, as the depiction of STM-4 is enough to impart the understanding of the principles; the interested readers can refer to the ITU-T standard G.707/Y.1322.

Review Questions

1. What were the limitations of PDH systems in developing higher data rates?

2. What were the factors that paved the way for development of SDH?

3. What are the advantages of network wide synchronization?

4. Illustrate with an example the key difference between direct multiplexing and hierarchical multiplexing.

5. Which basic concepts were applied to design the SDH?

6. What came first, SONET or SDH? When were they developed and what are their geographical jurisdictions?

7. Discuss in brief the advantages of SDH.

8. With the emergence of SDH, what happens to the existing PDH and other equipment?

9. Which were the prevailing PDH hierarchies in the world? Describe with the help of a block diagram.

10. Discuss in brief the compatibility issues tackled by SDH.

11. How does the SDH compare with PDH in O&M features?

12. Discuss in brief the disadvantages of SDH. Are they serious enough to impede the introduction of SDH?

13. Describe the concept of a "transport module" in the context of SDH.

14. Which is the smallest transport module defined in SDH? What is its capacity?

15. How are various transport modules of SDH and SONET matched? Make a comparative table showing the exact and commonly known data rates.

16. Does SDH need more bandwidth for the same payload as compared with PDH? If so, why?

17. How is the STM-1 formed from 63 E1s? Explain with the help of a diagram.

18. What is the frame rate of STM-1? What are the factors governing its choice?
19. How are the bit rates of higher-order STMs arrived at? What are they?
20. Describe the process of justification of E1 tributaries in STM-1.
21. What exactly is meant by positive and negative justification?
22. What is the role of justification control bits and justification opportunity bits?
23. What is the margin available for variation in the bit rates of E1 tributaries in the STM-1 justification process? Show through calculations.
24. What is the role of R and O bits in the 500-μs multiframe structure made of four E1s?
25. What do you understand about a "container" in the context of SDH? What role does it play in the formation of STM-1 and how?
26. What types of containers are used in SDH for various types of tributaries and how many of them are accommodated in STM-1?
27. Explain with illustration the concept of "mapping" in the context of SDH.
28. What is a virtual container? What is its need and how is it different from a container?
29. What is the role of overhead bytes in virtual containers? What are they called and why?
30. Calculate the bit rate change as we move to VC-12 from C-12.
31. Explain the concept of "path" in the context of SDH.
32. What is a regenerator section in SDH, what is its necessity?
33. Define a multiplex section in SDH. How is it different from regenerator section?
34. What do you understand about the terms RSOH, MSOH, SOH, and POH?
35. What are SDH layers? What is their need?
36. What is a tributary unit, and why it is called so?
37. What is the role of pointer in a tributary unit? What does it consists of?
38. Show the positions of various bytes of a TU in a diagram.
39. What are the limitations for higher orders of multiplexing beyond STM-256?
40. What are tributary unit groups? How are they formed?
41. With the help of schematic diagrams, explain the multiplexing of TUs into STM-1.
42. What is LKM numbering scheme of tributary units?
43. What higher-order path is defined in SDH? What is HO-POH?

44. Define administrative unit and AU pointer.
45. How is STM-1 formed from AU-4?
46. Draw the schematic of the complete multiplexing chain of STM-1, indicating the activity at each stage.
47. What do you understand about T-1, T-2, T-3, DS1, DS2, DS3, etc.?
48. Explain how the bit rates of T-1 and E1 tributaries are matched at higher levels of multiplexing in STM-1.
49. How is container C-3 formed?
50. Draw the complete multiplexing path for E3 and T-3 up to STM-1.
51. Which is the alternative path followed by C-3 for multiplexing of E3 or T-3?
52. How VC-3 can have a lower- or higher-order path?
53. Draw the multiplexing structure for E4 tributary.
54. Is a multiplexing route provided for E2 tributary in STM-1? If not, why?
55. What is meant by routes in the multiplexing structure of STM-1? What governs the choice of a particular route?
56. What is the significance of AUG?
57. Draw the complete multiplexing structure of different routes of STM-1 showing activities performed at each stage.
58. Show the bit rates at various stages in the above diagram.
59. What is the significance of numbers "9" and "270" in the STM-1 frame structure? Explain.
60. Draw the frame structure of STM-1 showing the details of overhead and payload bytes. Why is the frame structure represented in a matrix of 9 rows and 270 columns?
61. What is a pointer, and what is its role in STM?
62. Show the location of AU pointer in an AU-4 on a diagram.
63. Why exactly is the pointer needed in SDH? Explain using illustration.
64. What do you understand by floating payload and floating frame?
65. Explain the process of pointer interpretation and regeneration with the help of a block schematic.
66. How does the AU-4 pointer facilitate justification? How are positive and negative justifications achieved?
67. What is TU pointer and what is its role?
68. What are the advantages and disadvantages of the pointer?
69. What is the difference between AU-4 pointer and AU-3 pointer?
70. Explain the functions of 9 bytes of AU-4 pointer with an example.

71. Why is the pointer unit 3 bytes long?
72. Explain the process of justification by AU-4 pointer clearly indicating the role played by each bit.
73. What is NDF? Which bits make it up and what role does it play?
74. Calculate the bit rate adjustment range possible with AU-4 pointer. How does it compare with the requirement of an SDH node?
75. Is frequent pointer adjustment desirable? If not, why?
76. Why is the AU-4 pointer 3 bytes long?
77. What are the differences in AU-4 pointer and AU-3 pointer?
78. Explain the TU pointer, giving details of the functions performed by each byte.
79. How are the higher-order STMs formed from STM-1? Explain the multiplexing processes in AUG level and STM-1 level multiplexing.
80. What are the differences in the frame structures of higher-order STMs as compared with STM-1?

Critical Thinking Questions

1. What could have been the alternatives other than SDH/SONET for getting rid of the problems of low capacity and hierarchical multiplexing of PDH?
2. Differentiate clearly between synchronous and asynchronous multiplexing.
3. Work out the sizes of VC-12 and VC-2 assuming different hypothetical bit rates for E1, E2 and T-1, T-2.
4. Is the pointer necessary in SDH? Think of an alternative.
5. Calculate the percentage of overhead and payload bytes in cases of E4 and STM-1.
6. Calculate the payload to overhead ratio for E1, E2, E3, and E4 tributaries.
7. Compare the system of layers with that of the postal service.
8. Explain why tributary units are formed by grouping in the ratio of $3 \times 7 \times 3$.
9. An E2 tributary has not been catered to in SDH; what route could it have followed had it been facilitated?

Bibliography

1. Optical Transmission in Communication Networks, *Web Pro-Forum Tutorials*, International Engineering Consortium, 1998.
2. P. Moulton and J. Moulton, *The Telecommunications Survival Guide, Understanding and Applying Telecommunication Technologies to Save Money and Develop New Business*, Pearson Education, 2001.
3. S.V. Kartalopoulos, *SONET/SDH and ATM, Communications Networks for the Next Millennium*, IEEE Press, 2003.
4. P. Harikumar, *Teaching Notes on Synchronous Digital Hierarchy*, IRISET Secunderabad 2002.
5. J.C. Bellamy, *Digital Telephony*, John Wiley and Sons, Singapore, 2003.
6. G.S. Pandian, *IP and IP-Over SDH*, Training course, Railtel Corporation of India, 2003.
7. *TJ100MC-4 Multi-Service Access Node, Software User Guide*, Tejas Nerworks India, 2003.
8. J. John and S.K. Aggarwal, *Proceedings of the International Conference and Exposition on Communications and Computing IIT Kanpur*, February 2005.
9. P.V. Shreekanth, *Digital Transmission Hierarchies and Networks*, University Press, 2010.
10. NIIT, *Introduction to Digital Communication Systems*, Prentice-Hall, India, New Delhi 2004.
11. *Introduction to Synchronous Systems*, Training material, M/S Alcatel, 1994.
12. *In Service Course on SDH Systems*, Regional Telecom Training Centre, Hyderabad, 2000.
13. *Broadband Communications and SDH*, Training material, Tejas Networks India, 2002.
14. ITU-T Recommendation G.707/Y.1322 (2007) Amd-1, *Network Node Interface for the Synchronous Digital Hierarchy (SDH)*, International Telecommunication Union, Geneva, Switzerland.
15. ITU-T Recommendation G.736, *Characteristics of a Synchronous Digital Multiplex Equipment Operating at 2048 kbit/s*, International Telecommunication Union, Geneva, Switzerland.
16. ITU-T Recommendation G.774.02, *Synchronous Digital Hierarchy (SDH) Configuration of the Payload Structure for the Network Element View*, International Telecommunication Union, Geneva, Switzerland.

11

Operations and Maintenance in SDH

A very strong set of operations and maintenance features is one of the biggest advantages of the SDH over its predecessor technologies, second only to the huge capacity that it provides.

One of the most beautiful features of the digital technology is the ease of signaling. If we recall the kind of signaling associated with the analog technologies, the signaling functions were, by and large, restricted to the call setup and release, such as detecting the "on hook" and "off hook" conditions, extending the dial tone, seizure of trunk line, call terminations, etc. Adding any extra features to signaling called for increased complexities and, of course, increased costs. In sharp contrast to this, digital technology provides for a very large number of signaling functions. In fact, it will not be an exaggeration if we call it an unlimited number of functions. The secret of this quantum jump lies in the fact that the digital signal does not differentiate between signaling and speech when it comes to carrying it across the distances. Signaling information is just a few extra bits added to the speech data. Since these additional bits are not speech data, or payload as it is called, it is called the "overhead." However, the overhead is considerably insignificant in comparison to the total payload data, but what gives it a real outstanding status to digital systems is that the provision of this overhead is extremely easy and cost effective as compared with that of analog technology signaling. In fact, it was unimaginable in the analog era to provide for so many O&M functions in the system as in digital technology systems. The functionalities of the overhead bytes are defined in ITU-T recommendation G.707/Y1322 (2007), amendment I.

The scope of such O&M functions is so large as compared with the signaling functions of the analog era that the term signaling has been replaced by "operation administration and management/maintenance." The scope of OAM activities includes not only call processing, but a large set of activities including adding and dropping channels of varying capacities at various nodes in the link, remote configurations of the systems, performance monitoring, diagnostics and fault rectification, provisioning of channels and their protection paths, billing, generating various types of management reports, and so on. (The terms O&M, OAM, and OA&M are used synonymously. A stands for administration and M stands for either maintenance or management. Thus, OAM or OA&M becomes "operations administration and maintenance/management." The old telephone systems required only signaling, and the PDH standards improved greatly upon the fault detection and maintenance features, which were named O&M. The SDH standards have enormous features for management and administration of the network besides having vastly improvised features for fault detection and maintenance. It includes a large number of administrative and management functions such as provisioning of new channels, provision or disabling of protection channels, billing, generation of management reports, etc. Thus, all these features put together are termed as OAM; another variation for this nomenclature is OAM&P, which stands for operations administration and maintenance/management and provisioning. For further discussion in this chapter, we will use OAM to represent all these features.)

The overhead bytes provided in SDH incorporate all the perceivable activities as on date, and still a large number of bytes have been left unused and are allocated for future applications. Thus, SDH has a very vast OAM segment, which is far ahead of any of the past technologies or any of the contemporaries.

Broad categorizations of OAM activities may include:

(i) Performance monitoring
(ii) Fault diagnostic and rectification
(iii) Network management

We will discuss the subject under these categories in various sections of this chapter.

11.1 Performance Monitoring

Performance monitoring of an SDH link can be done in two ways:

(i) System out-of-service performance monitoring
(ii) In-circuit performance monitoring

11.1.1 Systems Out-of-Service Performance Monitoring

The best way to monitor the performance of a telecommunication link is to take it out of service and look for the bit error performance or bit error rate (BER), as it is popularly known. This test or monitoring is usually resorted to at the time of initial commissioning of the link or when the link completely fails and major repairs are carried out. A typical setup for BER testing is shown in Figure 11.1.

A fixed bit pattern is sent down the transmission link up to the last node. At the last node, the signal is "looped back," i.e., transmitted toward the

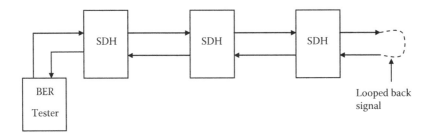

FIGURE 11.1
A typical BER test setup.

source. At the sending station, the same bit pattern is thus received, which is analyzed for bit errors by an equipment called a BER tester, which could be either stand-alone equipment or a part of multifunction test equipment.

The number of "errored bits" is counted over a period, which could be 1 h or so for a repaired link or could be 24 h or even longer for initial installations. With properly designed and executed optical fiber links BERs of 1 in 10^9 are easily achievable, while those of 1 in 10^{12} are not uncommon. The BERs of radio links are considerably lower and vary with the season due to various environmental factors.

11.1.2 In-Circuit Performance Monitoring

The BER test method, as brought out in the previous section, is the best way of knowing the performance of a digital telecommunication link. However, a better customer satisfaction can be achieved if the performance of the circuit can be monitored while it is in service, without any need to disrupt the services for carrying out a BER test.

With this approach, a number of performance monitoring features have been built into SDH. All this monitoring is achieved by dedicating specific bytes in the overhead portion. The monitoring is done at all the SDH layers, i.e., the regenerator section layer, the multiplex section layer, and the path layer. The layered SDH model is shown in Figure 11.2 for recapitulating the concept.

The overhead bytes, carrying out the functions of performance monitoring and fault detection and other management functions, are likewise divided into these three layers, which facilitate fast and proper localization of problems if any exist in the link. The frame structure of the SDH frame is shown in Figure 11.3, which gives the details of the overhead bytes allocated to these layers.

There are a total of $9 \times 3 = 27$ bytes reserved for the regenerator section overhead (RSOH), and there are $9 \times 5 = 35$ bytes reserved for the multiplex section overhead called MSOH. Of the first nine columns, only the fourth row does not pertain to the section overhead, as it is allocated to the AU pointer (we have seen the details of pointer and its functioning in Section 10.16).

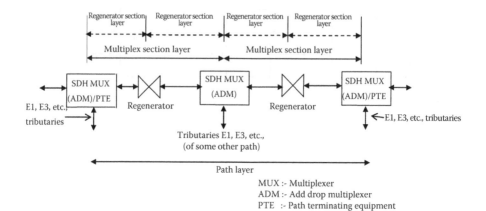

FIGURE 11.2
SDH layers.

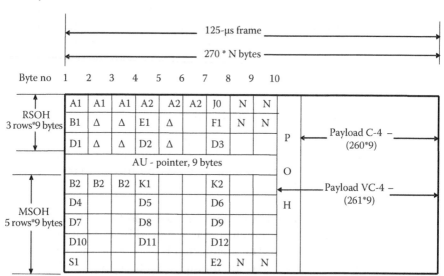

FIGURE 11.3
Overhead bytes in an STM-N frame.

Column 10 is allocated to the overhead bytes for the path layer; thus, the path layer has $9 \times 1 = 9$ bytes, and this is called path overhead (POH; in fact, higher-order POH [HO-POH], applicable to VC-4 payload; we will see the lower-order POH [LO-POH] later in this section, which will be applicable to VC-12 and other lower-order tributaries). Thus, we see that there is specific allocation of overhead bytes to all the three layers of SDH.

Now let us see how the performance monitoring is achieved with the help of these bytes.

11.1.2.1 Parity Check

One of the simplest and most popular methods of checking the "in-circuit" performance of a telecommunication link is parity check. Since this check is generally performed by bit interleaving of a particular set of bits, it is popularly called "bit-interleaved parity" (BIP).

The process of error detection through the use of parity check has been explained in detail in Section 6.7.2. However, it is repeated here for ease of reference, as we will need to refer to it frequently for the succeeding discussion.

The whole data of a frame is divided into a number of blocks. The number of blocks will be equal to the number of bits performing the parity check. Consider a data stream of 64 bits divided into 8 blocks for an 8-bit parity check.

As can be seen in Figure 11.4, the 1s in each column are added, and if the total of the number of 1s in the column is an odd number, then a 1 is placed in the result row, i.e., the parity bit. Similarly, if the number of 1s is even, then a 0 is placed as the parity bit. This ensures that the number of 1s in a column including, parity bit is "even." Thus, this is called "Even Parity." Conversely if we ensure the number of 1s to be odd in every column including the parity bit, then it will be called an "Odd Parity." In this example we can see that the total number of 1s in block number 1 are 5, i.e., "odd" number, thus a 1 is placed on the parity bit to make it "even." Similarly, the total number of 1s in block number 3 are 4, which is "even"; thus, a 0 is placed on the parity bit to keep it "even" and so on.

The parity bits so generated are transmitted along with the rest of the bits of the frame. At the receiver, the calculations are again performed in the

Block number →	1	2	3	4	5	6	7	8
	1	0	1	0	1	1	0	1
	1	1	0	1	0	0	1	0
	0	1	1	0	1	1	1	0
	1	0	1	1	1	1	0	1
	0	0	0	1	1	0	1	1
	1	1	0	1	0	1	0	0
	0	1	1	0	0	1	1	0
	1	1	0	0	0	0	1	1
Parity bits →	1	1	0	0	0	1	1	0

FIGURE 11.4
Parity bits checking of data.

same manner as with the transmitter. The parity bits received at the receiver from the transmitter and those generated locally are matched at the receiver. If both of them match correctly for all the bits independently for each parity bit, there is no bit error in the transmission, and if there is a mismatch, then there are blocks (a block is a set of a defined number of bits) with errors, in as many numbers as the number of mismatched parity bits.

Although the technique is very simple and straightforward, its accuracy is very limited. Suppose there are two bit errors in the block of data being checked by the parity bit. If both these errors are positive or both of them are negative, the parity will still detect an error, but if the bit errors are one positive and another negative, they will cancel each other and parity will show no error. Similarly, there could be a large number of bit errors, but as long as they balance themselves, no error will be detected by the parity check. Also, the number of bit errors is not known; all that is known is that there is some bit error in the block. Then why is this method deployed in all modern systems? The answer is, that despite its above-mentioned limitations, the error indications generated by the parity check are a good measure of transmission errors. More importantly, these indications enable performance monitoring without taking the circuit out of service.

In practical situations, we hardly ever need to know the exact number of bit errors happening in the system. All we are interested in is the general performance of the circuit. We need to know whether the errors are within the tolerable limits for a particular application or for a particular customer. The actual number (of BERs) also tells us the same story. Thus, in making a decision about the action to be taken, parity check gives a reasonably good indication of circuit performance. Generally, an accuracy of ≥90% is obtained in bit error monitoring by B1, B2, and B3 bytes (except in BIP-2 of VC-12 level). The functioning of B1, B2, B3, and BIP-2 will be clearer as you read through this section further.

In case of SDH, the parity check is carried out at all the three layers. The RSOH has been provided with byte B1 for the regenerator section layer, the MSOH has been provided with bytes B2 for the multiplex section layer (refer Figure 11.3), and a byte B3 has been provided for the path layer. B3 byte actually exists inside the VC-4, which is a payload for STM-1 and which in turn carries a payload of C-4 and one column of 9 bytes as the overhead. The overhead bytes in VC-4 are shown in Figure 11.5.

The parity bytes of all the three layers are shown as in Figure 11.6.

Nyte B1 provides the BIP-8 parity for the whole data between any two regenerator sections. For the purpose of calculating the parity, the whole of the STM-N frame is divided into eight blocks (or 8 columns as shown in Figure 11.4). Each block thus contains (STM-N)/8 bits, with each bit of the B1 byte denoting the parity for each such block. On parity mismatch in any one of them, the block is declared as errored block. This leads to the error performance being monitored in terms of errored blocks. The number of bits is only 8 for this parity check irrespective of the total number of bits

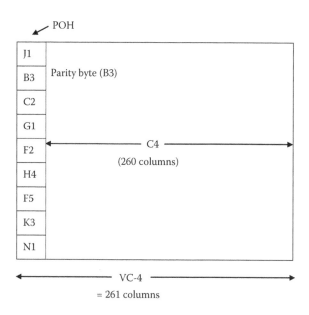

FIGURE 11.5
Parity byte in VC-4 POH.

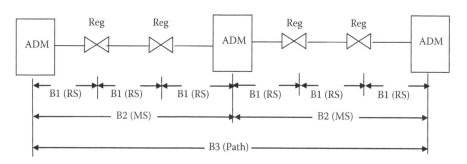

FIGURE 11.6
Distribution of parity bytes over SDH layers.

in the frame, which may vary from 19,440,000 for STM-1 to 311,040,000 for STM-16.

The parity for each of the eight blocks is calculated as shown in Figure 11.4. The result is transmitted by the transmitting station along with the next frame data. The receiving station separates the parity bits and compares them through an "exclusive OR" gate with the locally generated parity bits, applying the same principles. In case of a difference, the result of the "exclusive OR" operation will be a 1. Thus, the number 1s generated will indicate the number of errored blocks. (An "exclusive OR" operation is a logical operation in simple digital logic circuits. In case the reader is not familiar with digital logic circuits, he may refer to any text providing fundamentals of digital logic circuits or elements. For the time being and for the understanding of the current discussion, it is enough to assume that there is a device that produces the desired results on comparison.)

Byte B2 can be located in the MSOH section in Figure 11.3. There are three of them. These bytes provide the parity check for the multiplex sections. As they are three, they provide BIP-24 parity (8 bits × 3). Thus, each bit represents a column of (STM-1)/24 bits. Note that it is (STM-1)/24 and not (STM-N)/24 because these B2 bytes are provided in each STM-1 of the STM-N configuration. Thus, for each STM-1, there will be 24 error blocks in an STM-N. For example, STM-4 will have 4 sets of error blocks of 24 numbers each, and STM-16 will have 16 sets of error blocks of 24 numbers each.

The process adopted for B2 bytes is the same as that for B1, except that the B2 bytes are processed separately for each STM-1 in an STM-N, whereas the B1 byte is processed for the whole of the STM-N. Thus, there are 24 error blocks generated in each STM-1. The parities are calculated and transmitted and then compared at the receiver with locally generated parities and the block is declared having an error in case of a mismatch.

B3 bytes similarly provide for the parity check at the path level, i.e., for end-to-end user data. Separate parity at each layer of SDH provides quick localization of fault. It becomes easy to locate where the error blocks are generated (and hence the fault lies): between two regenerator sections or two multiplex sections or between the user entry and exit points.

The path level parity byte B3 indicated above is actually applicable for a higher level path, i.e., VC-4 (or VC-3 in some cases). As we know from the discussions in Section 10.9, there is a lower-order path, defined at the VC-12 level too. Figure 11.7 shows the overhead bytes of VC-12.

The first two bits of the byte V5 provide parity check; other bits have other functions for the VC-12 payload, which we will see shortly. The parity check is accordingly called BIP-2. The whole VC-12 is divided into 2 blocks, one consists of all odd-numbered bits and the second one consists of all even-numbered bits. The parity bit is then calculated in a way similar to the one explained for B1 or B2 in the preceding discussion.

11.1.2.2 Indications Generated by Parity Bytes

The parity bytes B1, B2, and B3 generate the error counts in the blocks of data as discussed in the preceding section. Now, let us see how they are being used by the system to generate useful indications for taking preventive maintenance actions.

11.1.2.2.1 B1

The B1 byte monitors the number of errored blocks between two regenerator sections. The information regarding the number of errored blocks received is not conveyed by the receiver back to the transmitter. This is classified as

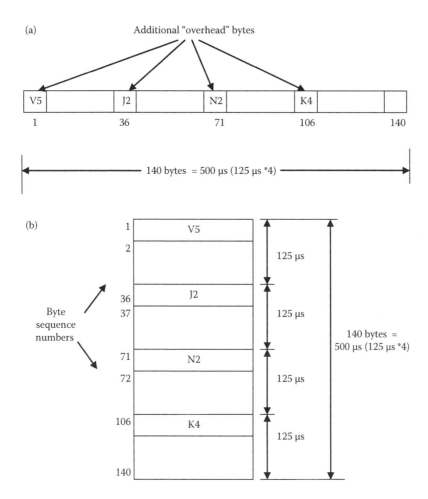

FIGURE 11.7
Overhead bytes of VC-12. Two types of representations are shown in (a) and (b).

"class B" performance parameters and is not mandatory to be deployed. This parameter is generally used while commissioning the system where it helps in monitoring the errored blocks between two adjacent nodes.

11.1.2.2.2 B2

The "B2" bytes monitor the number of errored blocks between two multiplex sections. As seen earlier, "B2" has a very positive feature where the number of these bytes per STM-1 is fixed, i.e., 3 bytes, thus when the number of STM-1s increases, these bytes also increase, maintaining the size of the block being monitored by each bit to be constant with changing configuration. The number of errored blocks counted by B2 bytes at the receiver are communicated back to the transmitter through the "M1" byte (refer to Figure

11.3) provided in the MSOH. This message containing the number of errored blocks, sent back by the receiver to the transmitter, is called "multiplex section remote error indications" (MS-REI).

B2 performance monitoring is classified as class 'A' performance parameter and is mandatory to be provided in the systems. It gives a fairly accurate approximation of number of bit errors in the multiplex section. The accuracy is considered to be up to 90% as compared with the actual BER. The information conveyed by B2 through the "M1" byte, i.e., MS-REI, is used in automatic protection switching of the link, depending upon the number of block errors. The number of block errors for switching the protection link through the APS (automatic protection switching) may be in the range of 1 in 10^5 to 1 in 10^9 depending upon the sensitivity of the application or the sensitivity of the customer.

11.1.2.2.3 B3

The parity byte "B3" provides for monitoring of end-to-end errors in the user data, on the higher-order path. Since the "B3" byte is associated with each VC-4, the number of bits in the error block remain constant (each block 261 × 9 × 8/8 = 2349 bits).

The information regarding the number of errored blocks detected by the "B3" byte is communicated back to the transmitter by use of the "G1" byte (refer to Figure 11.5). This message is called higher-order path remote error indications (HP-REI). The message is conveyed using the first 4 bits of the "G1" byte (other bits are used for some other indications, which we will see shortly).

This indication forms an important parameter for error monitoring and is used for end-to-end automatic protection switching (APS) of customer links (path level) depending upon the number of errored blocks.

11.1.2.2.4 V5 (Bits 1 and 2)

As discussed earlier in this section, the first two bits of the V5 byte of VC-12 payload (refer to Figure 11.7) are also parity bits, which provide for a parity check of VC-12 payload. The block errors detected by this BIP-2 are known as lower-order path background block errors (LP-BBE). Bit 3 of the same V5 byte is used for sending a message, i.e., a 1 is sent to the path originating node indicating an error and a 0 is sent in case of no error. This indication is called a lower-order path remote error indication (LP-REI). The former name for REI was FEBE (forward end block error).

Based on the number of errored blocks, reported by B1, B2, B3, and BIP-2, a number of performance parameters have been defined by ITU-T. These parameters enable an operator to assess the performance of the network elements (NEs) against the standards specified and to take a suitable remedial action if required. These parameters such as errored seconds (ES), severely errored seconds (SES), etc. will be discussed after the next section, where we will discuss other alarms and indications generated in SDH due to various other faults, as some of the performance parameters take them into account as well.

11.2 Fault Diagnostics and Restoration

Besides the degradation of performance as discussed above, there are several other causes due to which the NEs of SDH may start malfunctioning such as framing getting disturbed, signal not reaching the receiver at all, reaching partially, etc. SDH provides a very large number of alarms and indications for diagnostics of faults to a very precise extent, which takes the network manager quickly to the root of the problem and helps in speedy restoration.

11.2.1 Loss of Signal

When the optical power of the received signal at a receiver is zero, very low, or very high, an alarm is generated, which is known as loss of signal (LOS). The optical power should be so low or so high that it results in the degradation of BER to more than 10^{-3}. The situation may arise due to damage or cut to the OFC cable (in case of OFC, which is normally the case for SDH), malfunctioning of the laser optical source of the transmitter, heavy losses in the media due to poor/damaged joints, etc.

On detection of LOS (which is also known as R-LOS because the interruption has to be in between two regenerator sections, as all the multiplex nodes are also regenerator nodes in their functioning) alarm, an alarm indication signal (AIS) is generated by the receiver to inform all concerned. We will see about the AIS after learning about one more defect in the next section.

11.2.2 Loss of Framing

The SDH signal is a continuous flow of bits in a stream at the given rate of 155 Mbps for STM-1 or 10 Gbps for STM-16. If we are not able to mark the beginning and end of the individual STM-1 or STM-16 frames, we will not be able to put the whole data to any use. For separating individual frames from the bit stream, we use something called "framing" or "frame alignment techniques" (framing is discussed in Section 7.1.3).

To achieve the frame alignment, 6 "framing bytes" are provided in each STM-1 (refer to Figure 11.3). There are 3 A1 and 3 A2 bytes. In case of STM-4, STM-16, or STM-64, these bytes get multiplied in numbers by the factor N of STM-N. These bytes have a fixed pattern, which is recognized by the receiver of an SDH node and the frame alignment is performed if the A1 byte is 11110110 and A2 is 00101000.

If the A1 and A2 bytes (in proper sequence) are not received by the receiver for five consecutive frames (125 × 5 = 625 μs), which means that we are not able to achieve frame alignment for 625 μs, then the receiver enters into a fault mode called "out-of-frame" (OOF) state and generates an alarm called "OOF alarm."

If this out-of-frame status continues for 3 ms or more, the receiver will enter into a loss-of-frame (LOF) state and generate an LOF alarm. Naturally, the system cannot continue to work, and a service interruption occurs. Since this LOF is associated with the regenerator section, it is also called R-LOF or regenerator section loss of framing.

Now if we start receiving A1 and A2 bytes properly and the receiver is able to carry out the frame alignment properly and the condition stays for 1 ms or more, the LOF alarm will be discontinued and the receiver can assume normal status.

When the LOF alarm appears, the receiver sends an AIS signal to the downstream nodes. AIS is discussed in the next section.

11.2.3 Alarm Indication Signal

In the previous two sections, we have discussed LOS and LOF defects, which give rise to generation of AIS alarms. This AIS alarm is conveyed to the next node down the line to inform the system about the LOS or LOF problem. The AIS consists of the normal RSOH and all 1s, which means that the whole of the STM-1 or STM-N frame is filled with 1s, except the R-SOH, which is kept as it is. The situation is depicted in Figure 11.8.

The AIS is relayed by all 1s (+SOH) by all the downstream nodes until a multiplex node is reached, from where the AIS is conveyed by bit numbers 6 to 8 of K2 byte (refer to Figure 11.3). If bit numbers b6 to b8 of K2 byte are 111 for 3 frames or more, an AIS is indicated.

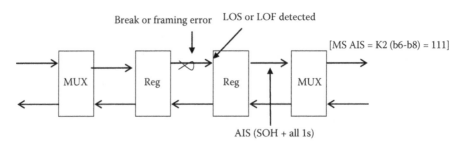

FIGURE 11.8
Detection of LOS or LOF and generation of AIS in SDH.

Thus, the next multiplex section node is informed about the problem in the upstream node. From any or all of these MS nodes, the management gets the information for initiating an appropriate remedial action.

There is no harm in filling the complete STM-N with all 1s in the transmission from all the regenerator section nodes downstream, as there is no dropping or adding of customer traffic from regenerator nodes and the traffic from an upstream multiplex node is already disrupted, which has generated AIS. In contrast to this, there may be useful traffic running between the first MS node downstream and subsequent MS nodes, thus filling the transmission from the MS nodes with all 1s will interrupt the traffic. Thus, first MS section onward the AIS is conveyed through the use of byte K2, while the normal traffic is carried in the STM-N bytes.

We have thus informed the downstream nodes about the problem. How about informing upstream nodes? Yes that needs to be done. The upstream nodes or the transmitter is informed about this by RDI. Let us see it in a little more detail.

11.2.4 Remote Defect Indication

As soon as the multiplex section receives an AIS from the upstream regenerator node, it regenerates two signals. First is the AIS signal to be conveyed to the next MS node, through 111 in b6 to b8 of the byte K2, and the second signal it generates is a remote defect indication (RDI) that is sent to the upstream MS node (and not to the RS node that sent the AIS indication; this is because any action whatsoever for remedial activities can be started by the management after getting the information about the fault through the MS section, and the RS node, even if informed by RDI, could not have initiated any action), through the use of the K2 byte again; this time, however, the value of bits 6 to 8 of the K2 byte will be 110 indicating the RDI.

Since the RDI is communicated between two multiplex sections, it is called MS-RDI. The situation is depicted in Figure 11.9.

If the break in the OFC had taken place between nodes C and D instead of B and C, the MS node D would have detected LOS or LOF signals instead of AIS. In that case, the MS node will generate the MS-RDI alarm as well, to be transmitted to the MS node A.

11.2.4.1 *Higher-Order Path Remote Defect Indication*

When an AU-AIS alarm, AU-LOP alarm, higher-order path unequipped (HP-UNEQ), or higher-order path trace identifier mismatch (HP-TIM) is detected by receiving equipment, it generates a higher-order path remote defect indication (HP-RDI) alarm to inform the upstream transmitter. This HP-RDI indication is sent by sending a 1 on the fifth bit of the G1 byte. The transmitter will recognize an HP-RDI alarm if it receives the 1 on the 5th bit

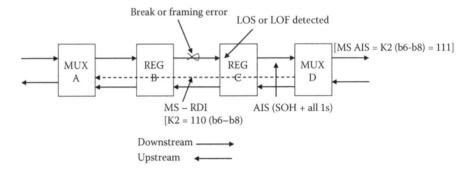

FIGURE 11.9
Remote defect indication.

of the G1 byte for 3 or more consecutive frames (we will see about all the unknown terms shortly).

11.2.5 Further Transmission of AIS to AU + TU Levels

For a proper and fast localization of fault, it is very useful to communicate the AIS alarm (which is generated per the discussions in the preceding sections) to the concerned administrative and tributary units. To visualize how AIS travels, let us consider the topology shown in Figure 11.10.

Three multiplex nodes that connect to many directions have been shown in Figure 11.10. Only in the direction of travel of AIS are the receiving and transmitting lines shown separately; in other directions, they have been shown as common lines for maintaining simplicity in presentation.

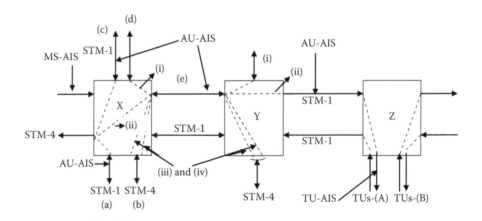

FIGURE 11.10
Transmission of MS-AIS to concerned AU-AIS and TU-AIS.

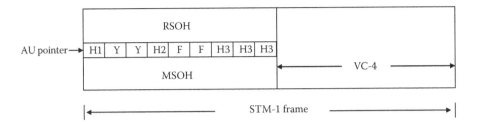

FIGURE 11.11
Location of AU pointer, i.e., H1YY, H2FF, and H3H3H3 bytes.

At node x, the MS-AIS is received from the previous multiplex node or an STM-4. This STM-4 diverges to various directions at node x with three directions, A, C, and E, connecting STM-1s to it. Thus, AU-AIS is sent on these three STM-1s. At node y, this AU-AIS is placed on the link connecting node z, as that is the only STM-1(ii) coming from the directions of AIS. At node z, TUs-(A) are connected to this STM-1, while TUs-(B) are connected to some other direction's STM-1. Thus, the AIS is passed on to TUs-(A), where it is called TU-AIS.

The AU-4 (VC-4 + pointer) belongs to the higher-order path whereas TU (VC-12 + pointer) belongs to the lower-order path (see Section 10.16). The AU-AIS is indicated by converting all the bits of the AU pointer, i.e., all the bits of the bytes H1YY, H2FF, H3H3H3 to 1s. Figure 11.11 shows the location of these bytes.

The TU-AIS is similarly indicated by converting all the bits of the TU pointer bytes V1, V2, and V3 to 1s.

Thus, an AIS indication or alarm will be passed on to all the TUs, which are ultimately traveling over the affected section.

However, the TUs receive the AIS alarm in many other cases of problems in addition to the service interruption type discussed above. We will see them as we proceed to look at the details of some more alarms and indications.

There are many other types of faults that may lead to the generation of AIS alarm at AU and TU levels. These are

AU-AIS: multiplex section excessive errors (MS-EXC) for AU-AIS and loss of AU pointer (AU-LOP)

TU-AIS: higher-order path unequipped (HP-UNEQ), higher-order path trace identifier mismatch (HP-TIM), and higher-order path signal label mismatch (HP-SLM)

We will discuss them briefly.

11.2.6 Multiplex Section Excessive Errors

We have seen in Section 11.1.2.2 of this chapter that the B2 bytes count the number of block errors in the multiplex section receiver and transmit the information to the transmitter about the number of block errors received through the M1 byte (called MS-REI). Now what the transmitter does out of this information?

The purpose of monitoring the number of errors is twofold:

(i) To check the system for possible faults and initiate corrective action before the complaint is received from the customer.
(ii) To take the circuit out of service if the quality is very poor and provide an alternative link to the customer.

Decision (ii) is taken if the number of errors crosses a preset threshold level. The threshold can be set during or after the installation through the management tools. A threshold of 10^{-3}, 10^{-4}, or 10^{-5} is generally set for this purpose. If the block errors are reported by B2 and this threshold (AU-AIS) is activated, the automatic protection switching is activated to provide the customer service on the protect path (if the protect path is programmed) and the errored circuit is taken out of service for rectification of fault.

11.2.7 Loss of Pointer (AU-LOP and TU-LOP)

The AU pointer locates the VC-4 payload in an STM-1. If the pointer is confused, then the payload cannot be located, leading to a failure (we have discussed about functioning of pointer in detail in Section 10.16). The AU-LOP indication appears because of two reasons:

(i) The value of the AU pointer is an invalid value for 8 consecutive frames
(ii) The new data flag (NDF) appears for 8 consecutive frames (for NDF details, see Sections 10.16.5 to 10.16.8)

The NDF appears (i.e., the first 4 bits of the first H1 byte, the AU pointer) when there is a change in the location of payload VC-4. The normal value of NDF is "0110," and a change (called the NDF appearance) is indicated by "1001." Once this change comes, NDF is supposed to have appeared. The value of NDF is normalized to "0110" in the next frame, and the NDF is not supposed to reappear at least for the next three frames.

The AU-LOP alarm is generated if NDF keeps on appearing for 8 consecutive frames.

The AU-LOP alarm is conveyed by converting all the bits of VC-4 to 1s. Detection of all this 1s signal generates the AU-AIS alarm.

Similarly, the TU-LOP alarm is generated if the value of the pointer is invalid or the NDF appears consecutively for 8 frames. In this case, the pointer bytes of lower-order overhead, i.e., V1 and V2 bytes, play the role, which is played by H1 bytes in the case of AU-LOP.

11.2.8 Higher-Order Path Unequipped and Lower-Order Path Unequipped

Byte C2 in the HO-POH (see Figure 11.5) is called a "signal label" byte because it represents the type of signal being carried by the VC-4 and the type of multiplexing structure used. The types of signals could be 2Mb PDH tributaries, Ethernet, FDDI, etc. The multiplexing structure could be any of the multiplexing routes per the ITU-T recommendations (Section 10.14).

It also indicates whether the VC-4 is equipped (presence of payload) or not. If the VC-4 is unequipped, the value of C2 byte is set to "00000000," i.e., all 0s. As soon as this condition is detected, a HP-UNEQ alarm is generated and a TU-AIS alarm is also generated by inserting 1s in all the bits of the VC-4 payload.

The signal label function is performed by bits 5 to 7 of the V5 byte in the LO-POH in exactly the same way. The "000" value of these bits indicates an unequipped VC-12 multiframe and consequently generates an alarm called LP-UNEQ. Other combinations of bits indicate the type of mapping such as asynchronous, floating, bit synchronous, etc. A "001" indicates that the path is equipped with nonspecific payload, and a "111" indicates and generates the VC-AIS alarm.

11.2.9 Trace Identifier Mismatch

SDH provides for checking the continued connection of a receiver with the intended transmitter. This is done at the section layer by transmitting and receiving a mutually agreed upon pattern called "section access point identifier" in the byte J0 of the RSOH (see Figure 11.3). At the higher-order path layer this function is achieved by the byte J1 of HO-POH (Figure 11.5) and at the lower-order path layer by the J2 byte (Figure 11.7).

In case of any outage of the system, these "trace identifier" bytes help in quick localization of faults. Any mismatch of J0 byte between the transmitter and receiver of a regenerator section results in an alarm called regenerator section trace identifier mismatch (RS-TIM).

The J0 byte also acts as an identifier of individual STM-1s in an STM-N frame.

Similarly, the J1 byte provides for the checking of continuity of connection between the transmitter and receiver at the higher-order path layer. Any mismatch between the transmitter and receiver at this layer results in the generation of an alarm called the higher-order path trace identifier mismatch (HP-TIM).

The J1 byte is also the first byte of the VC-4 payload, thus it marks the beginning of the higher-order path layer payload. The AU pointer always points toward the J1 byte.

At the lower-order path layer, it is the J2 byte (Figure 11.7) that carries out the function of trace identification between the intended transmitter and the receiver. In case of mismatch, the alarm generated is accordingly called lower-order path trace identifier mismatch (LP-TIM).

11.2.10 Signal Label Mismatch

The byte "C2" discussed in Section 11.2.8 performs one more function that is similar to the trace identification being performed by J1 byte. The configuration of "C2" byte has to match between the transmitter and receiver of the higher-order path layer. In case of a mismatch, an alarm is generated called HP-SLM.

A similar function is performed at the lower-order path layer by signal label bits, i.e., bits 5 to 7 of byte V5. The alarm generated is accordingly called lower-order path signal label mismatch (LP-SLM).

11.2.11 Tributary Unit Loss of Multiframe

We have seen in Section 10.6 to 10.11 that the individual E1s (2 Mbps PDH tributaries) are converted to C-12s, then to VC-12s, and then to TU-12s.

These TU-12s are arranged in a multiframe of 500 μs, consisting of 4 periods of 125 μs each. The LO-POH bytes are interleaved in the whole of this 500-μs period.

To be able to demultiplex the individual TU-12s and consequently the individual E1s from this multiframe, we need to know the location of each frame precisely within this multiframe. The basic frame (125-μs) sequence number is indicated by the byte H-4 (see Figure 11.5).

If the receiver of any SDH signal receives an invalid value of H4 (the valid values being 00000000 to 00000011), an alarm called tributary unit loss of multiframe (TU-LOM) is generated. This alarm is also generated if 2 to 10 numbers of multiframes are not in proper order.

11.3 Summary of SDH Alarms and Indication

We have seen the individual performance monitoring and fault diagnostic features of various overhead bytes of SDH at all the three layers, which are regeneration, multiplexing, and path layers. It will be quite useful to put all of them together in the form of a flow chart, which will not only act as a quick reference but will also facilitate clear understanding. The flow chart is drawn in Figure 11.12. A summary has also been tabulated in Table 11.1 for a quick reference.

11.3.1 SDH Performance Indicators

The performance of SDH equipment is generally categorized in terms of the indicators anomalies, defects, and failures. These terms are defined as below.

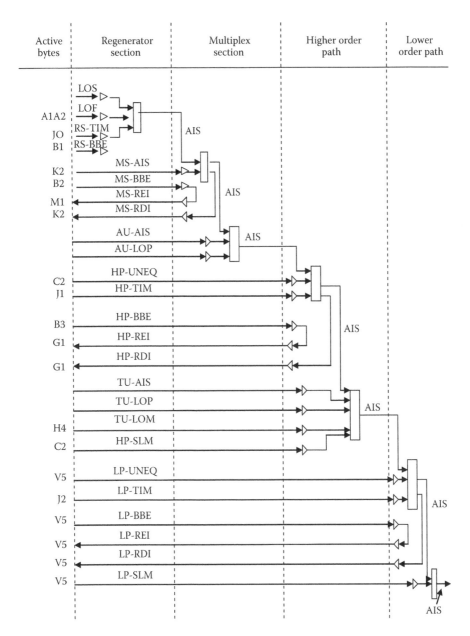

FIGURE 11.12
Flowchart summary of SDH alarms and indications.

TABLE 11.1

Summary of SDH Alarms and Indications

Defect/ Alarm	Expansion	Detection Criteria
LOS	Loss of signal	Absences of signal or drastic reduction in received power or very high number of bit errors
OOF	Out of framing	A1 and A2 bytes not received in proper sequence for 5 frames (625 µs)
LOF	Loss of framing	OOF remains for >3 ms
RS-TIM	Regenerator section trace identifier mismatch	J0 byte at the transmitter and receiver not matching
RS-BBE	Regenerator section back ground block error	Block errors are detected through the mismatch of receiver and locally computed B1 byte
MS-AIS	Multiplex section alarm indication signal	Bit numbers 6 to 8 of K2 byte are 111 for 3 or more frames
MS-BIP	Multiplex section bit interleaved parity	Bit errors detected through the mismatch of received and locally computed B2 bytes
MS-REI	Multiplex section remote error indication	Number of block errors MS-BIP conveyed through M1 byte
MS-RDI	Multiplex section remote defect indication	Bits 6 to 8 of K2 byte are 110 for more than 3 frames
AU-AIS	Administrative unit alarm indication signal	All the bits of the AU pointer H1, H2, and H3 are 1
AU-LOP	Administrative unit loss of pointer	Invalid AU pointer for 8 frames or NDF for 8 frames
HP-UNEQ	Higher-order path unequipped	Byte C2 contains all 0s
HP-TIM	Higher-order path trace identifier mismatch	Mismatch of J1 byte at the transmitter and receiver of higher-order path layer
HP-BBE	Higher-order path background block error	Block errors detected through the mismatch of received and locally computed B3 bytes
HP-REI	Higher-order path remote error indication	Number of block errors of higher-order path conveyed through first 4 bits of G1 byte
HP-RDI	Higher-order path remote defect indication	Fifth bit of G1 byte = 1 for >3 frames
TU-AIS	Tributary unit alarm indication signal	All bits of TU pointer bytes V1, V2, and V3 are 1s
TU-LOP	Tributary unit loss of pointer	Invalid pointer value or NDF for 8 or more frames
TU-LOM	Tributary unit loss of multiframe	Improper multiframe for 2 to 10 frames or invalid H4 byte
HP-SLM	Higher-order path signal label mismatch	Mismatch of "C2" byte in the transmitter and receiver
LP-UNEQ	Lower-order path unequipped	Bits 5 to 7 of V5 byte = 000 for more than 5 frames
LP-TIM	Lower-order path trace identifier mismatch	Mismatch of J2 byte at the transmitter and receiver of the lower-order path

(continued)

TABLE 11.1 (Continued)

Summary of SDH Alarms and Indications

Defect/ Alarm	Expansion	Detection Criteria
LP-BBE	Lower-order path background block error	Block errors detected through the received and locally computed first 2 bits of V5 byte
LP-REI	Lower-order path remote error indication	One or more block errors conveyed by 1 in the 3rd bit of V5 byte
LP-RDI	Lower-order path remote defect indication	Bit 8 of V5 is set to 1 for defect, else 0
LP-SLM	Lower-order path signal label mismatch	Mismatch of bits 5 to 7 of V5 byte in the transmitter and receiver

11.3.1.1 Anomaly

An anomaly is degradation in circuit performance, which does not cause an interruption of service. These are small deviations of the achieved circuit characteristics from the desired characteristics. The anomalies are indicated by bit interleaved parity (BIP), remote error indication (REI), etc.

11.3.1.2 Defect

A defect is the state when the circuit is not able to perform its designated function. A defect is detected when the persistent anomalies build up to exceed a maximum time limit. When a defect is detected, it calls for immediate attention of maintenance engineers to diagnose the cause, and isolate the faulty section or equipment to avoid interruption to the service. The indicators of defects are alarm indication signal (AIS), remote defect indication (RDI), loss of frame (LOF), out of frame (OOF), loss of pointer (LOP), loss of multiframe (LOM), etc.

11.3.1.3 Failure

When the defect persists for a period beyond a certain limit, it renders the equipment or the node unable to perform the required function. The situation is declared as a failure, and the circuit is treated as interrupted.

11.4 Performance Monitoring Parameters

To make the "in-circuit" performance monitoring quantitative and to enable the SDH system and the management to take decisions about the circuits,

ITU-T has defined a number of parameters. The decisions that need to be taken are

(i) Keep the circuit running as it is (if the errors are within tolerable limits).

(ii) Take the circuit out of service and switch the customer on protection path.

(iii) Keep the circuit under watch and switch over to step 2 if the performance does not improve.

The parameters defined in ITU-T G-826 and G-828 are as follows:

(i) *Errored block (EB):* EB is defined as a block where one or more bits are in error. (A block is defined as a set of bits. For example, say all the bits of STM-N frame have to be divided into 8 blocks for monitoring by B1 byte, with each bit of B1 monitoring STM-N/8 bits.)

(ii) *Errored second (ES):* ES is defined as the one second's period with one or more errored blocks, or at least one defect. (We have seen all types of defects earlier in this chapter.)

(iii) *Severely errored second (SES):* SES is defined as the 1-s period, which contains >30% errored blocks or at least one defect.

(iv) *Background block error (BBE):* BBE is defined as errored block not occurring as a part of SES.

(v) *Unavailable seconds (UAS):* UAS is defined as the time during which the signal is completely not available, due to alarms or errors that are more than SES for at least 10 s. The system unavailability situation is depicted in Figure 11.13.

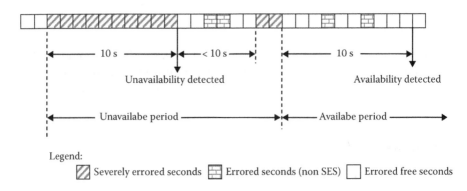

FIGURE 11.13
Unavailable and available periods (for a single direction).

The period of unavailability of the system is detected after 10 SES, but this period of 10 SES is also counted in the unavailability period. Similarly, the availability of the system is detected after 10 s of a "non-SES" period, but again, this 10-s period is also counted as an available period.

Figure 11.13 shows the system availability criteria for a single direction. For a bidirectional path, nonavailability of either of the paths will cause the system unavailability.

(i)   *Errored seconds ratio (ESR):* ESR is defined as the ratio of ES to total seconds in "available time" during a fixed measurement interval.

(ii) *Severely errored seconds ratio (SESR):* SESR is defined as the ratio of SES to total seconds in "available time" during a fixed measurement interval.

(iii) *Background block error ratio (BBER):* BBER is defined as the ratio of background block errors to total blocks in available time during a fixed measurement interval.

11.4.1 Error Performance Objectives

The above-mentioned parameters have to be within certain limits for the performance of the circuit to be called acceptable. The limits have been defined by ITU-T, with reference to a hypothetical reference path of 27,500 km, spanning international boundaries. (Without any length of the link being defined, the error parameters will not yield much meaning, as the smaller links will tend to have much smaller figures of errors as compared with longer links.) This hypothetical reference path has been assumed to cross four countries. The ratio of errors has also been distributed among national and international regions and media. The complete details are out of the scope of this text; however, the interested readers can refer to ITU-T recommendation G-826 to G-828 for further reading.

To get a feel of the limits, however, as an example, we can have a look at the parameters for STM-1, which are as follows:

Bits/block = 6000–20,000
ESR = 0.16
SESR = 0.002
BBER = 2×10^{-4}

These are called "end-to-end performance objectives." Since the above-mentioned limits are for a link of 27,500 km length, crossing four countries, the smaller links should correspondingly have fewer errors. It becomes a joint responsibility of the equipment manufacturers and service providers to ensure these limits.

In an international scenario, there could be equipment from several vendors, and multitudes of telecommunication operators could be involved in service provisioning, thus the percentage distribution of error limits by ITU-T plays a major role.

All these parameters are to be monitored at a frequency of 15 minutes causing 96 events in a day. A total period of 1 month of continuous monitoring (15 min each event) is required to produce the defined results.

In the course of our discussions on the performance monitoring and fault diagnostics of SDH, so far in this chapter, we have discussed the roles of many bytes in the "overheads" of regenerator section, multiplex section, and path. I am sure you must be worried about the roles of the remaining bytes. We will discuss the roles of the remaining overhead bytes now, and after that, we will have a comprehensive look at the total overhead activities.

11.5 Roles of Other Overhead Bytes

Many overhead bytes of each of the layers, regenerator section, multiplex section, and path are still left unexplained so far in this chapter. Let us see them layer wise.

11.5.1 SOH Bytes

11.5.1.1 Bytes E1 (RSOH) and E2 (MSOH)

These are known as "order wire bytes" (refer to Figure 11.3). Order wire is a maintenance facility provided in all transmission networks, whether analog or digital, radio or wire line, or OFC. The order wire is a direct telephone connection between different stations of the transmission system. The channel utilized for this facility is from within the capacity of the system itself. Using this dedicated communication facility, the staff carry out installation, maintenance, and repair activities in coordination with each other placed at different station/nodes.

Byte E1 is provided in RSOH and byte E2 is provided in MSOH. Each of them provides a 64-kbps speech channel. Since RSOH is processed in the regenerator stations, E1 provides the order wire phone between any two regenerator stations, whereas E2 provides the order wire phone between two multiplex sections, as the MSOH is processed at multiplex nodes only. Figure 11.14 depicts the situation.

Thus, the combination of E1 and E2 bytes provides the order wire communication between all nodes/stations.

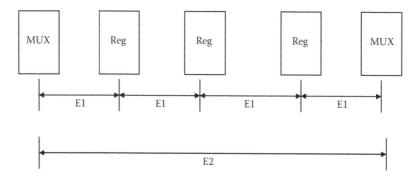

FIGURE 11.14
Order wire bytes in regenerator and multiplex sections.

11.5.1.2 Byte F1 (User Channel) (RSOH)

This byte is reserved for the user (who is generally the network operator or service provider) for any special applications. For example, it may be used by the operator to provide temporary order wire connection between the regenerator nodes by providing a 64-kbps speech/data channel.

11.5.1.3 Bytes K1 and K2 (MSOH)

These bytes perform one of the most important functions of SDH, called "automatic protection switching."

When the performance of a customer circuit falls below the acceptable levels or the signal is completely or substantially lost, the customer circuit is automatically switched over to "protection path." This switching action is achieved by K1 byte and the bits b1 to b5 of K2 byte. These bytes are provided in the multiplex section because no switching is required in the regenerator section as no customer links take off from the regenerator nodes. The automatic switching feature is also called a "self-healing" feature of the SDH.

The other three bits, i.e., bits 6 to 8 of K2 byte, are used for the MS-RDI alarm as discussed in Section 11.2.4 of this chapter.

11.5.1.4 Byte S1 (Synchronization Status Message Byte)

This byte is called synchronization status message (SSM) byte. Bits 5 to 8 of this byte continuously determine the quality of the receive clock in the multiplex section equipment. Different bit patterns have been defined for these bits, indicating different quality levels of the receive clock.

The information about the quality of the receive clock is used by the equipment to select the best quality of clock being received out of all the incoming signals with the help of this byte.

11.5.1.5 Δ Bytes (RSOH): Media-Dependent Byte

RSOH provides for 6 numbers of Δ bytes for carrying special information or commands for transmission media specific applications. For example, the excessive delay in satellite links will require special timing settings for many activities, or the direction of traffic in an OFC link may be identified by these bytes when using only one fiber for transmission and reception.

11.5.1.6 Bytes X (Reserved for National Use)

There are four (X) bytes in RSOH and two (X) bytes in MSOH, which are reserved for national use, i.e., for special features that are applicable within the country.

11.5.1.7 Bytes D1 to D12: Data Communication (DCC) Bytes

These 12 bytes, D1 to D3 (3 bytes) in RSOH, and D4 to D12 (9 bytes) in MSOH, are together called data communication (DCC) bytes. As we have discussed earlier, SDH has a very strong set of OAM features. We have also seen a number of error performance and fault diagnostics functions. All these functions are put to use by the operator through these data communication bytes. In addition, the DCC bytes provide a host of other management functionalities such as provisioning of customer links, making cross connects, providing "working" and "protection" paths, billing, maintenance activities, report generation, etc. The DCC bytes are also called an embedded control channel (ECC).

The data rate provided on this channel is $12 \times 8 \times 64$ kbps = 768 kbps. It is a very strong tool for the management of all the OAM activities of the SDH network.

11.5.1.8 Unmarked Bytes

There are as many as 4 unmarked bytes in RSOH and 25 unmarked bytes in MSOH. All these unmarked bytes are reserved for future use, clearly indicating a huge potential for improvement in the OAM of SDH networks.

11.6 Overhead Bytes Summary

Having seen the role of the overhead bytes of all the layers of SDH, it is time to have a comprehensive look at the functioning of each and every byte for quick reference and clarity of understanding. We will see this in the same order as that of the preceding discussion in the chapter.

11.6.1 Regenerator Section Overhead

The various bytes allocated to RSOH and their functioning is shown in Figure 11.15a and b.

Byte no.	1	2	3	4	5	6	7	8	9
	A1	A1	A1	A2	A2	A2	J0	X	X
	B1	Δ	Δ	E1	Δ		F1		
	D1	Δ	Δ	D2	Δ		D3		

(a)

Byte	Function
A1A1A1 A2A2A2	Framing bytes. Used for frame alignment of STM-N
J0	Regenerator section trace identifier
XX	Bytes reserved for national use
B1	Bit-interleaved parity check of whole STM-N
Δ Δ	Media-dependent bytes
E1	Order wire byte for regenerator section
F1	User channel (for the use of service provider)
D1D2D3	Data communication bytes (DCC)
Blanks	Byte reserved for future standardization

(b)

FIGURE 11.15
(a) RSOH bytes. (b) Functions of RSOH bytes.

11.6.2 Multiplex Section Overhead

The various bytes of MSOH and their functions are shown in Figure 11.16a and b.

11.6.3 Higher-Order Path Overhead

The HO-POH bytes and their functions are shown in Figure 11.17a and b.

B2	B2	B2	K1			K2		
D4			D5			D6		
D7			D8			D9		
D10			D11			D12		
S1					M1	E2		

(a)

Byte	Functions
B2B2B2	Bit-interleaved parity check for each STM-1
K1, K2 (b1 to b5)	Automatic protection switching
K2 (b6 to b8)	Multiplex section remote defect indication (MS-RDI)
D4 to D12	Data communication bytes
S1	Synchronization status message (SSM)
M1	Multiplex section remote error indication (MS-REI)
E2	Multiplex section order wire byte

(b)

FIGURE 11.16
(a) MSOH bytes. (b) Functions of MSOH bytes.

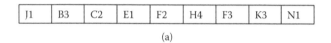

(a)

Byte	Bit numbers	Function
J1	All	Higher-order path trace identifier byte. First byte of HO-POH. AU pointer points to this byte for locating VC-4
B3	All	Bit interleaved parity check byte for VC-4
C2	00000000	Signal label byte, indicates an unequipped VC-4
	00000001	Indicates that VC-4 is equipped but is nonspecific
	00000010	Indicates multiplexing (TUG) structure
	00000011	Indicates that TU is locked
	00000100	Mapping is asynchronous. E3 or T3 into C3
	00010010	Mapping is asynchronous. E4 into C4
	00010011	ATM mapping
	00010100	MAN (DQDB) mapping
	00010101	Mapping of FDDI
	11111110	Q 181 test signal mapping
	11111111	VC-AIS
G1	—	Higher-order path status byte conveys the number of errors found by B3 byte to the path originating equipment (REI) and also conveys remote defect (RDI)
	b1, b2, b3, b4	Remote error indication
	b8	Spare
b5 to b7	000, 001, and 011	No remote defect
	010	Remote payload defect (LCD)
	100	Remote defect (AIS, LOP, TIM, UNEQ or SLM, LCD)
	101	Remote server defect (AIS, LOP)
	110	Remote connectivity defect (TIM, UNEQ)
	111	Remote defect (AIS, LOP, TIM, UNEQ or SLM, LCD)
F2 and F3	—	User (network provider) channels for higher-order path
H4		Identifies individual frames in a multiframe TU structure for dropping tributaries
K3	b1 to b4	Automatic protection switching for higher-order path
	b5 to b8	Reserved for future use
N1	—	Monitoring of higher-order tandem connection

(b)

FIGURE 11.17
(a) HO-POH bytes. (b) Functions of HO-POH bytes.

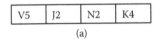

(a)

Byte	Bit numbers	Functions
V5	b1–b2	Bit interleaved parity check at lower-order path layer
	b3	Remote error indication (REI)
	b4	Remote failure indication (RFI)
	b5–b8	Signal label
	000	VC path unequipped
	001	Path equipped but nonspecific payload
	010	Asynchronous floating mapping
	011	Bit synchronous floating mapping
	100	Byte synchronous floating mapping
	101	Reserved for future use
	110	Test signal Q-181 mapping
	111	VC-AIS
	b-8	Remote defect indication (RDI)
J2	—	Lower-order path trace identifier
N2	—	Monitoring of lower-order tandem connection
K4	b1–b4	Automatic protection switching or lower-order path
	b5–b7	Enhanced remote defect indication
	b8	Spare

(b)

FIGURE 11.18
(a) LO-POH bytes. (b) Functions of LO-POH bytes.

11.6.4 Lower-Order Path Overhead

The bytes allocated to lower-order path and their functions are shown in Figure 11.18a and b.

11.7 Network Management in SDH

So far, in this chapter, we have seen the enormous number of operation and maintenance features of SDH. However, all these indications and alarms will be of little use if the management is not able to take quick action for restoration of faults and considered action for improvement of the performance or expansion of the network based upon the reports and statistics. In addition to fault management, network expansion, etc., a host of other activities are required to be performed by the management of the telecommunication network. Beginning with the initial configuration of the equipment of various nodes, provisioning of the customer links, making cross connects, setting up of parameters for various activities such as alarm, order wire, traps routing, disconnections and reactivations of the links, automatic protection switching, programming, billing, etc., are some of the other activities that the management needs to carry out on a regular basis.

All these functions are carried out with the help of a network management system (NMS).

The NMS in SDH is provided through the use of DCC bytes. There are three bytes in the regenerator sections D1 to D3, facilitating a data communication channel of bit rate 3×64 kbps = 192 kbps. This is called the DCC_R channel. Similarly, there are nine data communication bytes, i.e., D4 to D12 in the multiplex section, facilitating a data communication channel DCC_M at a bit rate of 9×64 kbps = 576 kbps. Thus, the total channel capacity of DCC including DCC_M and DCC_R is 192 + 576 = 768 kbps. Since the DCC_M bytes are not available in the regenerator section, the DCC_R bytes are used to connect the regenerator NEs or nodes. NE is a standard nomenclature used by ITU-T. Thus, the DCC_M channel forms a kind of backbone and the DCC_R forms kind of a LAN, connecting the local NEs to the backbone.

The Network Management System for SDH has been defined by ITU-T in their specification number G.784 and G.7710/Y.1701 (07/2007). The NMS is generally proprietary software of each SDH vendor, though it may operate on common networking protocols. We will consider here a simplified approach to develop the understanding of the concept.

11.7.1 Network Management System

In general, the NMS for SDH networks is a three stage management system. The first stage is the local management of the NEs (nodes) through local craft terminal (LCT), which could be a laptop PC, loaded with the requisite software if the NE is not preloaded with the software. The second stage is a regional management, and the third stage is the overall or national level management. Figure 11.19 shows this hierarchy.

The hierarchical structure of NMS helps in providing distributed control, distributed authorization, and customer-managed links.

11.7.1.1 Distributed Control

The control of the network management gets distributed between the regional NMS and the central or national NMS. The area under regional NMS may in fact be called a "subnetwork." In a big network spanning the whole country, it is practically impossible to control all the activities from a single NMS center. The regional NMSs are generally delegated authorities (through NMS software authentications) for day-to-day activities such as provisioning up to a certain capacity, fault rectification, maintenance activities, error performance monitoring, configuring, changing protection switching, etc. The central NMS keeps the overriding authority on all activities but may

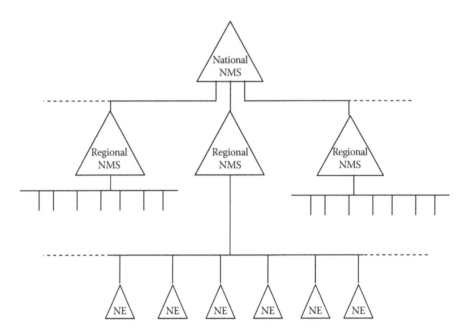

FIGURE 11.19
Hierarchical NMS of SDH.

generally perform the function of bulk provisioning, coordination between all regional NMSs, and make decisions based on error performance and other statistical reports for upgrades or network expansion, etc.

A module of the NMS software called an "agent" is normally loaded in the NEs. With the help of this agent, the NE can be managed locally through an LCT (a laptop) or remotely through the regional or national NMS. The regional NMSs could again be a "subnetwork management module," or they could just be the terminals of the national NMS, with restricted controls and authorities.

11.7.1.2 Distributed Authorization

Within a region, there may be different groups of people dealing with different activities such as provisioning, billing, performance monitoring, fault reporting, customer interface, etc.

There may even be different technical groups dealing with specific issues, say one group is expert on PDH, another group is expert on Ethernet or ATM, etc. All these groups or teams of people can be given authorities to manage only those activities on which they have expertise. Although all groups may have a common NMS, each of these groups will have authorization to access only the concerned controls. There may even be a distribution of one or two expert activities to each region.

11.7.1.3 Customer-Managed Links

At times some bulk customers may want some control on the part of the network that is provided to them or at the least they may want a monitoring facility. In such cases, a remote terminal of the NMS may be provided at customer premises with access to limited network resources such as monitoring and making small changes in configuration in the portion of the network under their use. Even if such controls are passed on to the customer, the overriding authority and the control over OAM features are with the network administrator.

11.7.2 Complete Telecommunications Network Management

A complete telecommunications network management model has been defined by ITU-T in its specification M-3000, which extends beyond the boundaries of SDH network. It includes the SDH and non-SDH portions of the managed network are per ITU-T G-784. The model is depicted in Figure 11.20.

We can see that, as compared with the model shown in Figure 11.19, the above model has one extra layer, i.e., of the TMN, and one layer less, i.e., of NEs. If NEs are added to this model, it will depict a complete model of the telecommunication management network.

As discussed in the beginning of the chapter, the NMS has to perform a large number of activities. In the next section, we will discuss the important ones briefly.

11.7.3 NMS Activities

All the NMSs of SDH are proprietary software of the particular vendor, which means that the NMS of one vendor will not work with NEs of any

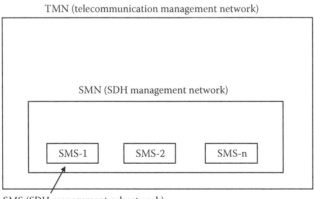

FIGURE 11.20
Telecommunication management network (TMN). (From the International Telecommunication Union, Fig. 5.3/G.784 (06/99), p. 10 of ITU-T recommendation no. G.784, June 1999. With permission.)

other vendor. Consequently, there is quite a lot of difference in the features provided by each vendor. However, there are many basic tasks that are to be facilitated by everyone. We will see these basic common activities of the NMS here.

The NMS activities can be broadly classified into the following categories:

   (i) Configuration

  (ii) Provisioning

 (iii) Performance monitoring, setting thresholds

 (iv) Fault diagnostics

  (v) Security

 (vi) Timing management

(vii) Maintenance operations

11.7.3.1 *Configuration*

When any NE is installed, it needs to be configured, before it can start functioning. The configuration may include defining several parameters such as

- Whether it is an add/drop multiplexer (ADM) or terminal multiplexer (TM) as the same equipment can function in either mode.
- What type of network application is it in? Point-to-point configuration, linear add and drop, ring add and drop, or mesh configuration?
- Synchronization: The NE may be connected to several other NEs. It needs to know from which NE it has to receive the synchronization clock. The preference for the synchronization clock has to be from a better and higher-order source. It also needs to be configured for activating the timing manager, which selects the best quality of clock during normal operation, from among various sources depending upon the SSM.
- Mapping of tributary signals to be done on various ports such as E1, E3, STM-1, Ethernet, etc. for incoming as well as outgoing signals from various directions.
- Creation of user groups is another activity to be taken up on initial configuration. There could be users with different privileges, logging on to the NE for different purposes. User could be administrator, user, operator, or diagnosis engineer. The administrator has complete control on the NEs for all purposes and also for setting the privileges of other users. The operator may have limited access for provisioning, setting the alarms and security, etc. The users may only be able to view the data and carry out limited provisioning. The diagnosis engineer may be authorized to diagnose the problem

if any and report to appropriate authorities or take necessary maintenance action.

- Time stamping is done by setting the local time correctly in the NE, so that when it is operational, it can send and receive meaningful data to/from NMS.

11.7.3.2 Provisioning

Provisioning of the services is the main purpose of establishing a telecommunication network. NMS provides an excellent tool for fast and efficient provisioning. This may include activities such as circuit provisioning, protection path provisioning, making of cross connects, and order wire provisioning.

11.7.3.2.1 Circuit Provisioning

Depending upon the hardware supported by a particular NE, any type of tributaries can be dropped or inserted remotely through NMS. The NMS logs on to the NE by inserting its IP address through a graphical user, menu-driven interface. The particulars of payload to be dropped or inserted are displayed according to the hardware available in the NE. The tributaries such as E1, E3, Ethernet, STM-1, etc. may be dropped or inserted from any direction on specific ports. The status of the NE is also available in the NMS, which gets updated with each activity relating to provisioning.

11.7.3.2.2 Point and Click Provisioning

Many NMSs provide graphical user interface for point and click provisioning. The end-to-end nodes are selected, the payload is defined, and NMS does the rest of the job, configuring all the NEs in the path for necessary payload provisioning.

11.7.3.2.3 Protection Path Provisioning

Whenever a circuit is provisioned between any two NEs, the NMS allows it to be provided on either a normal path called "work path" or a standby path called "protect path." Either only work path or both may be provided. In case both the paths are programmed (generally, the protect path on a physically different route, so that its chances of availability are more in the event of a failure of the route carrying the work path), the work path normally carries the traffic. The protect path takes over as soon as the work path fails. This switch over can be done manually or automatically, depending upon the programming done through NMS.

The automatic protection switching can also be programmed in "revertive," or "nonrevertive" modes. In revertive mode when the failure of the work path is restored, the traffic comes back to the work path after a predefined time called wait to restore (WTR). In the case of the nonrevertive mode, the traffic continues to be on the protect path, even after the failure of the work path is restored.

11.7.3.2.4 *Cross Connect*

For carrying the end-to-end traffic in the tributaries of level E1, E3, Ethernet, or STM-1s, the intermediate NEs need to just pass on the traffic from the receiver of one direction to the transmitter of another direction. This job is performed by "cross connecting" the required time slots. The tributaries for this purpose are called "time slots." There are 63 E1 time slots in an STM-1. Each of these time slots in the receiver of an NE can be mapped to any of the similar time slots in the transmitter of other directions. They can also be mapped to different directions in case of a multidirectional NE. All these cross connects can be made through the use of menu driven commands in the NMS, by addressing the required NE through its IP address.

11.7.3.2.5 *Order Wire Provisioning*

We have seen that the order wire is an important maintenance facility (see 11.5.1.1 of this chapter). E1 and E2 bytes in the RSOH and MSOH, respectively, provide the order wire phone. These order wires can be enabled or disabled with the help of NMS. Their ringing period and "calling number" (like normal telephone numbers) may also be programmed.

11.7.3.3 **Performance Monitoring**

Although the performance of SDH NE can, to a certain extent, be monitored through the alarms and indications provided on the equipment chassis, the real and complete monitoring is done only with the help of NMS. The NMS keeps a log of all events, including all the error alarms, failures, provisioning, systems availability, unavailability, etc. It also produces results and management information system (MIS) reports, which can help the management make proper short- and long-term decisions.

With the help of menu-driven commands, the threshold of each type of alarms are set by choosing an appropriate value from within the available range. These parameters include errored seconds (ES), severely errored seconds (SES), unavailable seconds (UAS), background block error (BBE), errored blocks (EB), error seconds ratio (ESR), severely errored seconds ratio (SESR), background block error ratios (BBER), etc. (All these parameters have been defined in Section 11.4.1.)

For the purpose of error monitoring, the time is divided into 15-min intervals, which means that all the events taking place are recorded by the system, and a 15-min summary is presented. The whole day (midnight to midnight) is thus divided into 96 such intervals.

NMS at any given time can display the performance of all the events mentioned above for any 15-min interval, be it current, previous, or any other previous interval. All these parameters are monitored for each of the layers, i.e., regenerator section, multiplex section, path, or at the tributary unit.

A 30-day summary is also presented by NMS for all the above parameters for each layer separately.

11.7.3.4 Fault Diagnostics and Management Alarms

A very large number of alarms are generated by SDH to exactly pinpoint the location and type of fault. Each alarm and the combination of alarms enable the network administrator or any other user as defined earlier to reach the root of the problem very quickly. All these alarms at any of the NEs can be seen on the NMS terminal screen and appropriate actions can be initiated. For the convenience of the staff, the alarms are graded into a few severity levels. The more severe the alarm, the faster attention it needs.

The levels are critical, major, and minor. The critical alarms are those that affect the service and need immediate attention. The major alarms too are service-affecting alarms but less severe than critical alarms, and minor alarms are those that may not affect the service immediately and can be attended to with ease. It is for the network operator to set the severity level that is facilitated by the NMS through dropdown menus.

Typically, a critical alarm could be "very high temperature," indicating overheating, suspected fire, etc., or "no message from peer node," indicating a link disruption or so. The major alarms could be failure alarms such as AIS, RDI, etc. and minor alarms could be a mismatch of software version, configuration, etc.

11.7.3.4.1 Traps

Depending upon the error thresholds set up by the operator, various in-circuit performance monitoring alarms are generated. In addition, a large number of failure or defect alarms are also generated by the system. All these events can be sent to many servers, where various teams of management are located. These could be customer-care teams, technical support teams, field maintenance teams, higher management, or equipment vendor. Such messages containing the events are called "traps." These traps can be viewed by any of the teams any time, and they can take appropriate action according to their assigned responsibility. The NMS allows these traps to be diverted to various servers, whose IP addresses can be fed through a menu driven command.

11.7.3.5 Security

To avoid the possibilities of intended or unintended (but undesirable) manipulations of the system parameters settings, records, etc., the NMS provides for various types of users to be defined. All these users can be defined by the user called administrator. Each of these users is given a limited access to the network resources. When a user logs in with his password, he gets only specific permissions, for example, the following types of users may be created.

11.7.3.5.1 Administrator

The administrator will have all the privileges for complete management of the network. He is also able to create new users and delete them. He is able to assign them specific authorities for limited access to the network.

11.7.3.5.2 Operator

The operator may have permissions for retrieving and limited provisioning. He may be allowed to set the severity levels of alarms and others.

11.7.3.5.3 User

A user may have all "read privileges" (i.e., he can see most of the happenings to his circuits, but is not able to take any action) and may be given permission for limited provisioning.

11.7.3.5.4 Diagnosis Engineer

He may be permitted to view the diagnostic information and support the technical teams in the field.

11.7.3.6 Timing Management

The SSM timing manager module in the NMS can configure the synchronization preferences of any node.

The quality level may be enabled or disabled. The quality level may be output manually or automatically. The quality level may be set to DNU, PRC, SSU, invalid, etc. The clock source may then be selected by the node automatically.

11.7.3.7 Maintenance Operations

11.7.3.7.1 Loop Back

One of the most important maintenance and test procedures is to send a signal from the near-end node of the path and "loop it back" from the far-end node. This gives us results of continuity and errors. The loop-back is done by the staff at the far end. However, the NMS provides a facility for software loop-back, i.e., the loop-back condition is created at the far end node by giving a command through the NMS. This saves a lot of time and effort in rectifying faults and initial installation.

The loop-back can be done at any level including STM, E3, E1, etc.

11.7.3.7.2 Power Management

The LOS alarm gives us an indication that the signal is completely lost, partially lost, or the bit errors are very high. The action required to be taken in each of these cases is different. For example, if there is no receive power, we may have to send the cable team to the field to localize a possible cut, but if the power is available but is low, or BER is high, an indoor team may be required to take remedial action. In such cases, if the exact power level is known, we can take either of the above actions accordingly. Some vendors offer the facility of power measurement on the receiver port, through NMS, to help the situation. If the receive power is low (but not 0), transferring the circuit on a spare fiber or cleaning of connectors may do the rectification.

These are some of the features of SDH and NMS. As stated earlier, NMSs are proprietary software of each vendor, and hence, they differ widely in features and facilities. However, the above discussion is enough to get a feel for what NMS means to the SDH network.

Review Questions

1. What is the fundamental difference in signaling in digital systems as compared with analog systems and what is its impact?

2. Distinguish among terms O&M, OAM, OA&M, and OAM&P.

3. List the main constituents of OAM activities in SDH.

4. Which are the main categories of performance monitoring in a telecommunication system?

5. Explain the concept of "system-out-of-service" type of performance monitoring with the help of a block diagram. What are the measurements carried out?

6. What are the problems with or limitations of the "out-of-service monitoring" system?

7. What is "in circuit monitoring?" Why it is considered better?

8. The performance monitoring in SDH is done in various layers. Elaborate. What are the benefits of layered monitoring?

9. How are the overhead bytes allocated to different layers of SDH?

10. What do you understand about BIP? Explain the parity check process with an example.

11. What do you understand about blocks?

12. What are the limitations of the parity check process? Why it is still considered important and deployed in modern systems?

13. Which are the layers of SDH where parity check is facilitated? Which are the bytes allocated for in these layers? Show the distribution of parity bytes over all the layers in a diagram.

14. Where is the parity check byte located in VC-4?

15. What are errored blocks in STM-1? How are they generated and monitored?

16. Describe the parity check provided in multiplex sections.

17. How is the path level parity provided for in a higher-order path? What is the purpose of providing a parity check at all the SDH layers?

18. How is parity check implemented at the lower-order path level? Why is it called BIP-2?

19. What are class A and class B performance parameters? How is the information provided by B2 bytes used by the system?

20. What are MS-REI and HP-REI? How are they communicated and used?

21. What are LP-BBE and LP-REI?

22. What is LOS? How it is detected?

23. What are OOF, LOF, and R-LOF? What are the criteria for generation and clearance of these alarms?

24. What are the criteria for detection of AIS? How is AIS transmitted?

25. How is MS-RDI detected and communicated?

26. What is HP-RDI? How is it detected and communicated?

27. Show with the help of a example topology the transmission of MS-AIS to the concerned AU-AIS and TU-AIS and explain.

28. How are AU-AIS and TU-AIS indicated?

29. What is the function of MS-EXC? When and how it is performed?

30. When the loss of pointer takes place, how are AU-LOP and TU-LOP detected and communicated?

31. How is it detected that the higher-order and lower-order paths are equipped or not? What are the alarms generated and how?

32. What is meant by "trace identifier mismatch"? How is it detected at different layers of SDH and what are the alarms generated?

33. What is "signal label mismatch"? What is its significance and how is it at various levels?

34. What are the indicators for "tributary unit loss of multiframe"? How are they generated?

35. Describe anomaly, defect, and failure in the context of performance indicators of SDH.

36. List and discuss various performance monitoring parameters defined by ITU-T.

37. What are the limits of end-to-end performance objectives of SDH systems as defined by ITU-T? On what length of circuit are they monitored and what is the frequency and period of measurement?

38. What is "order wire"? Which are the bytes used for this purpose?

39. What is "automatic protection switching?" Which are the bytes reserved for this application?

40. What is SSM? How is it worked?

41. What is the role of media-dependent bytes and bytes reserved for national use?

42. What is "embedded control channel?" How is it implemented and what purpose does it serve?

43. What is the role of NMS in SDH? Which bytes provide a channel for NMS at regenerator and multiples levels?

44. Draw a schematic diagram of hierarchical NMS as suggested by ITU-T.

45. Discuss in brief the features of "distributed control," "distributed authorization," and "customer managed links" with reference to NMS.

46. Draw the schematic of the complete NMS as defined by ITU-T M-3000.

47. List the important activities to be performed by NMS.

48. Describe briefly the parameters to be defined during configuration of an NE.

49. What are the privileges of administrator, user, operator, and diagnosis engineer in NMS?

50. Describe in brief various types of provisioning facilitated by the NMS.

51. How is the NMS used for performance monitoring of a network? What are the important parameters monitored?

52. What are the fault diagnostic capabilities of NMS? How are various alarms graded and what actions are taken on them? What are traps?

53. For the security of the network, which are the categories in which all the users of the NMS are classified? What are their functions and privileges?

54. Does the NMS help in timing management of the node? If yes, how?

55. How are loop-back test and power measurements performed by NMS and how do they help in the system maintenance?

Critical Thinking Questions

1. What is the utility of user channel? Think of applications other than mentioned in this chapter.

2. How many bytes are reserved for future use? What could be the future usages?

3. What is the need to provide the overhead bytes separately in different layers?

4. What is exclusive OR operation performed for? Could the same results be obtained through some other method of comparison?

5. Classify all the alarms and indications described in this chapter into critical, major, and minor alarms.

Bibliography

1. *Optical Transmission in Communication Networks, Web Pro-Forum Tutorials,* International Engineering Consortium, 1998.
2. S.V. Kartalopoulos, SONET/SDH and ATM, Communications Networks for the Next Millennium, IEEE Press, 2003.
3. P. Harikumar, *Teaching Notes on Synchronous Digital Hierarchy,* IRISET Secunderabad, 2002.
4. *In Service Course on SDH Systems,* Regional Telecom Training Centre, Hyderabad, 2000.
5. *Broadband Communications and SDH,* Training material, Tejas Networks India, 2002.
6. *TJ100MC-4 Multi-Service Access Node, Software User Guide, Tejas Networks India,* 2003.
7. G. Saundra Pandian, *IP and IP-Over SDH,* Training course, Railtel Corporation of India, 2003.
8. ITU-T Recommendation G.707/Y.1322 (2007) Amd-1, *Network Node Interface for the Synchronous Digital Hierarchy (SDH),* International Telecommunication Union, Geneva, Switzerland.
9. ITU-T Recommendation G.852.1, *Management of the Transport Network— Enterprise Viewpoint for Simple Sub-Network Connection Management,* International Telecommunication Union, Geneva, Switzerland.
10. ITU-T Recommendation G.774.01, *Synchronous Digital Hierarchy (SDH) Performance Monitoring for the Network Element View,* International Telecommunication Union, Geneva, Switzerland.
11. ITU-T Recommendation G.826, *End-to-End Error Performance Parameters and Objectives for International, Constant Bit-Rate Digital Paths and Connections,* International Telecommunication Union, Geneva, Switzerland.
12. ITU-T Recommendation G.827, *Availability Performance Parameters and Objectives for End-to-End International Constant Bit-Rate Digital Paths,* International Telecommunication Union, Geneva, Switzerland.
13. ITU-T Recommendation G.828, *Error Performance Parameters and Objectives for International, Constant Bit-Rate Synchronous Digital Paths,* International Telecommunication Union, Geneva, Switzerland.
14. ITU-T Recommendation G.784, *Synchronous Digital Hierarchy (SDH) Management,* International Telecommunication Union, Geneva, Switzerland.
15. ITU-T Recommendation M.3000, *Overview of TMN Recommendations,* International Telecommunication Union, Geneva, Switzerland.

12

SDH Architecture and Protection Mechanism

One of the main advantages of SDH/SONET over PDH is the "resilience" provided in the SDH networks, besides the direct drop/insert facility and enhanced OAM features. Resilience means that the networks are tolerant to faults, and the customer traffic is not affected by system breakdowns. Although it is never possible to provide a 100% fault tolerant design for any type of network, SDH provides for fault tolerance to a great extent. The electronics of today are by and large very reliable, thanks to the large-scale integrated circuits and much improved manufacturing facilities, with much harder specifications as compared to the past. This makes the equipment much less prone to faults as compared to the media (referring to the OFC or buried copper cables), which is extremely prone to damage. The buried or underground OFC is subjected to a very hostile environment because of a large number of activities in the vicinity involving digging by municipal corporations, electricity companies, telephone companies, and private parties. The result is that the cases of damages/cuts to OFC are very frequent, leading to complete disruption of telecommunication traffic.

SDH provides for switching over the traffic to an alternative healthy media, in such an eventuality of an OFC cut, through a mechanism called protection switching. The resilience achieved has an associated cost. A large amount of system capacity has to be kept reserved for providing alternative/protection channels. Depending upon the quality of resilience to be provided, the reserved capacity could be up to 100% of the actual capacity provided to the customer. The switchover might take place due to a variety of other cases of signal failures or signal degradations (for details of SDH failure alarms and degradations, see the previous chapter), to cater for an uninterrupted traffic to the customer to a great extent. We will discuss these details in this chapter.

To achieve the overall objectives of resilience, flexibility, and management, the SDH network elements (NEs, also called nodes) are arranged in categories called terminal multiplexer (TM), regenerator (REG), add/drop multiplexer (ADM), and digital cross connects (DXC). These elements are then arranged in the network in linear, point to point, ring, or mesh topologies depending upon the need and reliability requirements.

In this chapter, we will discuss these elements and architectures. We will also see about various types of ring architectures and their relative merits and the mechanisms of protection switching in each case. The role of K1 and

K2 bytes of the MSOH, which is very important for protections switching, will also be explored. The practical deployment scenarios are generally very different and complex from the scenarios considered for explaining the principles; however, they work on the same principles for achieving the objectives. We will have a look at practical deployment as well.

SDH can operate on any media, such as paired copper cables, coaxial cables, radio, or optical fiber cables; however, the bulk of the present SDH network is on OFC. To keep our focus on the subject of architecture and protection arrangements, we will presume the media to be OFC throughout this chapter.

12.1 SDH Network Elements

The SDH network elements can be arranged in the following categories:

(i) TM

(ii) REG

(iii) ADM

(iv) DXC

Let us see the features of each of them.

12.1.1 Terminal Multiplexer

A TM is a multiplexer equipment beyond which there is no transmission. It could be located at the end of the telecommunication operators' territory, may be at customer premises, etc. A TM is depicted in Figure 12.1.

Thus, a TM has the STM transmission and reception on one of its sides and tributaries getting inserted or dropped on the other side. There is no regeneration at TM.

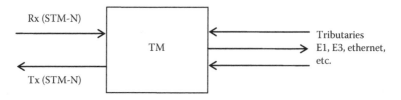

FIGURE 12.1
A terminal multiplexer.

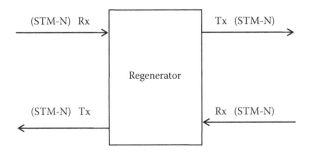

FIGURE 12.2
An REG node in an SDH network.

12.1.2 Regenerator

An REG node is an equipment which simply regenerates (or repeats) the signal received from either side and transmits it to the opposite side in an SDH network. Figure 12.2 shows an REG node.

There are no tributaries getting into or out of this equipment. An REG node is placed among two multiplex section nodes if the distance between them is more than what can be supported by the transmit power and the receiver threshold of the multiplex section nodes (typically 30–100 km). The REG nodes can be more in numbers if the distance is too large.

12.1.3 Add/Drop Multiplexer

At the ADM equipment, we regenerate the signal as well as add or drop tributaries. These are the nodes that form the multiplex section NE. An ADM is shown in Figure 12.3.

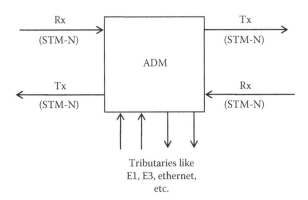

FIGURE 12.3
An ADM node.

12.1.4 Digital Cross Connect

DXC is basically a switching unit that facilitates the interconnection of any of the tributaries connected to it on one side to any of the tributaries connected on the other side of the equipment. It is just like a small automatic exchange interconnecting various subscribers.

In SDH, cross connects can be divided in two categories, namely,

(i) Higher-order cross connect

(ii) Lower-order cross connect

Let us see how they function.

12.1.4.1 Higher-Order Cross Connect

Higher-order cross connect switches the higher-order virtual container (VC-4) from one STM-1 to another STM-1. The access capacity in this cross connect is STM-1. Figure 12.4 shows this.

The STM-N is first broken into STM-1s (electrical signals), then any of the STM-1s on one side can be connected to any other STM-1s on the other side.

12.1.4.2 Lower-Order Cross Connect

The cross connections of lower-order payloads, i.e., VC-12, VC-2, etc., are required rather more frequently as compared to higher-order payload. This is owing to the fact that, as most of the user channels are tributaries, such as E1, E3, Ethernet, etc., the lower-order cross connects provide this switching. Figure 12.5 shows the concept.

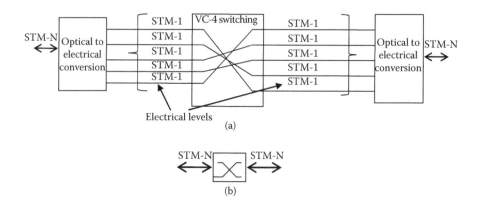

FIGURE 12.4

Higher-order cross connect (a) and its symbolic representation (b).

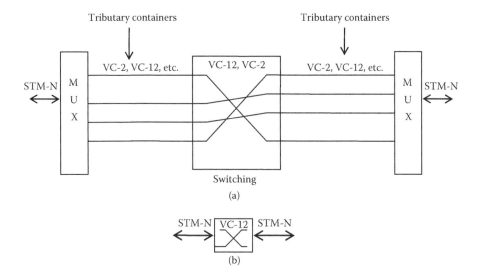

FIGURE 12.5
A lower-order cross connect (a) and its symbolic representation (b).

The STM-N signal is demultiplexed into tributaries and then any of the tributary containers or the lower-order payload on one side is connected to any of the tributary containers on the other side.

In practical situations, the operator may need to reroute a few STM-1s and drop and insert some tributaries. Hence, the ADMs will generally include cross connects of both higher- and lower-order levels, unless the network at the node is only of STM-1 capacity, in which case, only a lower-order cross connect will be required.

All these network elements are arranged in various types of topologies depending upon the need of the situation; these topologies are discussed in the next section.

12.2 SDH Network Topologies

The SDH network topology will basically depend upon the application. The simplest one could be point-to-point connection between two places, say, connecting two exchanges or two offices of a company. The more general application is the network of a telecommunication operator, which could be linear or a ring or a mesh network, again depending upon the geography of the nodes to be served and the requirement of reliability. These topologies are depicted in Figure 12.6a–d.

In all the above topologies, there is no difference in the functioning if direct provisioning is to be made to the customers. However, if a backup

(a)

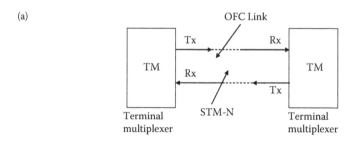

(b)

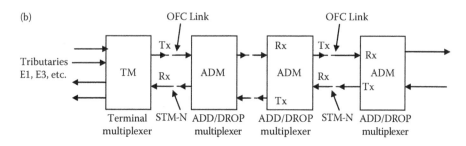

(c)

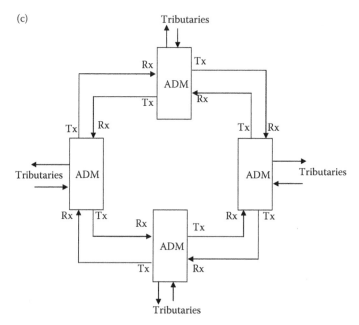

FIGURE 12.6
SDH network topologies. (a) A point SDH network formed by an OFC link connecting two
terminal MUXˢ. (b) A linear SDH network consisting of 4 or more nodes connected through
OFC links, with one of them being a terminal MUX and the rest being ADMs. The number of
nodes could be much larger. (c) A ring SDH network with four ADM nodes connected through
an OFC link. All the nodes are ADMs, as each one of them has to receive from one direction
and transmit to either direction besides adding and/or dropping tributaries.

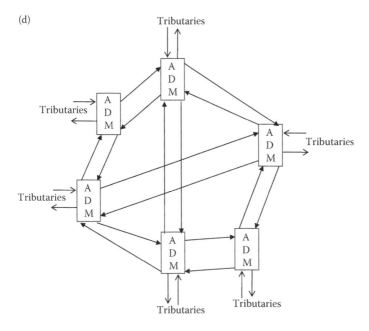

FIGURE 12.6 (Continued)
(d) A mesh SDH network consisting of 6 ADMs connected through OFC links in a ring, but 2 of them are interconnected directly to their opposite nodes. This creates multiple logical rings inside the main ring.

path is also to be provided for improving the reliability, the ring or mesh topologies are far better than the linear topology, and in many cases, they are the only answers for provisioning of a backup path. The backup path is also called a "protection path." Let us define these paths properly as they will appear frequently in our further discussion.

12.3 Work and Protect Paths

When a channel or circuit is provided to a customer, the normal config- ured channel for traffic carrying is called work path, working path, work- ing circuit, or working channel. The backup channel, which is provided for taking over the customer's traffic in case of the failure of the main link, is called the protect path, protection path, protection circuit, or protection channel. Figure 12.7 indicates the working and protection channels for a linear network.

Although the backup channels can be provided on a linear network as well, the ring topology is ideally suited to provide protection channels. Let us see how.

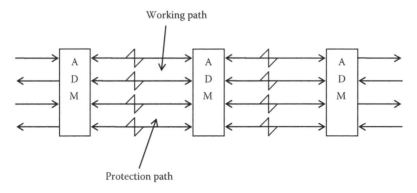

FIGURE 12.7
Work and protect paths in a linear network.

12.4 Advantage of a Ring

If the working and protection channels are provided on a linear network such as in Figure 12.7, either the working and protection circuits will have to be provided on different fibers of the same cable or two different optical fiber cables will have to be laid on different routes between adjacent nodes. The practical situations, however, are generally very different. For the project economies to make sense in the competitive world, no operator puts up two parallel cables on different routes only to connect the two nodes under consideration. Nevertheless, two such OFC routes may meet each other at two or more points due to the normal geography of the service area, and when it happens, a "ring" gets automatically formed. The ring formed automatically serves the purpose of two OFCs laid on divergent routes. The only requirement that such a ring may not satisfy is connecting each node by two different cables in the same direction. But when a ring is formed, the nonmeeting nodes on OFC can route their traffic through other nodes, in case of failure of working channel. Figure 12.8 depicts the situation.

The normal traffic between nodes A and C is via node B. When a cable cut occurs between nodes B and C, the traffic between nodes A and C can be diverted via nodes B→A→E→D→C because an "intact" OFC is still available connecting node A and C, although over a longer route.

Thus, without having a need to provide a parallel cable on a diversified route between each node, the purpose of provision of protection channels is served by the ring topology.

This was a simplified arrangement to demonstrate the advantage of ring for protection path provisioning. The actual ring implementations are little different, which are dealt within the subsequent sections; but before that, let us see the scheme of protection recommended by the International Telecommunication Union (ITU-T) in the next section.

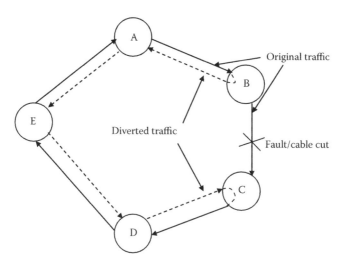

FIGURE 12.8
A ring network with five nodes.

12.5 Protection Switching Schemes

As we have seen in the preceding discussion, the protection channels are provided separately from the working channels in a network, to provide backup for customer traffic in case of failure of working channels. But providing a complete set of duplicate hardware, software, and media, which will normally remain idle and just wait for a failure of the working circuit to start their operations and again close shop on the restoration of the working circuit, is an example of very poor efficiency of the protection circuit and that of the system as a whole and thus is very expensive. ITU-T (G.841) has thus recommended schemes for a better utilization of this idle system, while maintaining the reliability of the network.

The schemes are

(i) 1 + 1 protections scheme
(ii) 1:1 protection scheme
(iii) 1:N protection scheme

Let us examine them in brief.

12.5.1 1 + 1 Protection Scheme

In this scheme of protection, the transmission of the signal is done simultaneously on working as well as protection path "always." In the case of a failure of the working path, the receiver switches over to the protection path. The concept is shown in Figure 12.9.

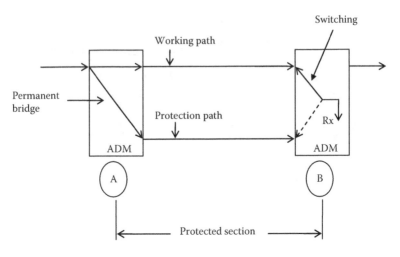

FIGURE 12.9
The 1 + 1 protection scheme between two nodes. Only the transmit path from A to B has been shown for clarity. Similar switching is carried out by the receiver of node A.

The permanent connection at the transmitter between working and protection paths is called a bridge. The switching is called "tail switching," and the circuit is called "tail switched" because the decision to the switchover is taken at the tail end of the signal. Figure 12.10 shows the scheme with both the transmitted and the received signals.

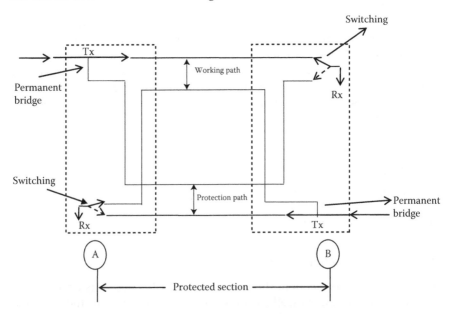

FIGURE 12.10
The 1 + 1 protection scheme, showing switching at both the nodes.

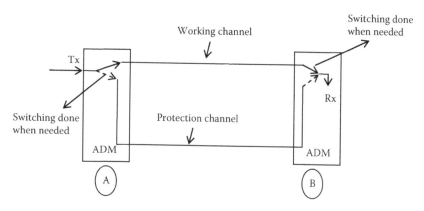

FIGURE 12.11

The 1:1 protection scheme. The transmitter as well as the receiver is connected to the "working channel" normally. In case of a failure of the working channel, the protection channel is chosen by switching action taking place at both the ends. Only the transmit path at node A and the receive path at node B is shown for clarity.

12.5.2 1:1 Protection Scheme

In contrast with the 1 + 1 scheme, this scheme does not provide a permanent bridge at the transmitter, which means that the transmission is not done in parallel on working and protect paths. The protection channel is kept reserved for providing the protection when needed. The channel is normally kept disconnected at both the transmitting and receiving ends, and the switchover is carried out at both the ends when needed in case of a failure of the working channel. Figure 12.11 depicts this situation. (Extra traffic: The protection channel is available completely idle when not being used for protection; thus, it can be used for carrying extra traffic. However, as soon as the need for protection arises, this extra traffic will get disconnected, and the channel will be used for protection of working channel traffic.)

Figure 12.12 shows the transmitting and reception paths for 1:1 protections.

In this case, the switching request is initiated by the transmitting end; thus, the protection is called "head switched" protection. (The transmit end being the "head" and receiving end being the "tail" of the signal.)

12.5.3 1:N Protection Scheme

The concept of 1:N protection scheme is the same as that of 1:1 scheme. The only difference is that the protection channel provides protection for N number of working channels. (Assuming that all of them may not need protection at the same time leads us to conclude that one channel may provide protection to more than one working channel.) The number N will depend upon the deployment scenario, but its maximum value has been fixed at 14 by the ITU-T. Figure 12.13 shows this protection scheme. (Automatic protection switching, or APS: The switching, as discussed above, from working to protection

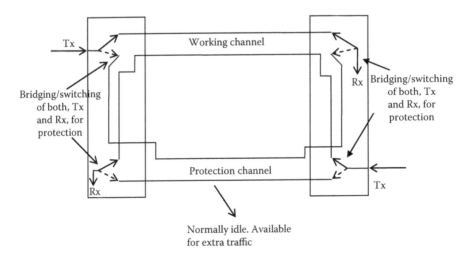

FIGURE 12.12
The 1:1 protection scheme, showing switching at both the ends.

paths takes place automatically in SDH; thus, it is called "automatic protection switching." This switching is carried out with the help of K1 and K2 bytes provided in the MSOH; refer to Chapter 11 for overhead details.)

We will see the details of functioning of K1 and K2 bytes during our discussion on various ring protection schemes in subsequent sections.

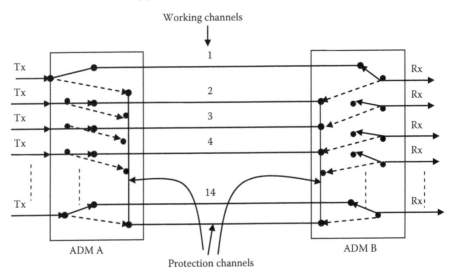

FIGURE 12.13
The 1:N protection scheme. One protection channel provides protection to 14 working channels. Whenever a working channel fails, it switches over to the protection channel. Switching takes place at both the ends, initiated at the transmitting end. Only transmission at A and receiving end at B has been shown for clarity.

12.6 Self-Healing Rings

We have seen the advantages that a ring offers, for the purposes of providing backup or protection channels to the customer traffic (see Section 12.4). With the help of APS, the traffic in the ring is automatically switched over to the protection path in case of a fault in the main or working path. Thus, the ring "heals" itself from the fault. Hence, it is called a self-healing ring (SHR).

These SHRs can be configured on two or four fibers. They could be dedicated or shared. They could be for multiplex sections or for end-to-end path of customers. All these details are the subject of our next section.

12.7 Types of Automatic Protection Switching

The automatic protection to the working traffic can be provided in many ways in SDH. The chart given in Figure 12.14 indicates these topologies.

You may notice that under the tree of ring protection, there are two main branches, viz. "MS shared ring" and "MS dedicated ring." MS stands for

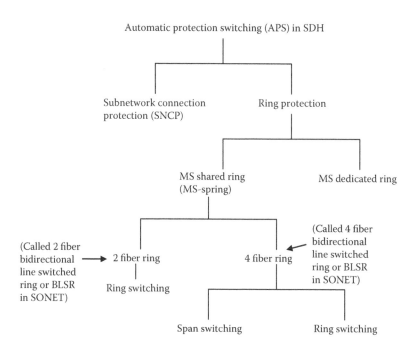

FIGURE 12.14
Types of APS mechanisms in SDH.

multiplex section. The protection is always provided between two multiplex sections, as there is no point in protecting an REG section because there is no traffic dropping or adding at the REG section. Thus, the rings providing protections for multiplex sections are called MS rings, dedicated and shared. We will see the details in the next sections.

Subnetwork connection protections (SNCP) is a path-level protection, and we will see it in detail after we finish our discussion on ring protection schemes. Other branches of the tree will become clear as we proceed.

12.8 MS Dedicated Ring Protection

This is a system of two counterrotating rings. One of them is dedicated for the working traffic, while the other one is dedicated to carry the protection traffic. Both the counterrotating rings are unidirectional. A dedicated protection ring is shown in Figure 12.15.

The capacity of the working ring and the protection ring has to be equal to each other. While the working ring transmits in a clockwise direction, the protection ring transmits in a counterclockwise direction.

When traffic has to be realized between any two nodes, say, nodes A and B, on the working ring, then A will transmit to B directly but will receive from B via nodes C–F, as there is no direct path on the working ring from B to A. This is because the scheme utilizes only one fiber for the working ring and another for the protection ring. Thus, the capacity utilized in any one section

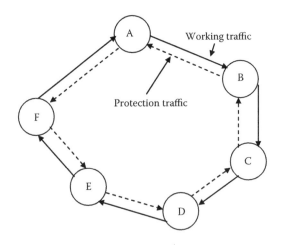

FIGURE 12.15
Dedicated protection ring.

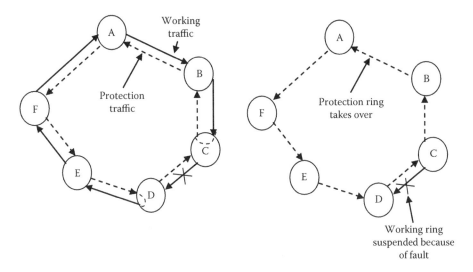

FIGURE 12.16
Fault condition on dedicated ring.

consumes the capacity of the whole ring. If the ring capacity is STM-16, then all the traffic links among various nodes together should not exceed STM-16 capacity.

In case of a fault in any of the sections, the whole of the working ring is suspended, and the traffic is transferred on the protection ring.

The situation is depicted in Figure 12.16. A fault between C and D nodes forces the entire working ring to be suspended because the whole ring is required for the traffic of each node, as seen above.

There are a number of unresolved issues related to the functioning of the dedicated ring, i.e., the time delay between any two nodes in the Tx and Rx path is different (because of unidirectional flow), the limited capacity, etc. Thus, this ring is still under study by ITU-T.

On the other hand, the MS shared ring offers much better performance and manageability, which is the preferred choice, at least for the time being.

12.9 MS Shared Ring Protection

This type of ring shares the working and the protection traffic, hence the name "shared ring." The protection is provided in the MS section, hence it is called MS Shared Protection Ring (in short, MS SPRING).

The switching action is performed at the section layer on the ring and does not involve path layer indications.

The MS SPRING can be realized using two or four fibers. Each type has its own merits; thus, both types are in use. We will discuss both these models and divide the subject into the following subheads:

   (i) Principles of working

  (ii) Switching mechanism

 (iii) Other important features

12.9.1.1 Principles of Working

12.9.1.1.1 Two-Fiber MS SPRING

A two-fiber MS SPRING is shown in Figure 12.17. Obviously the ring is formed using only two fibers.

Each of the fibers carries traffic that is meant for working and protection channels each in a 50/50 ratio. The direction of traffic on fibers 1 and 2 is opposite to each other. Thus, for the normal traffic, the transmission and reception between any two nodes is the same, leading to the same time delay.

In case of a fiber cut, the traffic of the working channel is carried on the protection channel by switching it back from the node, beyond which the fault occurs. The fault condition is shown in Figure 12.18.

As can be seen in Figure 12.18, in case of a fault between the nodes C and D, the traffic at C is switched back to reach D via nodes B, A, F, and E. Similarly, the traffic from node D to C, which was also direct, now takes a route via nodes E, F, A, and B. Thus, the working traffic is immediately restored on the capacity reserved for protection, providing 100% availability against a single fault.

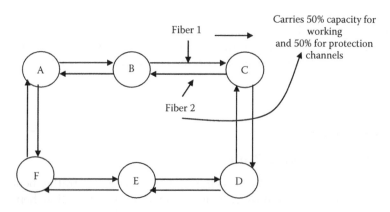

FIGURE 12.17
Two-fiber MS protection ring (normal operation).

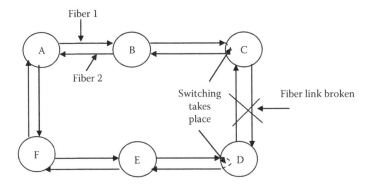

FIGURE 12.18
A two-fiber MS protection ring (fault condition).

12.9.1.1.2 Four-Fiber MS SPRING

The working of the four-fiber ring is similar to that of a two-fiber ring in many aspects, and it is different from it in many other aspects. Figure 12.19a shows the concept.

The working and protection channels are carried on separate pairs of fibers. Thus, the full capacity of the node is duplicated on the protection fibers. Generally, when an OFC cut takes place, the whole cable is cut and not a few fibers. In such a case, the likelihood of working fibers getting cut and protection fibers being intact is much lower. The four-fiber ring can give very good results only when the working and protection fibers are carried on separate cables and on different routes (Figure 12.19b).

A fault condition is shown in Figure 12.19b, where the working as well as protection fibers are cut between nodes C and D. In this case, the ring protection will come into force, where nodes C and D will transfer the Tx and Rx on the protection fibers toward nodes B and E, respectively, instead of sending the signal to each other directly. The signal of node C meant for node D will now be taken on protection fibers via nodes B, A, F, and E and vice versa for traffic from D to C. However, the fibers of working and protection channels may be taken on different routes, in which case only the working fibers may get damaged. Thus, there is no point in transferring the traffic on the protection fibers on the long path (for node A to F via BCDE and vice versa for node F) (see Figure 12.19c). The traffic in this case is transferred on the short path, i.e., on the protection fibers directly on the direct route between A to F. This is called "span switching" and is possible only on a four-fiber ring.

Thus, a four-fiber ring has an option of either span switching or ring switching. In APS protocol, the span switching has a higher priority, as it saves other spans for use by one other such case, leading to the possibility of multiple span switches being in operation at any given time.

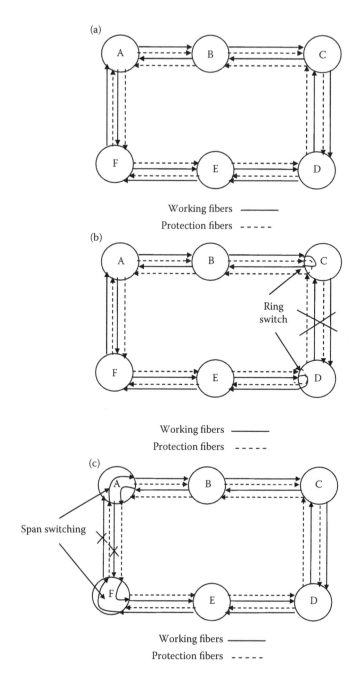

FIGURE 12.19
(a) A four-fiber ring with six nodes. (b) A fault on the whole cable in a four-fiber ring, causing a ring switch. (c) A fault on working fibers only, causing a span switch in a four-fiber ring.

The cost of four-fiber rings is naturally higher because the MUX has to interface four-fibers on either side. On the other hand, the two-fiber ring gives only half the capacity for operations as against full capacity of the four-fiber ring. The choice should depend upon the feasibility cum cost analysis, particularly in view of higher capacities being available at reasonably low costs.

Apart from the above-mentioned differences, the four-fiber ring has one more feature that is different from the two-fiber ring, which is

> When the traffic is working on the protection channel on the long path due to a fault on both the working and the protection fibers in a span, there is a possibility that the protection fibers in the span become available first. In such a case, the traffic will not switch over to the short path span protection but will continue to work on the long path ring protection. However, it will switch over to the span protection, in case the ring protection is required and command is given for restoration of traffic, which might have interrupted due to a fault in any other span.

12.9.1.2 Switching Mechanism

The protection switching is meant for diverting the traffic on the healthy system from the faulty system. The switching is thus automatically initiated in case of a fault of a predefined nature in the equipment or in the media. Various types of faults could include fiber cuts, signal failures due to various reasons, and signal degradations.

In addition to the automatically initiated switching, facility has been provided in SDH design to "force" a switchover from the working channel to the protection channel in case the operator desires to do so due to administrative reasons. Facilities of blocking the protection switching (called lock out) and carrying extra traffic during the nonutilization period of the protections channel are also provided.

Automatically initiated commands may include "signal fail" (SF), "signal degrade" (SD) etc., of various origins and types, while the externally initiated commands may include "lockout for protection," "forced switching," "manual switching," "exercise," etc.

The node includes a logical functional equipment called APS controller, which gets inputs from all the sources and makes a decision about switching. It sends the K1 and K2 bytes to the required nodes for necessary switching actions. Figure 12.20 depicts a schematic block diagram of an APS controller.

Let us consider the ring of Figure 12.18 again to understand the switching mechanism.

When the fiber link gets disrupted in this ring between nodes C and D (Figure 12.21), the loss of signal fault is detected at both these nodes. Node C sends this information through K1 and K2 bytes to node D via nodes B, A, F, and E, and node D also simultaneously sends this information to node C via nodes E, F, A, and B. On receipt of K1 and K2 bytes containing this

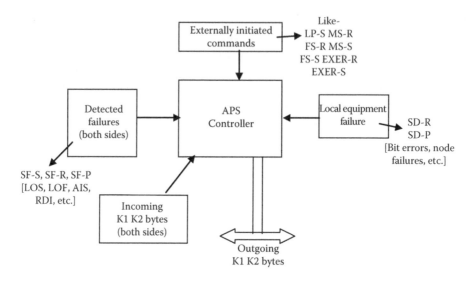

FIGURE 12.20
Schematic block diagram of an APS controller inside a node.

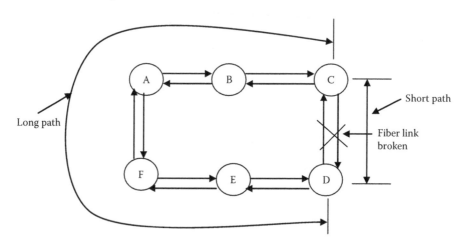

FIGURE 12.21
Two-fiber MS protection ring with a fault between nodes C and D.

information, node D switches over the working traffic to the protection link at node D, while node C carries out a similar activity at node C. Once the switchover is complete at both the ends, the K1 and K2 bytes are updated to carry this information to the other switching nodes. Intermediate nodes B, A, F, and E simply pass the K1 and K2 bytes as they are transmitted by nodes C and D. After both the ends switch over the traffic to the protection link, the K1 and K2 bytes resume the normal values and the circuits start functioning normally on the protection link.

After the fault is rectified, the K1 and K2 bytes of nodes C and D again change their values to convey to each other of the new development. This causes both the nodes to enter into a wait to restore (WTR) state. (WTR: When a link fails and the traffic is transferred on the protection link through APS, the traffic is not transferred back on the working link instantaneously on the restoration of the fault. This is because during the process of fault restoration, there may be intermittent small periods of "fault" and "no fault" situations due to rectification work in progress. Thus, a small period of nearly 900 s or so is allowed for stabilization of the signal after the intimation about fault restoration is available. This period is called WTR.)

The K1 and K2 bytes carry this WTR value for this period. After the expiry of the WTR period, K1 and K2 bytes again change values and restore the traffic on the working channel.

The flow diagram of activities and the values of K1 and K2 bytes are shown in Figure 12.22 for the above process.

Since the short path, i.e., node C to node D direct path, is what is considered faulty, only the long path has been shown in Figure 12.22. Normally the K1 and K2 bytes are addressed to the adjacent nodes on either side, but as soon as the K1 byte value is changed to a failure condition SF rings (SF-R), the K1 and K2 bytes get addressed to the destination node, which is supposed to activate the protection switching. The intermediate nodes B, A, F, and E simply pass through the K1 and K2 bytes as they are received from either direction.

The role of each individual bit of K1 and K2 bytes has been indicated in Figure 12.22. These roles have been defined by ITU-T in recommendation G.841. The full list of various values K1 and K2 bytes will be seen in this chapter shortly, but before that, let us see what happens when the fault gets rectified.

On rectification of the fault, a WTR state is entered by both the nodes. This is initiated by the APS controller by sending the SF-R signal on K1 and K2 bytes [K1 (b1–4) = 1011] on the long path from either node to each other, and a reverse request ring, or RR-R [K1 (b1–4) = 0001] on the short path by both of them to each other (after the restoration of the fault, the short path is also available). As soon as these crossing RR-R requests are received on the short path by the nodes, the WTR timing is started. On expiry of the WTR time, code is sent on K1 byte [(b1–4) = 0101] on the long path and RR-R as above on the short path from both the nodes. On receipt of WTR on the long path with RR-R on the short path, both nodes C and D switch over their receivers (drop the switches) to the normal working channel and send a no-request (NR) code [K1 (b1–4) = 0000] on byte K1 and bridged code [001 in bits 6–8 of K2] on byte K2 ("bridged" because the transmitters are still on protection path) to each other on the long path. On receipt of these codes on K1 and K2 bytes, both C and D nodes transfer their transmitters on the working path. Both the nodes now send NR code on K1 byte and "idle" code on K2 byte. The normal working is resumed.

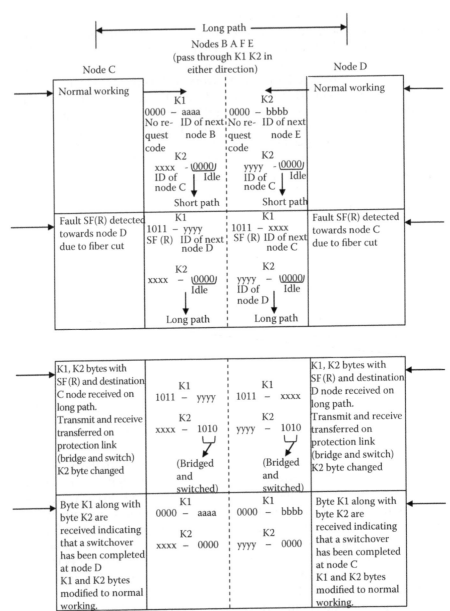

FIGURE 12.22
The protection switching process and role of K1 and K2 bytes.

This has been a specific case where both the fibers were cut, they were cut at the same time, and they were restored also at the same time. However, there may be many other situations where only one fiber gets disrupted, there is a difference in the timings of disruption of either fiber or a node failure, there could be a case of forced switchover due to administrative reasons, etc. In all such cases, the exact transactions of codes between the involved nodes will be a little different from what has been discussed above, with the deployment of various other codes described in ITU-T G.841 for K1 and K2 bytes for each type of signaling. Nevertheless, the mechanism of protection switching works on the principles similar to what has been discussed above. Discussion of each type of case is neither warranted for understanding the subject, nor is in the scope of this text. The interested readers can refer to the above-mentioned ITU-T recommendation for complete details.

The complete bit assignment tables for K1 and K2 bytes as per ITU-T G. 841 are given in Figure 12.23. (Note that the K1 and K2 bit assignment are differently defined in ITU-T G.783, which is meant for linear switching. The details are given here because its discussion is enough to understand the principles.)

K1 Byte (ITU-T G.841)			
Bridge request code (bits 1–4)			Destination node identification (bits 5–8)
Bit number	Command	Priority	
1 2 3 4			
1 1 1 1	Lockout of protection (span) LP-S or signal fail protection	Higher	The destination node ID is set to the value of the node for which that K1 byte is destined. The destination node ID is always that of an adjacent node (except for default APS bytes)
1 1 1 0	Forced switch (span) FS-S		
1 1 0 1	Forced switch (ring) FS-R		
1 1 0 0	Signal fail (span) SF-S		
1 0 1 1	Signal fail (ring) SF-R		
1 0 1 0	Signal degrade (protection) SD-P		
1 0 0 1	Signal degrade (span) SD-S		
1 0 0 0	Signal degrade (ring) SD-R		
0 1 1 1	Manual switch (span) MS-S		
0 1 1 0	Manual switch (ring) MS-R		
0 1 0 1 0 1	Wait to restore WTR		
0 1 0 0	Exerciser (span) EXER-S		
0 0 1 1	Exerciser (ring) EXER-R		
0 0 1 0	Reverse request (span) RR-S		
0 0 0 1	Reverse request (ring) RR-R		
0 0 0 0	No request NR	Lower	
Note: Reverse request assumes the priority of the bridge request to which it is responding.			

FIGURE 12.23
Assignment of K1 bytes for APS.

The whole of the APS procedure is carried out using these various commands conveyed by K1 and K2 bytes as far as MS-SPRING is concerned. Let us look at these commands in more detail to know the meaning of each.

During the preceding discussions, we have already seen NR, RR-R, WTR, and SF-R commands. The commands which have "(ring)" followed by the description are meant for a two- or a four-fiber ring, but the commands having "(span)" followed by the descriptions are applicable only to four-fiber rings because span protection is possible only in a four-fiber ring. (Span protection: In the case of a two-fiber ring, if the fibers get damaged or disrupted, the only path available for restoring the traffic is the long path on the ring (see Figure 12.21). However, in the case of a four-fiber ring, it is possible that only two fibers carrying the normal traffic get damaged or disrupted, leaving the protection fibers intact. In such a case, the protection is possible to be given on the same span (on the protection fibers) on which the other two, i.e., working fibers, got damaged. This is called "span protection." Thus, the commands for a four-fiber ring provide for span as well as ring protection.) Let us see the other commands.

12.9.2 Commands for Protection Switching

The commands for protection switching are communicated through K1 and K2 bytes as seen already. Let us see the functions of the commands activated through these bytes in brief.

12.9.2.1 Commands Activated through K1 Byte

12.9.2.1.1 Lockout for Protection (Span) (LP-S) or Signal Fail (Protection) (SF-P)

If due to any administrative or maintenance reason, it is desired that a particular span should not be used for protection for some time, SDH provides the facility in the form of this command. Using this command, not only the span switching is prevented in a particular span, but as a consequence, the switching is prevented in the entire ring (with one span locked out, the ring for protecting the failure in any other span can not get completed). The command thus prevents ring protection in the entire ring while a particular span is prevented for use of protection channels. In the balance spans, the span protection is possible. The command is also used to convey to adjacent nodes that the protection channel is in a failed state and should not be attempted for switching-over of traffic. The purpose served is the same as lockout; thus, the same command is used, and the only difference is that the LP-S is intentionally initiated, whereas SF-P is automatically initiated by the node.

12.9.2.1.2 Signal Fail-Ring (SF-R)

This command generates a request for "ring switching" in case of a signal failure. All signal failures of a two-fiber ring are protected using a ring protection. However, span protection feasibility also exists in the case of a four-fiber ring, as discussed earlier. Thus, in the case of a four-fiber ring, the span protection, if available, is used first, and ring protection is resorted to only if span protection is not possible due to failure or degradation of the protection channel in the particular span.

12.9.2.1.3 Signal Fail-Span (SF-S)

On detection of a fault on a four-fiber ring, this command generates a request for "span switching" of the working channel to the protection channel. The span switching can be initiated only if the protection channel fibers are intact in the same span as that of the failed span. The request is generated by the so-called tail end.

12.9.2.1.4 Forced Switch to Protection-Ring (FS-R) and Span (FS-S)

With the help of this command, the operator may forcibly switch the traffic between any two adjacent nodes from the working path to the protection path. The command is executed irrespective of the status of the protection channel, the only exception being that the protection channel is not in use for serving a higher priority request (for example, you can see from Figure 12.23 that LP-S is a higher priority request).

12.9.2.1.5 Signal Degrade-Protection (SD-P)

When a node detects an SD condition on one of the adjacent spans, on the protection channel, this command requests a switchover of that span protection to the ring protection. This can, however, be executed only if the ring is not serving any other high-priority request already. As this involves a protection path in the span, this command is applicable only to four-fiber rings.

12.9.2.1.6 Signal Degrade-Span (SD-S)

This command is used to transfer the working traffic on the protection channel on the "same span" in case of a degradation of signal in that span on the working channel.

12.9.2.1.7 Signal Degrade-Ring (SD-R)

This command requests a transfer of working traffic on the protection channel in the ring in case signal degradation is detected on working channel by the node. It happens in all cases in the case of a two-fiber ring. However, in a four-fiber ring, this transfer to "ring protection" takes place only if signal degradation of working as will as protection channel is detected in the same span.

12.9.2.1.8 Manual Switch to Protection-Ring (MS-R)

The function of this command is the same as that of SF-R. However, this command is not executed if either an SD condition is existing on the protection channel or it is not in use by a higher priority request (including the failure of protection channel). This command has a lower priority as compared to SF-R or FS-R commands (see Figure 12.23).

12.9.2.1.9 Manual Switch to Protection-Span (MS-S)

The function is the same as that of SF-S. However, this command is not executed if an SD condition is existing on the protection channel, or if the protection channel is in use by a higher priority request (including the failure of the protection channel). This command has a lower priority as compared to SF-S command (see Figure 12.23).

12.9.2.1.10 Exercise-Ring (EXER-R)

This command is for checking the functioning of the protection switching in the ring. The protection switching is exercised without actually completing the bridge and switch operations. This is like a mock drill of a fire fighting agency. Normal traffic is not affected. (APS is a very important facility of SDH. However, it is useful only if it works when required. In many cases, if the working channels keep functioning reliably, the protection channels may not be used for a long time. Thus, the system is not sure about their functioning—it may not work due to some reason in an emergency, when it is actually required. Thus, just to make sure that the protection switching is functional, this option has been provided.)

12.9.2.1.11 Exercise-Span (EXER-S)

Function is the same as EXER-R for span instead of ring. (In APS parlance, the head end is the node that switches over the transmitter portion of a channel— also called "bridging." The tail end is the node that switches over the receiver portion of a channel [also called "Switching"]. Since every traffic channel will have transmit and receive both functions at every node, each node is a "head end" for the other node [involved in switching] for the purpose of bridging, and it is also a "tail end" for the purpose of switching Figure 12.24).

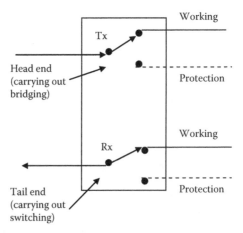

FIGURE 12.24
Head end and tail end of a node. The node that initiates a request for a bridge becomes a tail end, and the one that executes it becomes a head end.

12.9.2.1.12 Reverse Request Span (RR-S)

We have already seen the functioning of the switching mechanism of the two-fiber ring. The RR-S has a similar function of acknowledging the receipt of a "span bridge request" on the short path.

12.9.2.2 Commands Activated through K2 Byte

Refer to Figure 12.25 for K2 byte bit assignments.

12.9.2.2.1 Bits 1 to 4

These bits indicate the source ID, i.e., the ID of the node itself. (It is worth noting here that the source ID is of 4 bits. This means that there could be at the most 16 nodes in the ring. Similar restriction is imposed by the 4 bit destination ID of bits 5 to 8 of K1 byte. Thus, any SHR can not have more than 16 nodes. This limitation may appear to be a bottleneck, but actually it is not. Practical rings are generally of fewer than 16 nodes. Moreover, the distance covered by 16 nodes could be roughly 1600 km—assuming a distance of 50 km between each node and one REG node between any two MS nodes—which is quite a large distance for a single fault. The ring protection does not work in any case for double faults.)

12.9.2.2.2 Bit 5

This bit indicates a short path with a 0 and a long path with a 1. The short path is used for span protection and the long path for ring protection. The codes are also used during restoration, when both short and long paths communicate, as explained earlier in this section (see Figure 12.22).

K2 Byte (ITU-T G.841)			
Source node identification (bits 1–4)	Long/short path (Bit 5)	Status (bits 6–8)	
Source node ID is set to the node's own ID	0: Short path code (S)	**6 7 8**	**Command**
		1 1 1	MS-AIS
		1 1 0	MS-RDI
		1 0 1	Reserved for future use
		1 0 0	Reserved for future use
	1: Long path code (L)	0 1 1	Extra traffic on protection channels
		0 1 0	Bridged and switched (Br and Sw)
		0 0 1	Bridged (Br)
		0 0 0	Idle

FIGURE 12.25

Assignment of K2 bytes for APS. (From the International Telecommunication Union, Table 7.8/G.841 Byte K1 functions, p. 61 of ITU-T recommendation no G.841, October 1998. With permission.)

12.9.2.2.3 Bits 6 to 8

These bits have eight assignments, out of which two are MS-AIS and MS-RDI, two are reserved for the future, and the rest are as follows:

(i) *Extra traffic (ET) on protection channel:* A value of 011 on bits 6 to 8 of K2 byte indicates that the protection is carrying "extra traffic." This ET is a traffic that is carried on the protection channel when the protection channel is not used for protection. This traffic has a low priority and is disconnected when the protection channel is utilized automatically or manually for protection of the working channel. Naturally, the extra traffic is offered at much lower tariffs. The customers may not carry critical applications on this channel and may build their own reliability by hiring extra traffic channels from more than one operator.

(ii) *Bridged and switched (Br and Sw; 010):* When a request for transferring the working traffic on the protection channel is received by a node on bits 1 to 4 of K1 byte, it returns a Br and Sw code on K2 byte, after executing the request.

(iii) *Bridged (Br; 001):* Sometimes, only the bridging request is received by the node through the K1 byte. In such a case, only a bridge is executed, and accordingly, the K2 byte returns a Br code.

(iv) *Idle (000):* When everything is normal, the K2 byte carries this value.

12.10 Other Important Features of Protection Switching

Besides the features discussed above, there are some more important features of protection switching, which are as follows.

12.10.1 Switching Time

The APS in the ring has to be completed within a finite predefined time. This time has been defined to be less than 50 ms (for ring and span switching). However, this 50-ms requirement is subject to certain conditions, which are:

(i) The ring has no extra traffic.

(ii) All the nodes are in the idle state (meaning there are no detected failures, no active commands, automatic or external, and they are receiving only idle K1 and K2 bytes).

(iii) The total fiber length is less than 1200 km.

In case any of the above conditions are not met, the switching time is allowed to exceed 50 ms. Extra time is allowed for removing extra traffic or for negotiating and accommodating the existing switching commands.

12.10.2 Switching Initiation Time

The switching initiation time should be less than 10 ms for SF conditions caused by LOS, LOF, or AIS.

12.10.3 Operation Mode

Two modes of operation are possible, revertive and nonrevertive. In revertive mode of switching, the traffic is restored on the working channel automatically after the failure is restored after a WTR time. In nonrevertive mode of switching, the traffic continues to work on the protection channel even after the failure is restored.

12.10.4 Switching Protocol

The following switching protocol is followed:

(i) Up to 16 nodes should be accommodated on a ring (we have already seen this).
(ii) All spans on a ring should have equal priority.
(iii) Four-fiber rings should be provided with ring as well as span protection (we have seen this).
(iv) The state of a ring (i.e., normal or protected state) shall be known at each node. (This is ensured by the pass through operation of K1 and K2 bytes on all the nodes other than source and destination).
(v) Span bridge request should have a higher priority than a ring bridge request of the same type (priorities are per Figure 12.23).
(vi) Adding and deleting nodes in the ring should be possible.
(vii) It should be possible to split the ring into two separate segments with the ring protecting failures on two separate spans. This may be required when the ring is already working on the protection channel and a second failure of equal priority occurs.
(viii) Misconnection of traffic should be avoided by a deterministic process. AUG squelching should be done when needed.

12.10.5 Manual Controls

Manual controls such as lockout for protection, forced switching, manual switching, exerciser, etc., are possible.

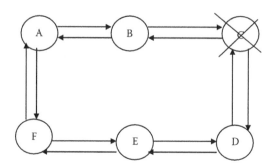

FIGURE 12.26
A node failure of node C causing a misconnection.

12.10.6 Misconnection

There are certain situations during the APS process when multiple activities start at the same time and clash and may cause a misconnection. A failure causing node isolation, a single point failure during the use of extra traffic, etc. are such situations.

Let us take an example of a failure of a node in a ring (Figure 12.26).

Node C is not accessible to node B; thus, it will redirect the traffic meant for node C via nodes A, F, E, and D on the protection path. Now node D is also not able to access the node C and therefore expecting a redirected traffic from node C via nodes B, A, F, and E. Nodes B and D both switch over to the protection path at the same time. If they happen to select a time slot on the protection path which belongs to the same AU-4, the traffic redirected from node B will get dropped at node D, whereas it was meant for node C. This causes a misconnection. Vice versa, i.e., from node D to node B, will also happen simultaneously in such a case.

The misconnection can be avoided by a process called "squelching."

12.10.7 Squelching

The process of squelching involves inserting AU-AIS (all 1s) in the signal required to be squelched.

In the above case, the node failure is recognized by the software and immediately an AU-AIS or squelched signal containing all 1s is inserted in the AU-4s meant for node C. Nodes B and D recognize this as an AIS signal and do not drop it.

12.10.8 Nonpreemptible Unprotected Traffic (NUT)

As is clear from the name these are the channels available in an MS-SPRING, which cannot be used for protection. Also, there is no protection provided to these channels. Thus, NUT is a kind of a reserve capacity in the MS-SPRING, which has to be used by the operator for exchanging the working and protection capacities.

When NUT has to be used, it has to be properly defined in all the nodes, such that the working channels, protection channels, and NUT channels are separately maintained by the system without any intermixing.

Note that the NUT is very different from extra traffic, which is a part of protection capacity and used only when protection is not in use.

12.10.9 DXC Byte Commands

Although most of the requirement of commands for protection switching is taken care of by the K1 and K2 bytes, some commands are executed through the use of data communication channel bytes. These commands are not time critical, i.e., the action need not be completed within 50 ms, some of them are as follows:

(i) *Clear:* This command clears WTR and all the externally initiated commands on the node.

(ii) *Lockout of protection (all spans):* We have seen earlier that a command generated through K1 byte for LP-S, which prevents the use of the "particular span" for any protection activity. However, there was no command on K1 byte for lockout for protection of the entire ring. This command, i.e., LP-S prevents the protection activities on all spans of the ring and thus on the whole ring. It causes all the traffic to be switched over to the working channel irrespective of the condition of the working channel.

Although the actual implementation of the ring protection will involve many more details, the whole of the working will revolve around the principles and features explained above.

Now it is time for us to have a look at the SNCP protection.

12.11 Subnetwork Connection Protection

Whatever type of protection we use in the SDH network, it will be of no use unless it serves the purpose of providing an alternative route for the customer traffic in case of failures to provide uninterrupted services to him.

The links provided to the customer, or to another telecommunication operator, by the main network owner of an SDH network, are called a "subnetwork." The protection provided to this subnetwork is called subnetwork protection, or SNCP. The subnetwork would be generally at the level of tributaries such as E1 or E3, which will require protection to be provided at the path layer, i.e., VC-12, VC-3, VC-4, etc.

Since the protection has to be provided on an end-to-end path of the sub-network connection, this protection scheme is basically a linear protection scheme. A dedicated unidirectional protection in 1 + 1 format is provided. (Although the scheme envisages a 1:1 protection as well, this is not yet in practice and has been taken up for further studies for defining the mechanism for directional switching that will be involved with 1:1 protection.)

We will examine the details of SNCP with respect to the principles of the working and switching mechanism. We will also have a look at other important features of the mechanism.

12.11.1 Principles of Working

The protection scheme provided is a dedicated 1 + 1 linear protection. At one end of the path, the transmission is permanently made parallel on the working as well as protection paths. The receiver is normally connected to the working path and switches over to the protection path in case of a fault on the working path. Figure 12.27a and b illustrate the principle.

While Figure 12.27a shows only the transmit path, Figure 12.27b shows both transmit and receive paths.

The working and protection channels have been shown in linear path. However, in practical situations, the two channels will have physically different routes, or else they may not actually be able to provide the intended protection, as both of them might get disrupted simultaneously, due to a common fault, say, a breakage in the OFC link. Thus, in practical situations, although the protection is linear, it may be implemented on a physical ring topology. Figure 12.28 shows such a ring.

Similarly, the SNCP protection can be provided on any other physical topology such as linear or mesh topologies.

12.11.2 Switching Mechanism

As discussed above, the transmit position of the path is always connected (bridged) to both, the working as well as the protection channel. In case of a fault, the receiver switches over to the protection path. The switching decision is taken by the receiver based on the local information. No APS protocol is required for the unidirectional 1 + 1 protection scheme.

The switching action is performed at the higher-order path (AU) layer or at the lower-order path (TU) layer. The switching does not involve the multiplex section layer indications.

The switching protocols followed are per ITU-T G.783 (vs. G.841, in case of ring protection).

If the protection switching only supports initiation through "defects," it is called "SNC(I)" or "SNC protection with inherent monitoring."

On the other hand, if the protection switching supports the initiation through "errors" (also called anomalies) in addition to the defects

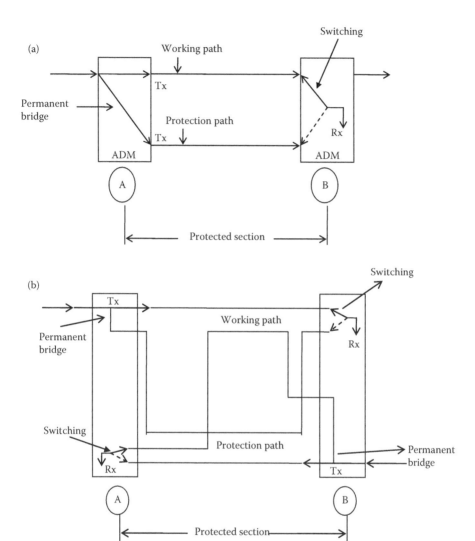

FIGURE 12.27
(a) The 1 + 1 protection scheme between two nodes. Only the transmit path from A to B has been shown for clarity. Similar switching is carried out by the receiver of node A. (b) The 1 + 1 protection scheme showing switching at both the nodes.

(or alarms), it is called "SNC(N)" or "SNC protection with nonintrusive monitoring."

The SNC protection can be used for complete "end-to-end path or termination connection points" or it can be used on a portion of the path. This makes it feasible to provide the end-to-end protection to the subnetwork, which may have intermediate MS-SPRING protections and unprotected multiplex sections in the path.

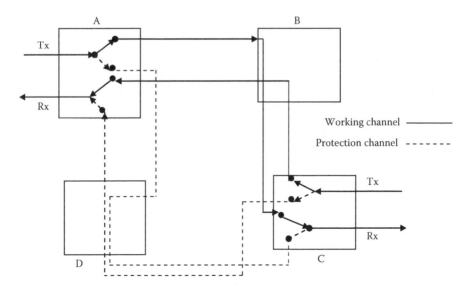

FIGURE 12.28
A 1 + 1 dedicated SNCP protection provided on a ring topology: SNCP ring.

12.11.3 Other Important Features

Some of the other important features of SNCP are as follows:

(i) *Switching time:* A switching time of 50 ms has been set as the target; however, the higher- and lower-order SNC protection should operate as fast as possible.

(ii) *Switching type:* At present only 1 + 1 unidirectional switching is supported. 1:1 bidirectional switching is being studied further.

(iii) *Holdoff time:* When a defect is declared in a VC, a timer starts. This timer is programmed for any period between 0 to 10 s, in steps of 100 ms. Switching takes place on the expiry of this time period. The purpose of this exercise is to initiate the protection of the path layer (SNCP) only if the fault does not get rectified automatically by some intermediate MS-SPRING protection.

In the next section, we will have a look at the comparative summary of the features of each type of protection.

12.12 Comparison of Various Protection Schemes

A comparative summary of the features of various protection schemes is given in Table 12.1.

TABLE 12.1

Comparative Summary of the Features of Various Protection Schemes

Feature	Two-Fiber MS-SPRING	Four-Fiber MS-SPRING	SNCP
Type of protection	Ring	Ring and span	Linear
Type of topology	Ring	Ring	Linear, ring, mesh
Protection layer	Multiplex section	Multiplex section	Path layer (HO and LO)
ITU-T recommendation	G.841	G.841	G.783
Dedicated/shared protection	Shared 1:1	Shared 1:1	Dedicated 1 + 1
External commands	Allowed	Allowed	Allowed
Manual control	Yes	Yes	Yes
Economy	Medium	Most expensive	Economical
Switching time	50 ms	50 ms	50 ms desirable
Holdoff time	Not defined	Not defined	0 to 10 s
Switch initiation commands	MS layer	MS layer	Path layer
Capacity for protection	50%	100%	50%
DXC bytes used for some functions	Yes	Yes	Yes
Possibility of misconnection	Yes	Yes	No
Unidirectional/bidirectional protection	Bidirectional	Bidirectional	Unidirectional

Practical networks are much more complex than what had been taken for illustration in the previous sections. Thus, the protection scheme implemented in practical situations is usually a mix of many of these schemes. Let us have a look at a hypothetical provisioning scenario in a typical network.

12.13 Deployment of Protection in a Network

Whatever scheme is used for protection by a telecommunication operator, it should result in uninterrupted (or close to it) service to the customer. A practical network is made per the demand of the business, which may or may not result in any ring or mesh formation on many locations, while on some other locations, there could be multiple rings. Thus, the end-to-end protection of the customer link may ultimately be a mix of all or some of the protection schemes that we have discussed in this chapter. While some sections of the customer link may be automatically protected due to ring protection existing on multiplex sections, some sections may have to be provided with linear protection.

A typical network shown in Figure 12.29 displays the protection provided on various sections to one of the customers.

In the example shown in Figure 12.29, the working path for the E1, originating at node K and terminating at node B, has been defined via nodes KJHGFECB. But the protection path is a linear protection between node K and node J. After that the protection path is a part of the ring JHGF, which has a MS-SPRING protection; thus, the protection path for the E1 of K node is not defined separately and the path protection will get automatically provided with MS-SPRING. Again, path GF is unprotected, path FEC is protected again by another MS-SPRING, i.e., CDFE, and finally, the portion CB is also unprotected. The overall protection provided to this link is at the path level and will be categorized as SNCP.

There could be several variations to this network depending upon the geographical reach of the optical fiber links. We have taken a customer whose subnetwork is from nodes K to B, which has unprotected sections. If the other end of the customer's link was at C, he would have had only one unprotected section and so on. In actual practice, it is generally possible to provide complete end-to-end protected paths to most of the customers, within the reach of the network, if due care is taken while designing the links.

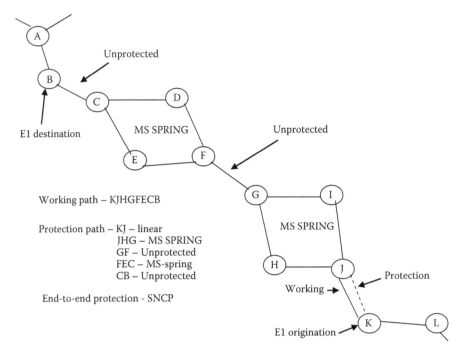

FIGURE 12.29
A typical deployment network with protection.

Review Questions

1. What is resilience in the context of SDH? What are the major factors responsible for disruptions in the network?
2. What kind of extra system capacities are required to provide resilience in the network?
3. Name the SDH network elements and explain their functioning briefly.
4. What is the difference between a TM and an REG?
5. Differentiate between lower- and higher-order cross connects. Which one is usually present in an SDH node?
6. What do you understand about network topologies of SDH? Which are the common topologies? Enumerate and explain with figures.
7. Which topologies can offer backup paths and why?
8. What are work and protect paths? What are protection path, protection channel, and protection circuit?
9. How do the rings get formed in service areas? What are the advantages of ring topology for the purpose of protection? How is the work and protect traffic routed?
10. What are the protection schemes recommended by ITU-T? Why are they required?
11. Explain the 1 + 1 protection scheme with the help of a schematic. What is the spare capacity required for this type of protection?
12. What are the differences between the 1 + 1 and 1:1 schemes of protection? What is extra traffic?
13. Distinguish between head switched and tail switched protections.
14. What are the advantages/pitfalls of a 1:N protection scheme? Show the switching scheme in a diagram.
15. What is APS? What do you understand about SHRs?
16. Show with the help of a chart the types of APS mechanisms in use in SDH.
17. What is a MS dedicated ring? Explain its working. What are its limitations?
18. What is MS SPRING? It works on how many fibers?
19. Explain the functioning and protection switching arrangement of two- and four-fiber MS protection rings.
20. What is the difference between ring switching and span switching? Compare their salient features.

21. What types of switching arrangements are provided for manual and automatic switching in the commands?

22. What is APS controller? Explain its working with the help of a block diagram.

23. How does protection switching take place with the help of K1 and K2 bytes? Explain the case of a two-fiber ring.

24. What is WTR? Why it is required?

25. Draw a flow diagram showing various activities and the values of K1 and K2 bytes during the process of protection switching to the protection path and explain the switching process.

26. What is a reverse request? What happens after a cable fault is rectified? Explain the process of resumption of normal working on the work path.

27. List the assignments of K1 and K2 bytes for APS.

28. What is the purpose of LP-S and SF-P commands? What is the difference between them?

29. Discuss SF-R and SF-S commands briefly and how they differ from FS-R and FS-S.

30. What is the application of higher or lower priorities in the APS commands?

31. What are EXER-R and EXER-S commands? Why are they required?

32. Explain the terms head end, tail end, bridging, and switching.

33. Why are the source and destination IDs of protection rings only 4-bits long?

34. When is the extra traffic carried by the channel and how is it indicted through K1 byte?

35. What are the terms "bridged and switched," "bridged," and "idle"?

36. Switching time can exceed 50 ms. Explain when and how.

37. What is the switching initiation time and what is the type of switching allowed in APS?

38. What is operation mode? What is meant by revertive and non-revertive modes?

39. Describe the switching protocol followed for APS.

40. Explain the process of misconnection through an example.

41. What is squelching? How does it help in avoiding misconnection?

42. What is NUT? How is it different from extra traffic?

43. Which are the commands executed through DXC bytes? Discuss briefly.

44. What is called a subnetwork and what is SNCP?

45. Explain the principles of SNCP with the help of sketches wherever necessary.

46. Describe the switching mechanism of SNCP. At what layer it is provided and what is the role of APS? What are the terms SNC(I) and SNC(N).

47. What are switching time, switching type, and holdoff time in the context of SNCP?

48. Compare the salient features of two-fiber MS SPRING, four-fiber MS SPRING, and SNCP.

49. How is the protection provided in practical networks? Explain with a hypothetical example. What type of protection is it?

Critical Thinking Questions

1. What is the most predominant media for SDH and why? What will be the effect on capacity and cost if the media is other than OFC?

2. Of the above, which type of protection scheme will be best suited to which type of application and why?

3. If we provide manual switchover of the services to the standby in case of disruption, what do you think about the capacity that should be kept reserved for being able to provide to all the customers at least 95% of the time? On what factors will this capacity depend?

4. Tabulate the comparative pros and cons of linear, ring, and mesh topologies.

5. The protection through APS could be revertive or nonrevertive— which one would you prefer and why?

6. What are the differences in the mechanism and features in the protection provided by SDH and RPR?

7. Compare the pros and cons of RPR provided on SDH with that of pure data RPR.

Bibliography

1. *Optical Transmission in Communication Networks*, Web pro-forum tutorials, International Engineering Consortium, 1998.
2. G. Saundra Pandian, *IP and IP-Over SDH*, Training course, Railtel Corporation of India, 2003.
3. *Broadband Communications and SDH*, Training material, Tejas Networks India, 2002.

4. *Introduction to Synchronous Systems*, Training Material, M/S Alcatel, 1994.
5. ITU-T Recommendation G.841, *Types and Characteristics of SDH Network Protection Architectures*, International Telecommunication Union, Geneva, Switzerland.
6. ITU-T Recommendation G.842, *Interworking of SDH Network Protection Architectures*, International Telecommunication Union, Geneva, Switzerland.
7. ITU-T Recommendation G.803, *Architecture of Transport Networks Based on the Synchronous Digital Hierarchy (SDH)*, International Telecommunication Union, Geneva, Switzerland.
8. ITU-T Recommendation G.783, *Characteristics of Synchronous Digital Hierarchy (SDH) Equipment Functional Blocks*, International Telecommunication Union, Geneva, Switzerland.

13

Data Over SDH

The telecommunication started with voice or speech of one person reaching another person at a distance that was beyond the range of the direct sound. The point-to-point local telephone was invented by Alexander Graham Bell in 1876. The need to talk to more people gave birth to manual telephone exchange, and subsequently, to automatic telephone exchanges, and our need to talk to the people in other cities gave birth to the long-distance communication.

Long-distance communication technologies have since made a long journey, from carrier communication of a few channels to the enormous capacity SDH networks. Although the media has gone through vast changes, beginning with copper cables, overhead aluminum conductor steel–reinforced wires, coaxial cables, microwaves radios, and finally reaching the optical fiber, the multiplexing technologies made a long journey as well, beginning with the analog FDM (frequency division multiplexing), PCM (24-channel digital) to PDH, and finally to the current SDH.

These developments facilitated people from one corner of the world to talk to those in another corner of the world at ease and at affordable costs. Although the speech quality for longer distances was poor in the initial analog days, it too improved greatly with the advent of the digital technology. In fact, the voice quality of the intercontinental calls today is almost the same as that of the local calls.

The focus of the telecommunication technology was, therefore, to improve upon the voice services in quantity and quality. Thus, all the telecommunication technologies of today, whether they are of switching or transmission, have been designed and optimized for the voice telephone traffic.

The telecommunication engineers were satisfied with the situation, until the computers arrived on the scene. The stand-alone computers did not need any telecommunication but needed to economically use the computers; another set of engineers invented technologies that facilitated the mainframe computer to be shared by a group of users. This development did not make much of a difference either to the health of the telecommunication engineers, but the computer users did not stop on just sharing the processing power of the mainframe computers, but also they wanted the computers to be able to "talk" to each other, to share information. This requirement promoted the development of "local area network," or LAN, as we know it today. When various businesses, institutes, and governments saw the power of the LANs, they wanted to share and use the information among LANs of their branches spread over long distances and in other cities. Since the computers talk to each other in digital language called "data," there was a need for long-distance "data communication."

Meanwhile, the world saw the development of another phenomenon in the information technology called "Internet." Another popular name given to Internet is "world wide web" signifying its worldwide reach. Internet is a data communication technology facilitating information exchange among computers spread around the world.

These developments forced the need to transport data from computers and LANs to distant computers and networks. However, as we have seen in previous chapters, the entire focus of the development of long-distance communication had been "voice," not only in analog era, but also in "digital era." There would not have been any problems if the design philosophies and the requirement of voice and data applications matched, but unfortunately, neither the design philosophy nor the functional requirement of the two matches each other.

The developments in the field of information transmission were so exciting that it was not possible for the engineers to wait and evolve an entirely new system for data communication over long distances, which led to the natural choice of "somehow using the existing long-distance infrastructure of voice telephony for the purpose of data communication."

Data has always been digital; thus, in the age of analog telephone lines, digital data was converted into analog form by devices called modems, and sent over the distances (for example, fax). The digital telephone technology of today is better suited to data communication as compared with analog lines, but there are still a large number of problems because, as mentioned earlier, telephone lines are only optimized for voice transmission.

A look at the prevailing telecommunication network of today reveals that almost the whole of the long-distance transmission deployment is on PDH or SDH technologies. Thus, we have no option but to carry the data traffic on this infrastructure. The problems and issues involved in transporting data over this infrastructure have long been the subject of global research of various telecommunications standardization bodies, resulting in a substantial number of standards for interworking of the two technologies. Currently, almost all the data traffic is being transported on the PDH or SDH (SONET in North America and Japan) infrastructure throughout and across the world.

In this chapter, we will discuss the problems of interfacing these two technologies, the solutions that evolved over time, and their limitations and shortcomings.

13.1 Problems in Interfacing Data and SDH

13.1.1 Difference in the Bit-Stream Structure

The bit-stream structures of data communication systems are widely different from those of SDH. The SDH consists of voice "channels," which are of fixed bandwidth, i.e., 64 kbps each. These channels are then multiplexed to form E1s (30 channels + 2 signaling channels), then 63 E1s combine to form an STM-1 transport module. These details have been discussed in detail in Chapter 10 on SDH principles.

On the other hand, the data bit stream consists of "packets" (or frames, when added with header, etc.). These packets in some systems are of fixed length (like in asynchronous transfer mode [ATM], they are 53 bytes long), and in some other systems, they are of variable lengths (like in Ethernet, they range from a few bytes to nearly 1500 bytes long). (ATM is one of the most popular data communication technologies; Ethernet is the most popular protocol for LANs.)

Thus, some mechanism had to be evolved which could allow the interface of these two different bits-stream structures to interoperate without any loss of user information.

13.1.2 Difference in Signaling (Protocol)

As is obvious from the history of development of either of these technologies, the signaling in SDH is very different from that of data communication.

SDH has a hierarchical signaling structure (E1s, E3s, etc.), tributaries have their own signaling (2 of 32 channels in an E1 are reserved for signaling and the rest, 30, carry user information [voice]), and when they are converted into containers, virtual containers (VCs), tributary units, administrative units, and finally STM-1s, they add signaling information called "overheads," at each and every stage. This signaling helps in proper multiplexing, de-multiplexing (adding and dropping), and management of tributaries (we have discussed this in detail in Chapter 10).

Again, the signaling in data communication, which is popularly known as "protocols," is very different from that of SDH. The data packets carry their own signaling information and are called frames. Each frame has some bytes allocated to signaling and some bytes to user data. The Ethernet frame is shown in Figure 13.1, for example.

This frame forms a basic unit of the bit stream and is the "only unit" of the bit stream; there is no further structure of signaling protocols. The first five and the seventh field carry the protocol information for successfully delivering the packet to the desired destination, whereas the sixth field carries the user data. (Figure 13.1 has been shown here only to emphasize the difference in the signaling structure of the data communication as compared with that

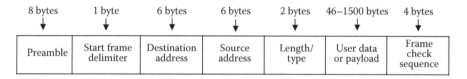

8 bytes	1 byte	6 bytes	6 bytes	2 bytes	46–1500 bytes	4 bytes
Preamble	Start frame delimiter	Destination address	Source address	Length/ type	User data or payload	Frame check sequence

FIGURE 13.1
Signaling protocol of an Ethernet frame.

of SDH; detailed discussion about its functioning is out of the scope of this work. The interested readers can find a good coverage in any text on networking or data communication.)

13.1.3 Difference in Throughput Rate

Although, as seen above, the data packets or frames could be of any length, ranging from a few bytes to 1500 bytes or so, their frequency of occurrence is also unpredictable. At any give time, there may be nil packets or there may be packets present at the data rate of 10 Mbps for Ethernet (or 100 Mbps, 1 Gbps, etc., depending on the system in use). Thus, the bit rate of Ethernet could vary from 0 to 10 Mbps.

In contrast, the bit rates for SDH are fixed, being different for different tributaries. For example, an E1 tributary has a bit rate of 2.048 Mbps, E3 has a bit rate of 34.368 Mbps, T1 has a bit rate of 1.544 Mbps, T2 has a bit rate of 6.312 Mbps, and so on (details on these have been discussed in Chapter 10).

Again, some mechanism for adaptation of these variable bit rates had to be developed before the data could be carried over the SDH systems.

As of today, the data is very much being carried over the SDH, as there is no other way to carry the data to long distances. This is how the Internet and the data applications are working. It is thus clear that the answers to the problems discussed above have been found and implemented. To solve the problem of different signaling/protocols, the data is being carried as payload in SDH, and to solve the problems of mismatch between the structured data streams of SDH and the continuous data streams of data communications, a procedure called "concatenation" is used. Let us have a look at them.

13.2 Data as Payload

The difference in signaling schemes or a protocol as brought out above in Section 13.1.2 is a very complex issue to resolve. A large variety of such protocols are used in data communication such as ATM, Ethernet, FDDI, HDLC, PPP, and many more, making the issue more complex. Thus, instead of breaking their heads to find a common signaling scheme and converting

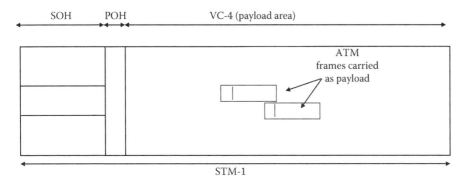

FIGURE 13.2
SDH carries the data communication frames as payload.

all data and SDH protocols into it, the engineers hit upon a very simple solutions, and that is to keep the SDH signaling as it is and treat all other data communication frames as "payloads" within SDH tributaries. The signaling or framing bytes and the user data bytes can together be treated as payload within the SDH tributaries and carried to the destination. At the destination, the frames are delivered to the respective data communication equipment, where the data communication equipment can process them as if they have been received from a neighboring data communication node. The SDH thus becomes a completely transparent carrier for the whole set of the data communication protocols. Figure 13.2 shows an STM-N frame, with ATM frames as payload.

This process solves our problems of handling different protocols for different types of data communication systems. The other two problems of different bit structure and different throughput are still unresolved. These problems are solved by a process known as concatenation. Let us see what this process of concatenation is.

13.3 Concatenation

Concatenation basically means "grouping." In our case of SDH, it applies to grouping of tributaries. To appreciate the need and application, let us consider an example of a 10 base T LAN, whose data rate is 10 Mbps (10 represents 10 Mbps; base T, a twisted pair cable media; LAN, local area network). The data generated by this LAN is to be carried over the SDH to a distant location, where it has to be delivered to another similar LAN. The data frames can be loaded into 5 E1s as payload. All the 5 E1s can be connected to an STM-1, and through STM-N, they can be transported to the other end. Figure 13.3 shows a schematic arrangement.

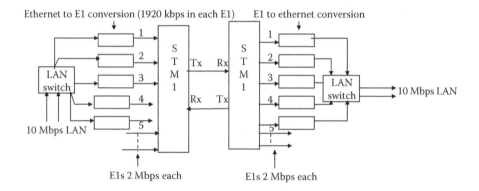

FIGURE 13.3
A 10-Mbps LAN traffic being carried on five separate E1s.

The diagram is simplified for demonstrating the principle. The Ethernet frames are actually "wrapped" into LAPS ("link access procedure," which is a protocol for loading Ethernet frames onto SDH). (In the field of data communication, the data is divided into packets of fixed or variable size, depending upon the protocol used. The data packet is then combined with a set of predefined bit pattern known as "header," which carries the information about the destination to which the packet is to be sent and some more information required for correct reception of the packet. The structure so formed is called a frame. Figure 13.1 shows an Ethernet frame with many fields in addition to data. Now if this Ethernet frame is to be transported on some other type of transmission backbone such as SDH, the whole of the Ethernet frame is treated as a payload by the new protocol, say LAP. The LAP adds a new header to the Ethernet frame, treating the Ethernet frame as a packet. At the destination, the LAP header is removed and the Ethernet frame is recovered in its original form, which can be transported on an Ethernet media. This procedure of treating the frames as packets or payload is known as "encapsulation" or "wrapping" and is the heart and soul of data communication. LAP is discussed in a later section in this chapter.)

As can be seen in Figure 13.3, the 10-Mbps LAN traffic has to be converted to an E1 payload and loaded onto 5 E1s, each of which carries 2-Mbps data. All the 5 E1s are then transported to the other end, where they are again combined into a 10-Mbps LAN after the E1s deliver them as payloads. Thus, the data has to be first broken into five different 2-Mbps streams and again combined at the other end. It would be much better if the data could be sent through a common bit stream of 10 Mbps, which would have automatically maintained the data sequence. Thus, if these 5 E1s could be grouped together to form a single data stream, the distribution of data at the sending end and again combining it at the receiving end would not be required. It will also avoid five separate E1 to Ethernet and Ethernet to E1 conversions at either end, and hence, the hardware cost will be reduced.

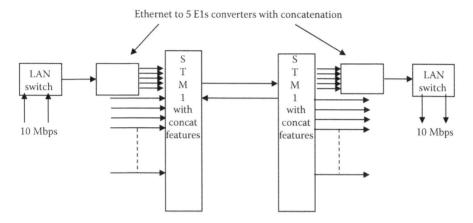

FIGURE 13.4
A 10-Mbps LAN traffic being carried on five concatenated E1s.

A mechanism for such grouping has therefore evolved, which is called concatenation. After concatenation, the scheme of Figure 13.3 would look like the one shown in Figure 13.4.

In this process of concatenation, the E1s (or any other size tributaries), which are contiguous to each other, are grouped or concatenated together, and hence, this is called "contiguous concatenation." We will see the procedure and some examples of contiguous concatenation in the next section.

13.4 Contiguous Concatenation

In our discussion in the previous section, we have taken the tributaries, such as E1, T1, E3, for the purposes of concatenation to enable easy understanding. However, in actual practice, the concatenation is done at the VC level. We have seen in detail in Sections 10.6 and 10.8 the formation of containers and virtual containers from the asynchronous tributaries. The asynchronous tributary (E1, T1, E3, etc.) is brought to a constant bit rate level by adding some justification and control bits. The unit formed is called a container. The container is added with path overhead bytes and is called a VC. The VC, when added to a pointer (tributary unit/administrative unit, or TU/AU), is called a tributary unit or administrative units. Figure 13.5 shows this process.

There are VCs of lower and higher levels pertaining to respective tributaries. The concatenation can be done at any of these levels, i.e., VC-4, VC-3, VC-2, VC-12, or VC-11, whenever the payload capacity required is more than the capacity of any of these containers.

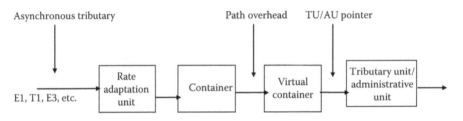

FIGURE 13.5
Process of forming container, virtual container, TU, and AU in SDH.

Let us take an example where the data communication payload capacity required is more than VC-4, and it is decided to concatenate a number of VC-4s, say "X" numbers.

The data communication payload is carried in the containers of the VC-4, whereas the "POH" portion of the VC-4s is filled with stuff bytes (which are discarded at the receiving end) except that of the first "VC-4." The AU-4 pointer, which points to the location of the first byte J1 of the VC-4, is retained, as it is for the first AU-4, whereas the rest of the AU-4's pointers are filled with bits, which indicate that this VC-4 has been concatenated. The concatenated payload structure so formed is called VC-4-X_c, where X is the number of VC-4s that have been concatenated and subscript c indicates "contiguous" concatenation. Figure 13.6 shows the arrangement.

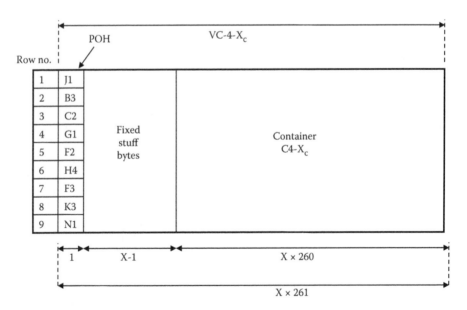

FIGURE 13.6
Contiguous concatenation of VC-4s, where X = 4, 16, 64, or 256.

The payload capacity of this concatenated container is X × 260 × 9 bytes. The value of X could be 4, 16, 64, or 256, depending upon the value of N in STM-N (N = 4, 16, 64, or 256). The pointer values of the concatenated AU-4s are shown below.

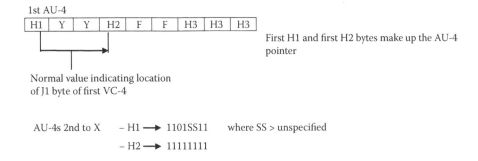

1st AU-4

| H1 | Y | Y | H2 | F | F | H3 | H3 | H3 |

First H1 and first H2 bytes make up the AU-4 pointer

Normal value indicating location of J1 byte of first VC-4

AU-4s 2nd to X   – H1 ⟶ 1101SS11   where SS > unspecified

                – H2 ⟶ 11111111

STM-1 has only one VC-4, STM-4 has 4-VC-4s, and STM-16 has 16 VC-4s. The respective data communication capacities of these concatenations will be (maximum capacities if X = N)

$$\text{STM-4 :– VC-4-4c} = 599.040 \text{ Mbps } (4 \times 260 \times 9 \times 8 \times 8000)$$

$$\text{STM-16 :– VC-4-16c} = 2396.160 \text{ Mbps } (16 \times 260 \times 9 \times 8 \times 8000)$$

The concatenation indication tells the receiver that this particular VC-4 has been concatenated, and thus it should not look for J1 byte, but continue to receive the bit stream. After reception of the bits, the knowledge of locations of stuff bytes is used in discarding them.

Thus, treating the data communication bit stream as "payload" and using "concatenation" allows us to use SDH for the transport of data communication services.

The process of arranging the data communication frames into the SDH payload area, in a systematic manner, is called "mapping." The international Telecommunication Union, Telecom Standards group (ITU-T) has defined the mapping of various types of data signals into SDH in their recommendation G.707. We will have a look at some of them in the next few sections to develop our understanding further.

13.5 Mapping of ATM Frames

ATM is one of the most popular data communication protocols. An ATM frame (also called a "cell"; in fact, cell is a more popular name in the case of

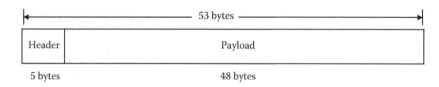

FIGURE 13.7
An ATM cell.

ATM instead of a frame) consists of 53 bytes. Of 53 bytes, 5 bytes are over-head bytes and 48 bytes are payload or data bytes. The ATM cell is shown in Figure 13.7.

Since the ATM cell is only 53 bytes long, it can be mapped into any size of payload of SDH, i.e., VC-12, VC-11, VC-2, VC-3, or VC-4. We will see here an example of mapping in two types of VCs, one in which the VC area is not an integer multiple of cell size and another in which it is.

13.5.1 ATM Mapping in VC-4

The ATM cells are arranged from the beginning of the payload area, one after the other, until the full payload area of VC-4, i.e., C-4 is filled (Figure 13.8).

Since the ATM cell size is 53 bytes and the VC-4 payload area size is $260 \times 9 = 2360$ bytes, the VC-4 is not an integer multiple of ATM cells. Thus, the boundary at the end of VC-4 will not align with the boundary of the

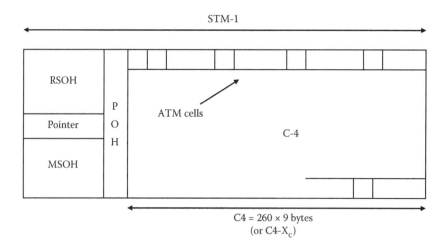

FIGURE 13.8
Mapping of ATM cells into VC-4 or VC-4-X_c.

ATM cell. A part of the last ATM cells is thus allowed to get into the next VC-4, or in other words, it is allowed to cross the boundary of the container VC-4.

Although the example shown above is for a single VC-4, similar mapping will be done for a concatenated VC-4-X_c. The mapping in the case of VC-3 or VC-3-X_c will also be similar.

13.5.2 ATM Mapping in VC-2

The process of mapping the ATM cells into VC-2 remains the same as that used for VC-4. The VC-2 structure is a four frame multiframe of 428 bytes, of which 4 bytes are overhead bytes and the remaining 424 bytes are data bytes. The mapping is shown in Figure 13.9.

We can see that each individual frame of the multiframe has an area of 106 bytes as the payload area. It is a great coincidence that it can accommodate just two ATM cells of 53 bytes each, completely aligning the boundaries at the beginning and at the end.

Similar results will be produced for mapping of ATM cells into VC-2-X_c, i.e., concatenated VC-2 containers as well. The only difference being that the POH of all the VC-2s, except the first one, will be filled with stuff bytes and the pointers will carry the concatenation indication as explained in Section 13.4.

The ATM mapping can be done in all other types of tributaries of SDH in a manner similar to the above two methods.

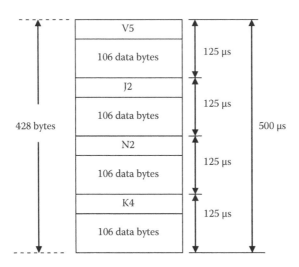

FIGURE 13.9
Mapping of ATM cells into VC-2.

13.6 Mapping of HDLC, PPP, Ethernet, IP, and LAPS

HDLC (high-level data link control) is a data communication protocol for data link layer which was developed by the International Standard Organization (ISO).

ITU-T modified HDLC and called it link access procedure (LAP). LAP was subsequently modified to LAP-B and finally to LAP for SDH. Point-to-point protocol (PPP) is a protocol that was especially developed for the Internet. Ethernet and IP are carried over one of these protocols.

All these protocols, i.e., HDLC, PPP, and LAPS, have a similar format, with some differences from one another, incorporated by different standardization organizations.

However, as far as the mapping over SDH is concerned, their exact shape and size and the contents do not matter; they have to be carried over the SDH as a payload. Figures 13.10a and 13.10b show the format of HDLC and LAPS frames, respectively, for example.

A detailed explanation of the functionality of these protocols is out of the scope of this text, but that knowledge is not necessary for understanding of the concept being dealt with.

The Ethernet and Internet Protocol (IP) are carried by one of these protocols. Figure 13.11 shows an Ethernet frame being carried over a LAPS frame.

As can be seen in Figure 13.11, the whole of the Ethernet frame is accommodated in the "data" area of a LAPS frame, except the "preamble" and "start of frame" bytes, which are discarded, as these functions are performed by SDH and LAPS. These bytes are again added at the receiver end to form proper Ethernet frames.

The data over SDH is also popularly known as packet over SONET (POS). Another name to the process applicable for Ethernet is Ethernet over SDH or SONET (EOS).

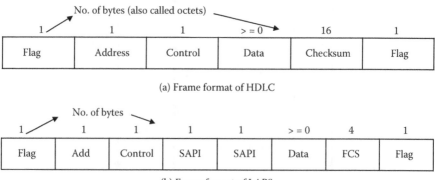

(a) Frame format of HDLC

(b) Frame format of LAPS

FIGURE 13.10
Frame format of (a) HDLC and (b) LAPS.

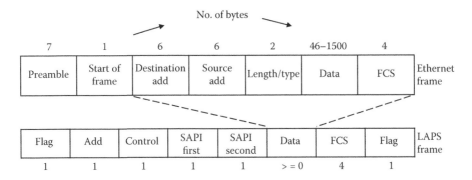

FIGURE 13.11
Ethernet frame being carried over a LAPS frame.

Now, coming back to the mapping of these frames over SDH, they are simply filled in the VCs or VC-X$_c$s, aligning the first frame with the beginning of the VC payload area. Since the total length of these frames is variable (the data field could be an arbitrary length for HDLC frames, whereas it could be 46–1500 bytes for LAPS frames; see Figure 13.11), their boundaries are not likely to match with the boundaries of VCs. Hence, the frames are allowed to cross the VC payload boundaries, with a part of the frame being loaded in subsequent containers. They can be mapped into any of the VCs, such as VC-4, VC-3, VC-2, VC-12, and VC-11, depending upon the requirement of the throughput. The mapping can again be either in a single container or in concatenated containers depending upon the requirement.

The developments so far discussed in this chapter facilitated the much-needed transport of the data communication bit stream over long distances. However, these procedures have some limitations and shortcomings, which are the subject of the next section.

13.7 Shortcomings of Data over SDH

When the data over SDH was made possible, it was a wonderful achievement, but as the time passed and we got used to it, we started realizing that things could be better than what they are. A number of shortcomings started surfacing, which limited the capacity and efficiency of the system. The important ones are as follows.

13.7.1 Requirement of Concatenation Feature

The concatenation feature is required to be available at all the nodes throughout the link on which the facility is required. This makes the deployment of concatenated services very difficult on the existing infrastructure, which may not have this feature.

13.7.2 Inefficient Capacity Utilization

The capacity utilization of the SDH, when used for data communication, remains poor, which results in poor overall circuit efficiency. For example, if a 100-Mbps traffic (100 base T Ethernet) is to be carried over the SDH, the nearest suitable capacity is approximately 150 Mbps in the container C-4 in an STM-1, leaving about 50 Mbps of the data capacity unutilized, leading to an efficiency of 66%. If a 1000-Mbps (GB Ethernet) is to be transported, the nearest SDH level is 2.5 Gbps (STM-4), resulting in only 40% or so efficiency.

13.7.3 Stuffing Bytes Requirement

Protocols such as HDLC, PPP, LAPS, etc. have a frame marker or flag with a value 011111110 in their frames. This bit pattern can occur in the data bit stream as well, in which case the system may consider the frame to begin at a wrong location. To avoid this situation, such patterns in the data stream are replaced by "stuff" bytes, which are discarded at the receiver. Sometimes, the number of such bytes could be very large, giving rise to poor efficiency.

13.7.4 Handling of Multiple Protocols

The SDH equipment has to handle a large number of existing data communication protocols such as ATM, HDLC, PPP, LAPS, FDDI, and many more, which increases the complexity of the equipment.

We have seen a brief description of the shortcomings above. All these shortcomings have been overcome by new developments in the field of data over SDH. So much so that the new mechanisms developed to tackle these problems have given rise to a new name to the SDH that is "next-generation SDH." We will discuss all these problems and their solutions in our next chapter on this subject of "Emerging Systems and Future of SDH."

Review Questions

1. How did the need of data communication arise?
2. Were there any problems in long-distance data communication on the telephone lines? If yes, please elaborate.
3. What had been the developments in the media and technologies since the invention of the telephone?

4. List the problems in interfacing data with SDH.

5. Bring out the difference in the bit-stream structures of data as compared with SDH.

6. What are the differences in signaling in the case of data communication and SDH; illustrate with an example.

7. How do the throughput rates of SDH and data communication differ?

8. Can data packets be carried over SDH as payload? Explain with an illustration.

9. Give an illustration of a 10-Mbps LAN being carried over 5 E1s on SDH.

10. Explain in brief the terms *packet, frame, encapsulation*, and *wrapping*.

11. What do you understand about concatenation? What are its advantages in the given example of carrying 10-Mbps data on 5 E1s?

12. What is contiguous concatenation? Illustrate the contiguous concatenation of VC-4s.

13. What is a concatenation indication? What happens to the POH of various VC-4s in case of contiguous concatenation?

14. Work out the data carrying capacities of STM-4, STM-16, and STM-64 in the case of contiguous concatenation of VC-4s.

15. What is called mapping with reference to carrying data over SDH? Illustrate the mapping of ATM cells into VC-4 and VC-2.

16. What do you understand about POS and EOS?

17. Discuss various aspects of mapping of Ethernet and LAPS protocols into VCs.

18. Discuss in brief the shortcomings of SDH for data communication.

Critical Thinking Questions

1. What problems could occur due to the above differences in data communication systems and SDH when we wish to carry data over SDH?

2. How can these shortcomings be overcome? Please try to work out before reading the next chapter.

3. Historically speaking, what came first, data communication or voice communication? When and in what form?

4. Show the mapping of LAPS frames in VC-4.

Bibliography

1. W. Stalllings, *Data and Computer Communications*, Seventh Edition, Prentice-Hall India, New Delhi, 2003.
2. A.S. Tanenbaum, *Computer Networks*, Fourth Edition, Pearson Education, 2003.
3. D.E. Comer, *Computer Networks and Internets*, Second Edition, Pearson Education Asia, 2000.
4. J. John and S.K. Aggarwal, *Proceedings of the International Conference and Exposition on Communications and Computing IIT Kanpur*, February 2005.
5. P.V. Shreekanth, *Digital Transmission Hierarchies and Networks*, University Press, 2010.
6. *Introduction to Synchronous Systems*, Training material, M/S Alcatel, 1994.
7. G. Saundra Pandian, *IP and IP-Over SDH*, Training course, Railtel Corporation, India, 2003.
8. ITU-T Recommendation G.707/Y.1322 (2007) Amd-1, *Network Node Interface for the Synchronous Digital Hierarchy (SDH)*, International Telecommunication Union, Geneva, Switzerland.
9. ITU-T Recommendation G.7041/Y.1303 (10/2008), *Generic Framing Procedure (GFP)*, International Telecommunication Union, Geneva, Switzerland.
10. ITU-T Recommendation X.85, *IP Over SDH Using LAPS*, International Telecommunication Union, Geneva, Switzerland.
11. ITU-T Recommendation X.86, *Ethernet Over LAPS*, International Telecommunication Union, Geneva, Switzerland.

14

Emerging Systems and the Future of SDH

When the demand for long-distance data traffic arose, it was carried on the available telecommunication backbone, which was designed and optimized for voice traffic. This backbone was initially built over analog lines on copper pairs, coaxial cables, microwaves, etc. It initially migrated to the digital technology without any change in the media with PDH multiplexing and transmission. The backbone finally switched over to the SDH multiplexing with optical transmission of lasers on optical fiber cables. With the huge capacity of optical fibers, the advancement of technology and scales of economy with high demand, the cost of equipment continued to slide, which further boosted the demand. Data communications proved excellent in solving complex social problems; hence, a substantial demand for additional telecommunication traffic came from the data segment.

With this growing demand of data traffic, the efficiency of the backbone telecommunication carrier came under a scanner, and the once-"wonderful" SDH system was faced with a number of allegations. All these allegations about the shortcomings of SDH systems were true. We have discussed these shortcomings of the SDH system with respect to data communications in the last section of the previous chapter in brief. All these shortcomings are basically pointers of poor efficiency of SDH for data communication. The SDH equipment, being precise and robust, also started loosing on cost comparison with their competitor, the data communication equipment.

This scenario started pushing for a case for a pure data backbone (or a pure Internet Protocol [IP] backbone, as the bulk of the data traffic today is on IP), which, in addition to providing an efficient backbone, would also eliminate the conversion equipment from data to SDH and vice versa at the interface points, further reducing the overall operational cost.

It was, thus, time for SDH to fight back. The standards developing agencies along with vendors started working toward making the SDH more data-friendly, so as to make it an efficient data transport backbone. The driving forces behind this were the large-scale deployment of SDH by operators, which was already in place, and the manufacturing facilities of vendors, which could not be replaced overnight, thus sinking the huge investments. All the shortcomings of the SDH have since been overcome, and the SDH so reformed is called the "next-generation SDH."

The war over the "Ethernet or SDH" backbone is, however, far from over. New developments in pure IP-based carrier/backbone services, and further enhancements in SDH capabilities, are still keeping the experts guessing on the issue of the final technology that will prevail in the future. Developments such as the resilient packet ring (RPR) and optical transport networks (OTN) have added more fuel to the fire. It appears that the "time," which is the final master, will only be able to answer this question. However, looking at the present scenario in which the SDH (next generation) is being widely deployed currently, we can safely infer that the NG-SDH is going to stay for at least 10 to 15 years more. This period of 10 to 15 years is required (a) for the operators to get proper returns on their investments, which they are making today, including the replacement of the equipment when the capacity gets exhausted, and (b) the trend is likely to continue for at least a few more years, until the alternatives are ready for large-scale deployment.

In this chapter, we will discuss the case for the Ethernet backbone and the challenge of SDH in the form of next-generation SDH. In the NGSDH, the technological improvements which have really made it the next-generation SDH such as virtual concatenations (V-CAT), the link capacity adjustment scheme (LCAS), and the generic framing procedure (GFP) will be introduced. We will also discuss in brief the deployment modes of NGSDH such as the multiservices provisioning platform (MSPP), the multiservices switching platform (MSSP), etc. The RPR and some other emerging technologies such as OTN and others will also be dealt with in brief. Finally, we will have a comprehensive look at the overall scenario.

14.1 Case for Ethernet Backbone

The phenomenal growth of the data communication traffic has been almost synonymous with the popularity of the IP. The majority of applications today use IP as the choicest protocol. A little discussion about the data communication applications will be worthwhile here.

On the enterprise front, the data communication has found tremendous success on account of the success of applications running with the networking of the computers. The glowing examples are airline and railways reservations, banking transactions and credit/debit card applications, and sharing of customer database and resources information by all the offices and staff of a company whose offices are spread worldwide. The enterprises mostly work on leased lines or virtual private networks (VPNs), on which they find it very convenient and economical to use voice-over IP (VOIP) services as well. Video conferencing is another application that added to the smoother functioning of the enterprise. The enterprises are, therefore, inclined to go for high bandwidth enterprise systems communication (ESCON). Fiber channel is another important application for enterprises for their storage area networks (SANs).

The internet is a revolutionary application of data communication, which has totally transformed the way the world works. From executives to students and from housewives to senior citizens, nobody remains untouched from the Internet and its benefits. The users of the Internet are continually growing, as well as their demand for higher and higher bandwidth. In the beginning, even a 64-kbps line was enough to run a cyber-kiosk with 8–10 computer terminals, but now even the individual customers are not happy with 256-kbps or even 2-Mbps speeds. The moment the speed is jacked up, the aspirations take a jump as well. While the initial applications were limited to e-mail, file sharing, etc., current applications include online gaming, video chatting, streaming video, etc., which are highly bandwidth-hungry applications.

Another application which is catching up very fast is IP-TV. The TV broadcast, which was through satellite and cable until now, is being switched over to DTH (direct to home) from the satellite and to the PSTN subscriber lines with the use of DSL technology. The minimum requirement of bandwidth per user is nearly 20 Mbps for this application.

Social networking has added another dimension to Internet usage. Individuals not only download information but also upload sizable amount of data to their profiles, which include pictures and videos. This has to some extent balanced the skewed upload-to-download ratio of the Internet.

Looking into all these data-centric applications, a natural question keeps popping up in the minds of the engineers: why transport the data on a system that is designed for voice and thus is inefficient for data communication? Why not have a whole IP backbone for carrying data traffic more efficiently and probably with fewer complexities?

The factors that go in favor of IP backbone are

(i) A high percentage of telecommunication traffic today is data (IP-centric).

(ii) The data communication devices are cheaper than those of SDH.

(iii) The networkwide clock synchronization is not required, thus timing requirements are less stringent.

(iv) Conversion from data to SDH and back will not be required. The integration with LANs will be much smoother.

(v) Even voice can be transported as data (VOIP).

(vi) The efficiency is much better than SDH because the backbone is optimized for data.

(vii) The growth is giving rise to further growth, bringing down the cost of the equipment because of large scale production.

(viii) Provisioning of new links and configurations become simple.

With so many factors in favor of IP backbone, let us see whether the SDH can stand against the odds.

14.2 SDH's Fight

The factors that go in favor of SDH and against the IP backbone are

 (i) SDH is already widely deployed and, currently, it is the only media available for almost all of the backbone traffic.
 (ii) TDM voice traffic still makes a significant and most important portion of the total traffic.
(iii) SDH has overcome the efficiency and interface problems to a great extent.
(iv) OAM features of SDH are very strong, whereas IP management backbone OAM is not yet fully developed.
 (v) Ethernet equipment available so far are only up to 10 Gbps, whereas SDH is available up to 40 Gbps.
(vi) A well-trained workforce is available for maintenance.
(vii) Automatic protection switching makes the SDH much more robust.

If you leave aside the improvements made by SDH in the so-called NGSDH, out of the above, the strongest factor that keeps the operators from switching over to IP core or backbone is the large-scale deployment of SDH, which is not only in place, but is also continuing to grow at a great pace. This is because of the demand from mobile telephony and, of course, from data services. If some enterprise desires to have a countrywide VPN or a leased lines network for their operations, can they be denied the service, or can they be asked to wait until we develop a pure data network? Certainly, it is not done. They will have to be provided the service from the existing deployment by adding a little more equipment or interfaces, wherever required. In other words, it is very difficult for the IP core to establish itself at the pace of the demand. This situation is leading to higher and higher SDH proliferation.

As mentioned above, the shortcomings of SDH with respect to data transport have been overcome to a great extent by recent developments. In the following few sections, we will have a look at all these features of SDH.

14.3 Next-Generation SDH

To facilitate a more efficient data transport on SDH, a number of improvements have been made in the procedures dealing with the transport of data traffic. With these improvements, the SDH is again being tipped as not only the technology of today, but also the technology of the future, which was once being regarded as a legacy technology. These improvements are as follows:

(i) V-CAT

(ii) LCAS

(iii) GFP

Let us discuss them in brief.

14.3.1 Virtual Concatenation

We have seen "contiguous concatenations" of payloads in the previous chapter.
Contiguous concatenation has the following drawbacks:

(i) The feature of "contiguous concatenation" is required to be provided on all the nodes of the link over which data transport is intended. This will involve the replacement of hardware/software at many nodes, before the provision of this facility can be made. This poses a great problem in practical implementations, as the replacement of equipment involves costs and down times which cannot be easily afforded; this has been the primary reason for "contiguous concatenation" being seldom deployed.

(ii) The capacity utilization of payload containers remains inefficient because of mismatch of the data rates of data communication application with those of SDH data rates.

Both these problems have been overcome by the use "V-CAT," which you will see next.

14.3.1.1 V-CAT Procedure

Problem (a) above (that of providing a concatenation feature in all the intermediate nodes) can be solved if we do not provide for contiguous concatenations but divide the payload into the required number of VCs (virtual containers) and just know their locations, so that at the receiving end, we can rejoin the bitstream in the original order. This is exactly what is done, and the process is known as "V-CAT." Since we do not need to do anything to the VCs (stripping of POH, stuffing of bytes, etc., as required in contiguous concatenation) except knowing their locations, nothing needs to be done at the intermediate nodes where they will transport the payloads as the normal SDH payload. The splitting of data is required only at the originating node and recombining is required at the terminating node. Thus, the V-CAT feature is required only at the terminal nodes between which the traffic is proposed. The knowledge

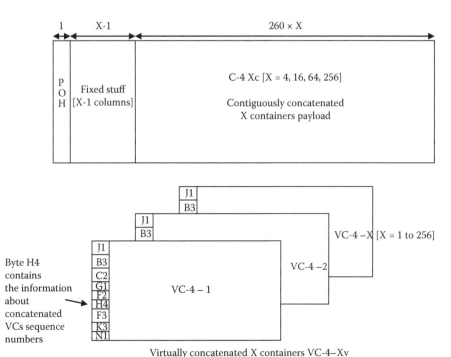

FIGURE 14.1
V-CAT of X VC-4 virtual containers.

of the VCs carrying the concatenated payload can be carried by any of the overhead bytes available to spare. The byte H4 of the POH is used for carrying this information. The V-CAT of X VC-4s is shown in Figure 14.1.

The VC-4s carry their POH as usual and travel normally on the network. At the receiving end, the information contained in the byte H4 facilitates the original data stream to be reconstructed by taking out the data in proper sequence from the concatenated VC-4s. The group of virtually concatenated payloads is called a VCG or a "V-CAT group."

Incidentally, the above procedure solves our second problem of inefficient capacity utilization as well. Let us see how it is done.

14.3.1.2 Mapping of 10-Mbps Ethernet

Let us take a case of a 10-Mbps Ethernet bit stream to be carried by SDH. The VC-12 (2 Mbps) or VC-2 (6.3 Mbps) will not be able to carry this, so the next higher level container, i.e., VC-3 with a capacity of 51 Mbps, will have to be chosen. Thus, a 10-Mbps data will be carried on a 51-Mbps container, giving only 20% efficiency. This is a case of a single container.

Let us take the case of a 1000-Mbps gigabit Ethernet. This data rate exceeds even the largest SDH container, i.e., VC-4 (150-Mbps payload). The contiguous

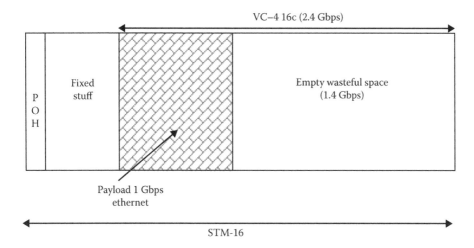

FIGURE 14.2
Mapping of 1-Gbps Ethernet over concatenated VC-4-16c.

concatenation VC-4-4c will give a capacity of only about 600 Mbps, thus we will have to go for the next higher level of concatenations, i.e., VC-4-16c (VC-4s with STM-16), which has an approximate capacity of 2500 Mbps, thus the efficiency is again only 40%. Figure 14.2 shows the gap for the Gb Ethernet.

14.3.1.3 Efficiencies of Other Services

The efficiencies of some more service payloads in the case of contiguous concatenation are shown in Figure 14.3. We can see that the efficiencies range from nearly 20% to 67%.

In sharp contrast to this, the efficiencies improve greatly in case of V-CAT because in V-CAT, we do not physically join or concatenate the VCs but keep them wherever they are and use an identifier to find their locations. In V-CAT, we can group any arbitrary number of VCs in a VCG (group of VCs), and mark them with a V-CAT marker and a sequence number. The marking is carried out at the transmitting end, and is deciphered at the receiving end.

Service	Bit rate (Mbps)	SDH Level	SDH Payload bit rate (Mbps)	Efficiency (%)
Ethernet	10	STM-0	48.384	20.66
Fast Ethernet	100	STM-1	149.760	66.77
Fiber Channel	200	STM-4	599.040	33.38
Gigabit Ethernet	1000	STM-16	2396.160	41.73

FIGURE 14.3
Efficiencies of various service payloads in contiguous concatenations.

For example, in the case of contiguous concatenation of VC-4s, we can have only VC-4-4c, VC-4-16c, VC-4-64c, and VC-4-256c sizes of the grouped container, but in the case of V-CAT, we can have any number of VC-4s grouped together, say VC-4-3c, VC-4-7c, or any other number from 1 to 256.

This flexibility allows a lot more precision in the mapping of various payloads giving rise to highly improved efficiencies. For example, a 10-Mbps Ethernet can be mapped to 5VC-12s to give an overall bit rate of 10.875 Mbps. Figure 14.4 gives the comparative efficiency for the payloads of Figure 14.3 with V-CAT.

As can be seen in Figure 14.4, by using any odd combination of VC-X with any number of them for grouping, we can achieve data carrying efficiencies near 100%. When we use the higher-order containers such as VC-4 and VC-3, it is called "higher-order V-CAT" or HO-V-CAT, and when we use lower-order VCs such as VC-11, VC-12, or VC-2, the procedure is called "lower-order V-CAT" or LO-V-CAT. The use of LO-VCs allows a very fine granularity of 1.5 Mbps in case of VC-11 and that of 2 Mbps in case of VC-12.

These two features together, i.e., ability to operate V-CAT from the path terminating nodes only and a very high efficiency, have given a tremendous boost to the SDH for the transport of data traffic.

In the above case, we have tried to match the service design capacity with the V-CAT container size. In practice, carriers may choose to create containers of sizes of only average bandwidth, or with margins for better or tolerable QOS requirements. They may also vary the actual container size as per the load requirement through proper monitoring.

In the example shown in Figure 14.4, we can choose to virtually concatenate the contiguous VCs or discrete VCs. In fact, the chosen VCs may be located anywhere in the STM-N. This feature offers a great practical advantage. An operator will generally establish his network with a certain capacity, which can sustain the demand for the next one to two years. He keeps on providing the services as per the demand of different capacities by different customers. Thus, at any given point in time, the operator may have a number of empty VCs, which may not be contiguous. In the case of a contiguous concatenation procedure, it will require a lot of effort and down time to reorganize the

Service	Bit rate (Mbps)	SDH Level	SDH Payload bit rate (Mbps)	Efficiency (%)
Ethernet	10	VC-12-5$_v$	10.875 (2.175 × 5)	91.95
Fast Ethernet	100	VC-3-2$_v$	96.768 (48.384 × 2)	100
Fiber channel	200	VC-3-4$_v$	195.536 (48.384 × 4)	100
Gigabit Ethernet	1000	VC-4-7$_v$	1048.32 (149.76 × 7)	95.39

Note: The subscript in case of V-CAT is "v," i.e., VC-4-7$_v$ and so forth.

FIGURE 14.4
Efficiencies of various service payloads in V-CAT.

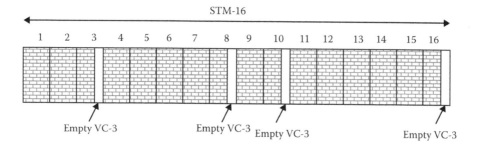

FIGURE 14.5
An STM-16 has four VC-3s empty at different locations, which could be utilized for additional service (POH and others are not shown to maintain clarity).

traffic of all the existing customers and to accommodate the traffic demand of the new customers. In many cases it may just not be feasible because of the granularity limitations in contiguous concatenations. However, in the case of V-CAT, the empty VCs are simply loaded with the customer traffic, irrespective of their location within the STM-N. This adds to the capacity utilization efficiency of the network as well as facilitates quick provisioning. Figure 14.5 depicts the situation.

Let us say there is a demand for an additional 100-Mbps payload. On examination of our STM-16 utilization, we find that one VC-3 each is empty in 3rd, 8th, 10th, and 16th VC-4s. Since we will need only two VC-3s, any two of these VC-3s can be directly configured for this additional demand through V-CAT. It would have been very difficult to accommodate this additional demand using contiguous concatenations.

14.3.1.4 Resilience through V-CAT

In addition to the three wonderful features above, V-CAT allows us to forward the individual VCs of a concatenated payload or VCG via different routes to the destination. This adds to the resilience or reliability of service, as in the case of a failure of a particular link, all the VCs are directed toward the remaining healthy links. The situation is illustrated in Figure 14.6 for a payload of GB Ethernet.

14.3.1.5 Payload Identification and Realignment

The different paths from source to destination may not be of equal length (Figure 14.6), thereby causing different propagation delays to different VCs. This difference in the propagation delay needs to be compensated to correctly reassemble the VCs of the VCG in proper order. The V-CAT procedure provides a means for this compensation as well. A mechanism involving frame number and sequence number of VCs facilitates correct concatenation at the receiving end.

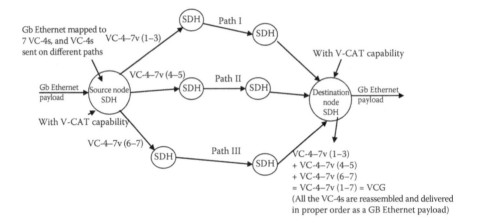

FIGURE 14.6
Individual VCs of the VCG routed to destination via different paths.

Let us consider the case of a gigabit Ethernet to be transported on the SDH network using V-CAT, as shown in Figure 14.6. The mapping details are depicted in Figure 14.7. The Gb Ethernet payload is mapped into 7 VC-4s. Their locations, as per availability in the STM-16 frame are 4th, 6th, 7th, 9th, 10th, 11th, and 15th. All these VC-4s are transported to the destination node as independent entities on the available SDH infrastructure. Each VC-4 is added to one of the SDH/ADM for transmission. At the receiving end the corresponding time slots are dropped. Because of different path length on each of the three paths (Figure 14.8), the individual VC-4s may reach the destination at different times.

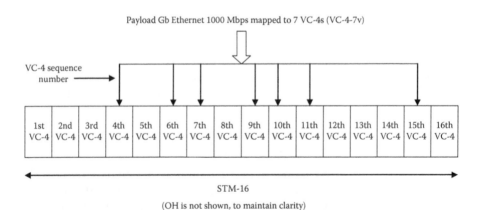

FIGURE 14.7
The mapping of 1 Gbps Ethernet payload into 7 VC-4s of STM-16.

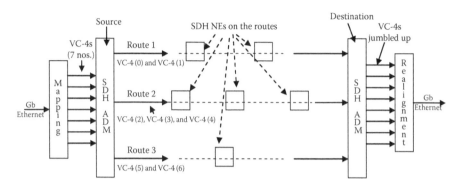

FIGURE 14.8
VC-4s containing the Gb Ethernet are transported independently via different routes.

Thus, the VC-4s belonging to the first frame (see Figure 14.7) means the payload for the first frame of "1/8000th second" may reach after the VC-4 carrying the payload for the 2nd frame of "1/8000th second" or may be after the 3rd or fourth frame. In such a situation, how do you identify and realign them at the receiver? We need to find the exact sequence of VC-4s, which was constructed out of the GB Ethernet payload before transmission.

To identify the VCs, each group of VCs falling in one frame is given a "frame number," and within the group, each VC is given a sequence number. Thus, if we check the content of H4 byte of the received VC-4, we can get the frame number and the VC-4 sequence number. We can first bring together all the VC-4s bearing the same frame number, and then arrange all the VC-4s in order of their sequence numbers. The third step is to arrange all the groups in order of their frame numbers, and the original payload structure gets reconstructed.

Now the question is, what are the limits of frame numbers and VC sequence numbers?

It has been considered adequate to cater for a maximum delay of 512 ms. Thus, the number of frames, which have to carry a unique frame number, would be 512 ms/125 µs = 4095 frames. The frame number will repeat after counting from 0 to 4095, but it is unlikely to create any duplication, as the delay caused by 4095 frames, i.e., 512 ms, is more than enough for the normal differences in the divergent route lengths.

The VC-4 sequence number is from 0 to 255, as per the V-CAT procedure. We are not likely to have more than 256 VCs in the higher-order path for which this sequence number is meant (STM-64 can have 64 VC-4s or 192 VC-3s).

Coming back to the frame number, it is actually called a "multiframe number" because the number is generated by a two-stage counter operated by the bits of H4 byte made up of multiframes. The bit allocation for this multiframe indicator is shown in Figure 14.9a and b.

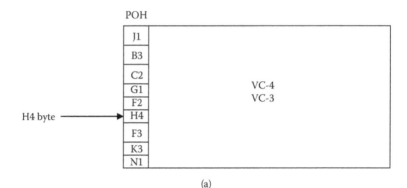

(a)

Bits 1–4 of H4 byte of first frame of the second multiframe (MFI-II)				Bits 1–4 of H4 byte of second frame of the second multiframe (MFI-II)				Bits 5–8 of H4 byte of all the frames of the first multiframe (MFI-I)				Frame number
0	0	0	0	0	0	0	0	0	0	0	0	0
0	0	0	0	0	0	0	0	0	0	0	1	1
0	0	0	0	0	0	0	0	0	0	1	0	2
–	–	–	–	–	–	–	–	–	–	–	–	–
1	1	1	1	1	1	1	1	1	1	1	1	4095

(b)

FIGURE 14.9
The frame number (multiframe number) generation through H4 byte. (a) Location of H4-byte in the POH of HO-VCs of VC-4 and VC-3. (b) The 12-bit frame number counter.

The generation of 4096 frames (0 to 4095) is done by a 12-bit counter that consists of the last four bits of byte H4 (last 4 bits, called MFI-I) and the first four bits of the "second frame of the second multiframe of 16 frames," as the least significant bits of MFI-II, and again the first four bits of the "first frame of the second multiframe of 16 frames" as the most significant bits of MFI-II.

The VC-4 (or VC-3) sequence number is generated by the use of the first four bits of the 15th and 16th frames of the multiframe (total 8 bits giving 0 to 255, i.e., 256 count). The bit assignment of the H4 byte is shown in Figure 14.10.

Thus, each VC-4 carries with itself two indicators, i.e., "multiframe indicator" and "sequence number" in its H4 byte, which allows the proper realignment of the payload at the receiving end. For the payload of Figure 14.7, the VC-4s of the STM-16 will have the numbers shown in Figure 14.11.

Some of the VC-4s having the multiframe number 0 may reach after those leaving the MFI number 1 or 2 or even more, while some might reach at other times. The receiver may, therefore, have a totally random sequence of VCs. However, with the help of the MFI number and VC sequence number, all of them are arranged in a proper contiguous order and delivered to the customer.

H4 Byte								First multiframe number	Second multiframe number
Bit 1	Bit 2	Bit 3	Bit 4	Bit 5	Bit 6	Bit 7	Bit 8		
				First multiframe indicator MFI-I (bits 1–4)					
Sequence indicator MSB (bits 1–4)				1	1	1	0	14	$n-1$
Sequence indicator LSB (bits 5–8)				1	1	1	1	15	
Second multiframe indicator MFI–(II)–(MSB bits 1–4)				0	0	0	0	0	n
Second multiframe indicator MFI–(II)–(LSB bits 5–8)				0	0	0	1	1	
Reserved "0000"				0	0	1	0	2	
"				0	0	1	1	3	
"				0	1	0	0	4	
"				0	1	0	1	5	
"				0	1	1	0	6	
"				0	1	1	1	7	
"				1	0	0	0	8	
"				1	0	0	1	9	
"				1	0	1	0	10	
"				1	0	1	1	11	
"				1	1	0	0	12	
"				1	1	0	1	13	
Sequence indicator SQ (MSB bits 1–4)				1	1	1	0	14	
Sequence indicator SQ (LSB bits 5–8)				1	1	1	1	15	
Second multiframe indicator MFI2 MSB (bits 1–4)				0	0	0	0	0	$n+1$
Second multiframe indicator MFI2 LSB (bits 5–8)				0	0	0	1	1	
Reserved ("0000")				0	0	1	0	2	

FIGURE 14.10
Bit assignment of the H4 byte of the HO-POH for V-CAT.

14.3.1.6 Payload Identification and Realignment in LO-V-CAT

The details we have discussed so far pertain to V-CATing of higher-order tributaries (HO-V-CAT) of SDH, i.e., VC-3 and VC-4, although in the beginning, we have seen that V-CAT is possible for lower-order tributaries such as VC-2, VC-12, and VC-11 as well. Let us see how the job is performed for lower-order tributaries. The process is called LO-V-CAT.

The process is exactly the same in the case of LO-V-CAT as that of HO-V-CAT. The only difference is that the functions of the MFI number and sequence number are performed by bit number 2 of K4 byte of LO-POH. The single bit

	VC-4 Serial number in STM-16 (time slot)	Information in H-4 byte	
		Multiframe number MFI	VC-4 Sequences number SQ
First frame of 125 μs	4th	0	0
	6th		1
	7th		2
	9th		3
	10th		4
	11th		5
	15th		6
Second frame of 125 μs	4th	1	0
	6th		1
	7th		2
	9th		3
	10th		4
	11th		5
	15th		6
And so on up to 4096 frames	Do	And so on up to 4095	Do

FIGURE 14.11
VC-4 identifiers.

is used in a multiframe of 32 frames, thus providing a 32-bit sequence, of which the first 5 bits are used for an MFI number and the next 6 bits are used for the sequence number. Note that MFI is only 5 bits long, as compared to that of HO-POH, which was 12 bits long (and provided a delay margin of up to 512 ms, after which the sequence repeated). The lower-order tributaries are not discriminated against those of higher order by the transmission media, which causes the same amount of delay in them as in those of higher-order tributaries. Let us see how this 5-bit MFI provides a 512-ms delay. The details of LO-POH and this multiframe are indicated in Figure 14.12.

The next important feature of a next-generation SDH network is the line capacity adjustment scheme (LCAS), which is the subject of our next section.

14.3.2 Link Capacity Adjustment Scheme

As is obvious from the name, LCAS is a scheme that helps in adjusting the link capacity in SDH networks. The capacity adjustment is possible in the normal (or legacy) SDH networks as well, but the process is very slow and

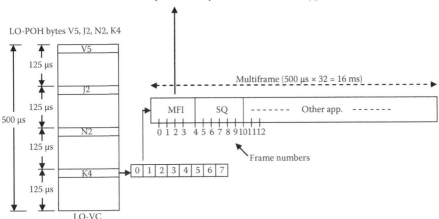

FIGURE 14.12
The second bit of K4 byte in LO-POH provides the MFI and sequence numbers for the identification of transmitted tributaries. The two stage multiframing process gives the desired delay margin of 512 ms (the LO-VC is itself in a multiframe format; thus, this becomes three-stage multiframe).

inefficient. The LCAS removes this bottleneck and provides a very fast and efficient mechanism for adjusting the link capacity.

LCAS is basically a control mechanism of V-CAT. In other words, the V-CAT mechanism is put to its proper use through the LCAS mechanism.

LCAS also helps in improving the reliability of the link. Various applications of LCAS are as follows.

14.3.2.1 *Improving the Link Reliability (Resilience)*

The LCAS provides for commands, through which the payload can be reconfigured into fewer or more numbers of VCs. Thus, when the normal traffic is routed via many routes, and one of the routes fail, the whole of the payload is reconfigured into the remaining VCs, which were traveling via the healthy paths. This process reduces the throughput, but the link is not totally disrupted. Let us consider the payload of Figure 14.6 (redrawn here as Figure 14.13).

14.3.2.2 *Automatic Removal of Failed Members*

Of all the members of the VCG (VCG consists of VCs [HO or LO], which are called by its members, e.g., in VC-4-7$_v$, there are 7 VC-4s in the VCG), if any one starts giving errors, LCAS can remove that member from the group to save the whole link from giving errors. The VCG is reconfigured accordingly.

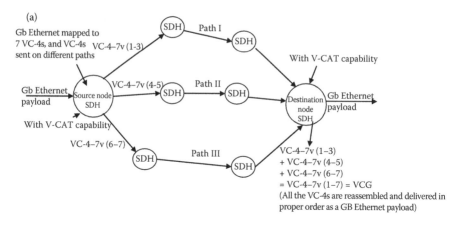

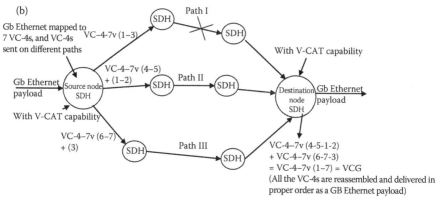

FIGURE 14.13
LCAS reconfigures the payload into the VCs traveling on healthy paths to maintain the traffic on the link. (a) Shows normal distribution of VCs, whereas (b) shows their distribution when path I has failed.

When the member becomes OK, it is again included in the group. This enables the VCG to continue to carry error-free data but at a reduced rate.

14.3.2.3 In-Service Resizing of Bandwidth

Most data networks are transmit data in bursts, where the bandwidth keeps on varying. In normal SDH, the bandwidth, once configured, cannot be changed without a sizable effort, and most probably involves a down time as well. Even though the V-CAT by itself does not provide the facility to dynamically change the bandwidth, LCAS provides a means for dynamically changing the bit rate or the bandwidth of the customer service by adding or deleting members to or from the group, while the service keeps running. Thus, it facilitates features such as bandwidth on demand. A customer can opt for a high bandwidth for a limited number of hours, while the rest of the day, he might manage with a certain minimum bandwidth. LCAS cannot, however, automatically adjust the VCG size according to the traffic pattern at any given time.

14.3.2.4 Provisioning of Asymmetric Bandwidth

LCAS provides a unidirectional protocol for the control of V-CAT, thus allowing an asymmetric bandwidth provisioning, which is a more prevalent pattern of data traffic as compared to symmetric bandwidth.

14.3.2.5 Working with Non-LCAS Nodes

The LCAS transmitters can interwork with non-LCAS receivers and vice versa. This improves flexibility of the overall network.

14.3.2.6 LCAS Operation (Higher Order)

The LCAS operations are performed by using the spare bits in the multiframes of H4 and K4 bytes. For HO-V-CAT, the H4 byte is used, and for LO-V-CAT, the K4 byte is used. The revised bit allocation of the H4 byte is called a "high-order control packet" and that of the K4 byte is called a "low-order control packet."

Let us see the revised distribution of duties of various bits of H4 byte, which includes the LCAS commands. They are shown in the table of Figure Figure 14.14.

The 16 frames block, counting from frame number 8 of the first multiframe to frame number 7 of the second multiframe is called a "higher-order control packet." These frames include all the commands for LCAS. The last two frames of this block include 8-bit CRC check (CRC-8) for the control packet error checking. Other commands and indicators are as follows.

14.3.2.6.1 Member Status

As shown in Figure 14.14, 8 bits have been provided for this purpose. Member status (MST) means the status of the members of the VCG group. Each bit indicates the status of one VCG member, which is given below

Bit Value		Status
0	—	OK
1	—	FAIL

The value 0 indicates that the member is healthy and 1 indicates a defective member. As we have seen earlier, there are 256 members in a VCG. Since each multiframe carries the status of 8 members, a second stage multiframe of 32 such multiframes will carry the status of all the members of a HO-VCG

H4 Byte								First multiframe number	Second multiframe number
Bit 1	Bit 2	Bit 3	Bit 4	Bit 5	Bit 6	Bit 7	Bit 8		
				First multiframe indicator MFI-I (bits 1–4)					
Sequence indicator MSB (bits 1–4)				1	1	1	0	14	$n-1$
Sequence indicator LSB (bits 5–8)				1	1	1	1	15	
Second multiframe indicator MFI–(II)–(MSB bits 1–4)				0	0	0	0	0	n
Second multiframe indicator MFI –(II)–(LSB bits 5–8)				0	0	0	1	1	
CTRL				0	0	1	0	2	
GID ("000X")				0	0	1	1	3	
Reserved (0000)				0	1	0	0	4	
"				0	1	0	1	5	
CRC-8				0	1	1	0	6	
CRC-8				0	1	1	1	7	
Member status MST				1	0	0	0	8	
Member status MST				1	0	0	1	9	
0	0	0	RC-Ack	1	0	1	0	10	
Reserved (0000)				1	0	1	1	11	
"				1	1	0	0	12	
"				1	1	0	1	13	
Sequence indicator SQ (MSB bits 1–4)				1	1	1	0	14	
Sequence indicator SQ (LSB bits 5–8)				1	1	1	1	15	
Second multiframe indicator MFI–(II)–(MSB bits 1–4)				0	0	0	0	0	$n+1$
Second multiframe indicator MFI–(II)–(LSB bits 5–8)				0	0	0	1	1	
CTRL				0	0	1	0	2	
0	0	0	GID	0	0	1	1	3	
Reserved (0000)				0	1	0	0	4	
"				0	1	0	1	5	
C1	C2	C3	C4	0	1	1	0	6	
C5	C6	C7	C8	0	1	1	1	7	
Member status MST				1	0	0	0	8	

FIGURE 14.14

Revised distribution of duties of H4 byte, including commands for LCAS.

(from 0 to 255). The status of the members will thus be refreshed after every 64 ms ($16 \times 125 \times 32$ µs).

14.3.2.6.2 Resequence Acknowledged

This signal is carried in the 4th bit of the 10th frame. Whenever any member is added or deleted from the VCG, the sequence changes, which is acknowledged through this bit by the receiving node.

14.3.2.6.3 SQ Sequence Indicator

We have already seen the details in the previous section.

14.3.2.6.4 Second Multiframe Indication

We have already seen the details in the previous section.

14.3.2.6.5 Group Identifications (Bit 4 of Frame Number 3)

This is a group identification (GID) bit, which is the same for all members of a VCG.

14.3.2.6.6 Control (Bits 1–4 of the Second Frame)

The assigned controls for various values are

0000: FIXED (no LCAS nodes, LCAS is not supported)

0001: ADD (a member is added to the group of an existing V-CAT channel)

0010: NORM (normal LCAS transmission)

0011: EOS (end of sequence. The highest number in the VCG group)

0101: IDLE (member is not in the VCG group or will be removed)

1111: DNU (do not use, MST fail status is reported by the receive side member to be removed from the group)

14.3.2.6.7 CRC-8 (Bits 1–4 of the Sixth and Seventh Frames)

These bits carry out an 8-bit CRC check on the control packet, improving its reliability. (CRC was discussed in Section 6.7.)

14.3.2.7 Lower-Order LCAS

We left some spare bits in the multiframe of K4 bit 2, in Figure 14.12, which showed the allocation of bit duties for V-CAT. These bits are used for the LCAS operations of the lower order. The revised bit assignment is shown in Figure 14.15.

The roles of all the fields are the same as those specified for a higher-order control packet.

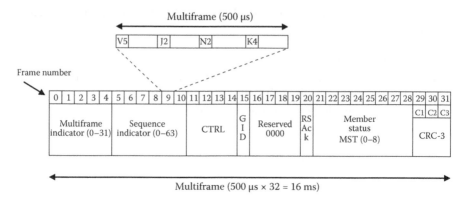

FIGURE 14.15
Bit assignment of "lower-order control packet."

The member status of 8 members of the VCG is indicated by one multi-frame; thus, 8 multiframes are required to carry the status of all the 64 (0–63) members of LO-V-CAT-VCG.

The LCAS operations are exactly the same as those performed for higher-order LCAS, explained in the previous sections.

Besides V-CAT and LCAS, GFP is the third major improvement that has been made in the next-generation SDH, as compared to the legacy SDH. Let us see about GFP in our next sections.

14.3.3 Generic Framing Procedure

As the data communication transport was facilitated on SDH network, a number of mapping procedures were developed to accommodate all the pre-vailing data communication standards and protocols such as HDLC, PPP, ATM, FDDI, etc. Special protocols such as LAPS were developed (ITU-T rec-ommendations X.85 and X.86) for carrying Ethernet and IP over the SDH. The overall scenario led to a large number of protocols and standards to be made available on the SDH devices. Many of these protocols had problems of uncertainty of bandwidth, which affected the planning and efficiency. The processing of ATM was quite complex and inefficient due to a large amount of overhead bytes.

A need was therefore felt to have a common protocol that can carry all types of user data signals as payload on the SDH and that also minimizes, if not completely removes, problems of flexibility, inefficiency, and reliability that are associated with many protocols.

Taking all these factors into account, ITU-T developed a new standard for data communication framing procedures on SDH, called GFP, which is defined in their recommendation number G.7041.

14.3.3.1 What Is GFP?

GFP is an "encapsulation mechanism" for transport of data signals on SDH. "Encapsulation" is a process in data communication in which a particular data packet (or even a frame) is added with a header of the "encapsulating protocol" to form a new frame. The new frame thus contains a header and the old frame. When this "encapsulated" frame is received by the sink, the new header is processed first. Figure 14.16 shows the concept.

Similarly, the frames pertaining to any of the user data protocol are encapsulated in the GFP frame, transmitted over SDH, received by SDH sink, and the user data is delivered to the other end with their own protocol as it was sent. A GFP frame can also be depicted similar to that shown above. It is shown in Figure 14.17.

GFP has offered many advantages to the transport of data over SDH. Let us see some of the important ones.

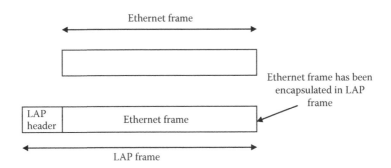

FIGURE 14.16
Encapsulation of Ethernet frame in LAP frame.

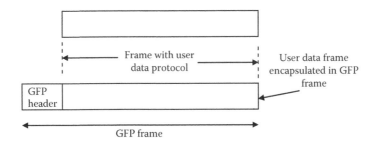

FIGURE 14.17
Encapsulation of user data frames in GFP frame.

14.3.3.2 Advantages of GFP

14.3.3.2.1 Common Encapsulation Protocol

GFP provides a common encapsulation protocol for all types of data protocols to be carried over SDH. This minimizes the complexities due to protocol
specific processing. For example, the processing of ATM, mapped on SDH, is
quite complex, which is an undue burden for simple data connections. This
gets avoided in GFP, as ATM cells (just like all other protocols such as HDLC,
PPP, Ethernet, IP, etc.; see Section 13.6 for mapping of PPP and HDLC in SDH)
get encapsulated in GFP, and the SDH is required to process only GFP frame.

14.3.3.2.2 Predictable Bandwidth

In case of transmissions using the protocols such as HDLC or PPP, there is a
strong possibility of usage of extra bandwidth without any gain, and sometimes, this extra bandwidth may become quite large, resulting in nonpredict-
ability of bandwidth requirement and poor efficiency. This is explained as
follows.

Protocols such as HDLC, PPP, etc. use a "flag," which is a bit sequence in
the beginning, and at the end of a frame; it is also called "frame delineator."
A generally used sequence is "01111110." The HDLC/PPP frame is shown in
Figure 14.18.

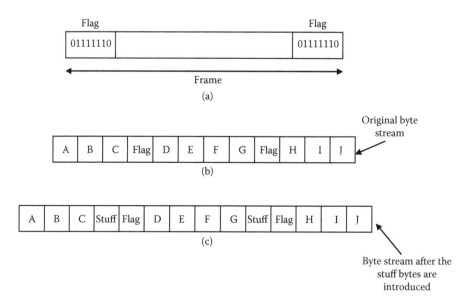

FIGURE 14.18
(a) HDLC/PPP frame with flags at either end. (b) Original byte stream. (c) Byte stream after
the stuffing.

When these frames are used for data transmission, there is a possibility that there may be a bit sequence in the data that matches with the "flag." The receiver might confuse these data bits with frame and might accept their occurrences as beginning or end of the frame. To avoid this possibility the data stream is checked at the transmitting end, and in case such a data sequence is encountered, a special byte called an "escape character or a stuff byte" with predetermined bit pattern is introduced "before" such a flag pattern in the data. The receiver understands the stuff byte and knows that the following flag such as a pattern is actually data and not a flag and discards the stuff bytes.

We can see that these stuff bytes increase the bit rate for no useful purpose. If the stuff bytes become too many (which can happen in the case of a malicious user), not only the efficiency of the network will go down drastically but also the actual requirement of the bandwidth cannot be correctly assessed.

This problem is solved by the GFP by having an "error control mechanism" for the header, which ensures that the probability of errors in the header (the header is similar to the flag of HDLC/PPP) is made extremely low through the use of CRC (see Section 6.7 for CRC principles). The header and only the header alone is always detected as a header, eliminating the need of stuff bytes.

14.3.3.2.3 Efficient Transport

GFP uses a very small header, with a total of 8 bytes. Some of these bytes are compensated because original headers of user data protocols are removed. Thus, the bulk of the GFP frame is the user payload, giving a very high efficiency. Moreover, as discussed in Section 14.3.3.2.2 above, the need of stuff bytes is eliminated, which again adds to the efficiency.

14.3.3.2.4 Optimized for Specific Services

It uses two types of framing procedures, one called "frame mapped GFP" (GFP-F) and the other one called "transparent mapped GFP" (GFP-T). While the first one is optimized for throughput, for protocols such as Ethernet, PPP, HDLC, etc., the transparent mapped GFP is optimized for time sensitive (or isochronous) types of applications such as voice or video. Thus, GFP provides an optimized transport service for all types of data traffic.

14.3.3.2.5 Supports Different Network Topologies

The optional "extension header" supports various types of topologies in the network. While the "null" extension header supports channelized point to point network, the "linear" extension header is for "flow aggregation" on point-to-point networks, and the "ring" extension header is for RPR topology.

In all, the GFP provides uniformity and efficiency in the transport of data over SDH, in addition to providing reliability and flexibility.

The frame structure of GFP and its mapping procedures are described in the next section.

14.3.3.3 GFP Frame Structure

The frame structure of GFP is shown in Figure 14.19.

As can be seen, the frame consists mainly of four portions, firstly the core header, followed by a payload header, user payload information, and finally a frame check sequence or CRC (CRC for the payload is optional).

The core header consists of two fields, the payload length indicator (PLI), which is of 2 bytes, and the header error check (cHEC), which is also of 2 bytes. This core header is the main strength of the GFP frame format. With the help of the core header, the GFP is able to correctly identify the beginning and end of a frame, instead of depending upon the "flags" as used by the other protocols. The PLI field indicates, in 16 bits, the length of the user payload. This 16-bit PLI sequence is passed through a CRC-16 process, and the CRC checksum is loaded into the next 2 bytes in the field cHEC and sent followed by the PLI field.

The receiver looks for this unique 32-bit-long bit stream to mark the beginning of the frame. A similar pattern will occur at the beginning of the second frame; thus, the receiver is correctly able to identify the beginning and end of the frame. The problem of stuff bytes as discussed earlier is not faced by the GFP header, as the probability of occurrence of the core header pattern in user data is 2^{-32}.

In addition to the core header, the GFP also has a payload header. The payload header is again divided into two fields, the payload "type" field, which

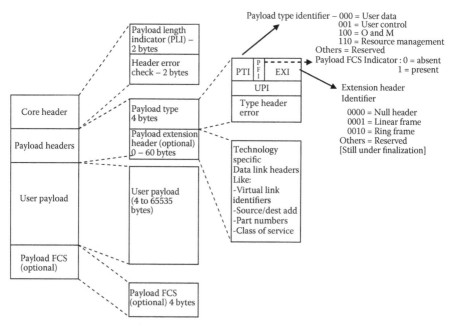

FIGURE 14.19
General structure of a GFP frame.

Payload type	UPI Value
Ethernet	0000 0001
PPP	0000 0010
Fiber channel	0000 1011
IP-V4	0001 0000
IP-V6	0001 0001
MAPOS	0000 1000
MPLS (direct)	0000 1101
MPLS (multicast)	0000 1110
RPR	0000 1010

FIGURE 14.20
GFP(F) payload types and their UPI values.

is 4 bytes long and the payload "extension" header, which is 0–60 bytes long, but it is optional. The payload "type" header is again divided into five fields. The first field consists of the first three bits of the first byte and is called "payload type indicator" (PTI). This field indicates whether the user payload is a data information or management information. The fourth bit of the first byte is called payload FCS indicator (PFI) and indicates whether FCS is present or not for the payload (the FCS field is optional for the payload). The last 4 bits of the first byte are called "extension header identifier" and identify whether there is an extension header or not (extension header is optional) and the type of extension header. The second byte of the payload type indicator is UPI or user payload indicator and indicates the type of user payload. A large number of user payload types have been defined; some of them are shown in Figure 14.20. Many other types are under consideration.

Followed by the UPI, there is a cHEC or the error check of the payload type of 2 bytes length.

14.3.3.4 GFP Mapping

Two types of mappings are permitted in GFP, and the frames are accordingly called "frame-mapped GFP" [GFP(F)] and "transparent mapped GFP." Because of these two types of framing procedures, GFP is able to offer optimized services for all types of data applications, whether they be throughput- or time-sensitive. The frame format used in both the cases is the same; however, the data carrying philosophies are widely different. We will see about these mappings in the next subsections.

14.3.3.5 Frame-Mapped GFP

This is the normal GFP protocol. It is just like any other encapsulation mechanism. The (F) had to be added to the GFP because of the development of another format of GFP called GFP(T) (transparent). Complete user frames are mapped

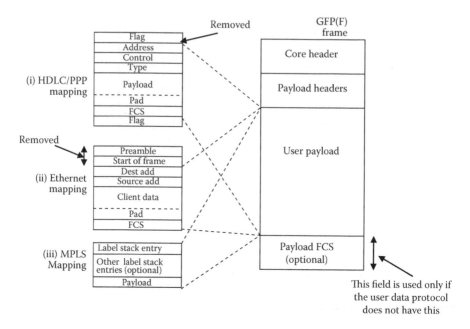

FIGURE 14.21
Mapping of some of the protocols on GFP(F).

as a whole in each of the GFP(F) frames, whereas GFP(T) frames can map a part of the user frame as well. The user data along with its overhead protocols is treated as a payload for the GFP(F) framing. However, to improve efficiency, some unnecessary overhead bytes such as "preamble," "frame delimiters," interframe gaps, etc. are removed by the GFP procedure. These removed protocols are added again while delivering the payload to the user at the receiving end.

The mapping of Ethernet, HDLC/PPP, and MPLS frames in GFP(F) is shown in Figure 14.21. It is very clear from the depiction, how simple it is to map any protocol into GFP(F). Thus, GFP(F) provides a very good amount of flexibility, along with efficiency.

14.3.3.6 Transparent Mapped GFP

The transparent GFP mapping is a very special type of mapping. It is very different from the normal mapping procedures. Although the GFP frame format of GFP(T) is the same as that of the GFP(F), the mapping procedure is entirely different.

GFP(T) has been developed specially for delay sensitive isochronous types of applications, such as digital video broadcast (DVB), fiber channel for SANs, etc. It is applicable for only those applications that use 8B/10B as a line code for transmission. (8B/10B is a line code that is used specially in optical fiber

Payload type	UPI Value
Fiber channel	0000 0011
FICON	0000 0100
ESCON	0000 0101
Gb Ethernet	0000 0110
DVB ASI	0000 1001

FIGURE 14.22
GFP(T) payload types and their UPI values.

transmissions. These types of line codes have very good timing content in the signal. The 8 bits of the data are transmitted using selected sequences from a 10 bit code; this allows selection of the best codes with respect to maximum transition from 0 to 1 to 1 to 0 for improving the timing content of the signal.)

The data signal is completely decoded, and the line code 8B/10B is converted into a line code of 64B/65B by the GFP process. The data is then transmitted at a constant rate. The GFP frame size is fixed irrespective of the frame size of the incoming data signal. The process does not wait for the complete arrival of the frame. The frame is broken as per the constant rate required for GFP(T). This ensures that the latency is minimized with fast transmission of parts of the data packets in separate GFP(T) frames. Similar to the case of GFP(F), the entire user data, including protocols, is treated as payload and mapped into the user payload area of the frame (Figure 14.21). The only difference in GFP(T) mapping as compared to GFP(F) is that the preamble, SOF, interframe gaps, etc. are not removed.

GFP(T) does not differentiate between various fields of the incoming data frames, the preamble, start of header or any other overhead bytes, and even the interframe gaps are treated as data. They are all converted to 64B/65B code and transmitted. It thus transparently transports the user data to the destination. GFP(T) thus actually becomes a layer 1 transport mechanism.

A list of some of the user payloads currently supported on the GFP(T) is shown in Figure 14.22. Although GFP(T) is very good for transmission of delay sensitive applications, it suffers from poor efficiency on account of continuing to transmit even when the user data is not present.

14.3.3.7 *Comparison of GFP(F) and GFP(T)*

As discussed above, GFP(F) and GFP(T) have been optimized for different types of data services. Both of them together cover almost all types of data applications with optimum efficiency. More applications are being added as the standard is still evolving. A comparison of GFP(F) and GFP(T) is given in Figure 14.23.

Thus, the GFP provides a uniform protocol for flexible, efficient, and reliable data transport over SDH, while the V-CAT and LCAS together provide for very high bandwidth utilization and resilience features. These developments

SN	Feature	GFP(F)	GFP(T)
1	Efficiency	High	Low
2	Frame length	Variable	Fixed
3	Optimized for payloads	Ethernet, HDLC, IP	DVB, SAN, GB Ethernet
4	Topologies permitted	Point to point, RPR	Point to point
5	Encapsulation layer	Layer 2	Layer 1
6	Payload contend carried	Preamble, SOF, gaps, etc. stripped off	Data, controls, and gaps all carried
7	Transparency of protocol	Low	High
8	Error correction	No	Single error correction per 64 characters
9	Alarms and defect communication to far end	Not clear	Defined

FIGURE 14.23
Comparison of salient features of GFP(F) and GFP(T).

have made the SDH, or the next-generation SDH as it is called, a truly data-centric transport platform.

One more development in the field of data transport is making waves, which is called RPR. We will have a small introduction of this new technology in the next section.

14.4 Resilient Packet Ring

RPR is a dedicated data network in a ring topology. It has been defined in IEEE 802.17 W/G. The need for RPR was felt because of the growing share of data traffic over the voice traffic and the development of technologies such as VOIP, which can carry even the voiceover data. Thus, as of today, all the telecommunication traffic, be it voice, data, or video, can be carried over a data or packet network.

RPR is a two-fiber ring, consisting of two counterrotating rings (each fiber-carrying signal in one direction throughout, i.e., one fiber in clockwise direction and the other one in counterclockwise direction) and supports a number of nodes. The distance covered by the ring depends upon the number of nodes and the capacity of the laser source. Figure 14.24 illustrates the concept.

The RPR can be formed on an independent physical media as a pure data network or it can be run over an SDH network as a concatenated payload.

Both the fibers carry the working traffic; there is no capacity reserved for protection.

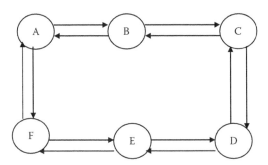

FIGURE 14.24
Two counterrotating packet rings form an RPR.

RPR is a layer two "media access control" (MAC) technology. Like Ethernet, each node knows the topology of the ring and routes the data meant for any other node optimally. Generally, the shortest path is chosen. The MAC decides how to avoid congestion and collisions, and optimizes the use of media amongst all the nodes. The bandwidth control between any two nodes is decided by a fairness control mechanism.

Some of the important features of RPR are as follows.

14.4.1 Classes of Service in RPR

To decide traffic priorities and bandwidth allocations amongst the nodes, RPR defines three classes of service. They are the following:

(i) *Class A service (highest quality):* Class A service is a committed information rate (CIR) service, which means that the provided bandwidth is guaranteed to the customer. The packet delivery also ensures minimum latency and jitter. The applications for this quality of service are normally the latency sensitive services such as voice and video.

(ii) *Class B service (medium quality):* Even the class B services offer a committed information rate (CIR), providing a guaranteed bandwidth, but the requirements of latency and jitter are less stringent. This service is generally provided to on line transactions type of applications.

(iii) *Class C service (lower quality):* This type of service is called a best effort delivery service. The information rate is not committed (No CIR), and the bandwidth is provided depending upon availability. The bandwidth is allocated to this class of service when the requirement of A and B class customers is less. Typical applications for this quality of service are ISPs or businesses requiring the bandwidth for off line activities.

14.4.2 Fairness Control

In the cases of congestion in the network, the nodes communicate to each other regarding bandwidth allocation using the "fairness control algorithm." A fairness control unit in the nodes decides the allocation of bandwidth.

14.4.3 Protection in RPR

In RPR, two paths are available for each packet to travel. In case of a disruption of traffic in any span, the packets automatically decide to take the other route. No bandwidth is reserved for protection; thus, in case of a failed span, the throughput will reduce but the nodes will keep on working. All the nodes are advised about the failed span by a mechanism called "STEERING." All the nodes keep on diverting their packets on the other route, until they are advised that the failed span has become functional.

However, in case of a "node failure," the traffic that was passing through the failed node to the other nodes is "wrapped" back via the other direction. This process is called "wrapping."

14.4.4 RPR Management

RPR, being a packet network, is quite easy to manage. A new node can be simply added or removed from the ring, it will get automatically detected or deleted, and there is no requirement of manual provisioning. The topology discovery and protection activation are automatic. Not much of the management is required through the operator.

Nevertheless, RPR has OAM functions such as configuration, performance monitoring, and fault management.

RPR has several advantages over SDH, as far as the data traffic is concerned, particularly when it is a packet based transport platform (and not configured on SDH). The capacity utilization is much better because there is no band width reserved for protection purposes as against the SDH, where 50% bandwidth is kept reserved for protection (in case 1 + 1 protection is provided). Being a packet network, statistical multiplexing allows it to over provision the bandwidth to an extent of 3–4 times. The automatic topology discovery makes the RPR a plug-and-play type of system, resulting in very fast provisioning as compared to SDH. In case of span failure, the bandwidth used by RPR is much less than that by SDH because the packets are rerouted from the source itself rather than traveling to the failed span and returning back.

The SDH in its new generation on the other hand is competing for data traffic on the strength of its deployed force.

Now, having seen the next-generation SDH and RPR, before we conclude this chapter, let us have a look at a few of the important terms that are frequently encountered when dealing with next-generation SDH and other new transport technologies.

14.5 New-Generation Network Elements

14.5.1 Multiservice Provisioning Platform

This is the basic element of a next-generation SDH network. In the discussion so far in this chapter, we had many times referred to the SDH node having V-CAT, LCAS, and GFP capabilities. All such nodes are called MSPPs. In addition to the SDH with features such as V-CAT, LCAS, and GFP, the MSPP, in keeping with its name, provides services such as Ethernet, packet multiplexing and switching, and sometimes has WDM capability as well. MSPPs are generally robust carrier class equipment.

14.5.2 Multiservice Switching Platform

When multiservice switching platform is provided with a large capacity for TDM switching it is called a MSSP.

14.5.3 Multiservice Transport Node

MSPPs with feature rich packet switching are called multiservice transport nodes (MSTNs).

14.5.4 Wave Length Division Multiplexing

The SDH multiplexing streams are carried on one pair of optical fiber on one optical wavelength. It is possible to multiplex several such SDH streams each working on a different wavelength, on a single pair of fibers. This increases the transport capacity of the optical fiber manyfold (see Section 15.3.2 for more details).

14.5.5 Optical Transport Network

Combining the transport backbone to include the high bit rate transport of SDH, i.e., STM-64 and STM-256, and 10-Gb Ethernet and DWM services, has been defined in the new standards of ITU-T in their recommendation G.709 and G.872. This technology provides a common platform for all the services, which presently exist on separate physical boxes and separate management planes. OTN is being perceived as the transport technology of the future.

14.5.6 Carrier Ethernet

Most of the data traffic initiates and terminates in Ethernet protocol as almost all the offices in the world deploy Ethernet for their LANs. But as it comes out of the office premises, it is converted to many other protocols for being taken to other cities. After reaching the destination city and the office, it is again converted to Ethernet. Sizable efforts go into this conversion at either end, which costs heavily besides reducing the efficiency of the data delivery. However, the conversion has been unavoidable due to historical reasons. The long-distance transport is only available on the voice optimized links of SDH/SONET because of it being the legacy requirement. With the ability to carry the voice, video, etc., which are time-sensitive services on the data protocols, the demand to build the backbone carrier service on Ethernet has been growing steadily over the years.

Ethernet has been developing itself to meet the changing requirements of the users. From 10 Mbps, it moved to 100 Mbps and then to 1 Gbps. It still, however, remained confined to the office LAN. The development of Metro Ethernet brought it out of the office, and it was able to build its own backbone, although for short distances. The development of 10-Gbps Ethernet has been a real boon for carrier Ethernet services. Ethernet backbone has the phenomenal advantage that an Ethernet packet may start its journey from a 10-Mbps LAN, traverse as an Ethernet packet itself throughout the world through many routers and finally get into its destination all without any conversion. With the development of standards for OAM features, the case of carrier Ethernet is becoming stronger, although presently, it is only scantily deployed.

After having seen the future trends in SDH, its data carrying enhancements, its capability with packet protocols, and an introduction to RPR, we are back to the question, whether the future belongs to SDH or pure packet data networks. Let us try to analyze.

14.6 What Is the Future, SDH or Packet Networks?

SDH was considered a technological marvel in the field of telecommunication transport when it was developed, and it is still considered so. However, since it was designed and optimized for voice, and today a bulk of the traffic is data, we keep reminding ourselves as to why rely on something that is optimized for some other application. SDH has more negatives than this alone. The cost is high because of precision electronics needed, synchronization rules being very stringent, and the management being complex. The data-minded engineers just want to get rid of these problems and have a pure packet network that will be able to carry real-time services such as voice and video.

SDH naturally had a very tough time in the discussion forms, but on the field, it is perhaps the only option. Realizing the needs of data implementations, SDH readily accommodated data by various mapping procedures such as LAPS, ATM mapping, etc., and when the inefficiencies were discovered, SDH developed into a new generation called next-generation SDH. We have seen the features such as V-CAT, LCAS, and GFP incorporated by SDH for efficiently carrying all types of data protocols.

But the concept of pure packet network was still attracting the vendors and operators because of the SDH's negative characteristics mentioned above. Some vendors tried to implement a pure packet network, which met with a heavy success at the enterprise level, but when it was tried for carrier-level services, it was not able to match the performance of SDH on the operation and management front. The SDH's robustness with data friendly features again gave an edge to it. The MSPP provided the TDM as well as the data services.

Meanwhile, RPR was developed as a pure data carrier transport service, which again had several advantages over SDH as brought out earlier in this chapter. RPR could operate as a separate packet network at the transport level or it could be configured on SDH. All the advantages of RPR can be availed even when it is configured as a concatenated payload on the SDH.

The biggest advantage SDH has over all its rivals is its already deployed network. On any given day, if an operator has to provide any additional service, he would prefer to use the existing infrastructure as far as possible with whatever additional interfaces required. Since various interfaces keep getting developed for SDH, they get deployed as well and add to the already deployed capacity of SDH. If the operator desires to provision a new service using an RPR ring, he may need to install a large number of RPR nodes to cover the geography of the customers. Such a deployment will have prohibitive costs, which may not make a business sense in a competitive environment. However, on the other hand, the operator may plan to establish an RPR network, keeping the future demand of 1 to 2 years in mind, and get returns on his investment during this period. In the world of uncertainties, he would definitely prefer to go for a safer option, which may be a mix of the two deployment philosophies.

A new platform recommended by ITU-T, which has already started getting manufactured and deployed, called OTN, is designed to optimize the transport of SDH-based services aside from Ethernet and DWDM services. This platform is likely to further boost the chances of long-term survival of SDH.

From all the above variations of technologies and market forces, it is very difficult to predict the future technology that will survive with time. However, most experts predict that the next-generation SDH is going to dominate the market for a long time to come.

The final scenario that appears to be emerging today is that the OTN deployment may slowly phase out other variations. The boundaries of TDM and data may fade with further development within the OTN. However, we will have to only wait and watch to really be the witness to what finally happens.

Review Questions

1. What were the driving forces for the next-generation SDH?

2. Which are the factors that will keep the SDH going for at least 10–15 years even if the backbone networks start functioning on pure data systems?

3. What are the socioeconomic and technological factors leading to a demand of pure IP backbone network?

4. List the apparent advantages of pure IP backbone.

5. What are the strengths of SDH that are keeping it going?

6. List the special features/improvements done in SDH to make it the next-generation SDH.

7. What are the limitations of contiguous concatenation for efficient data transport?

8. Describe the process of V-CAT with an illustration of VC-4s concatenation. What are the benefits of V-CAT over contiguous concatenation?

9. Which nodes in the network are required to have a feature of V-CAT and why? What is a VCG?

10. Explain with calculations how the efficiency of 1-Gbps Ethernet payload is adversely affected in contiguous concatenation.

11. Compare and tabulate the efficiencies of some other data protocols in contiguous concatenation.

12. Compare and tabulate the efficiencies of these protocols in case of V-CAT. What are the possible groupings of containers in V-CAT?

13. What are HO-V-CAT and LO-V-CAT? What is the advantage of LO-V-CAT?

14. How does V-CAT help in quick provisioning and improve capacity utilization efficiency? Illustrate with an example.

15. Does V-CAT help in improving the service resilience? If yes, illustrate with an example.

16. Explain the process of payload identification and realignment in V-CAT operations by illustrating an example of HO-V-CAT.

17. What do you understand about frame numbers and sequence numbers in V-CAT operation? What are the limits and how are they decided? How are these numbers carried over the network?

18. What is LCAS and what is its role in next-generation SDH?

19. Illustrate with an example how LCAS is able to reconfigure the payload for improving the reliability of service.

20. How is the service quality improved by LCAS by removing failed members?

21. What are the advantages of dynamically changing the bandwidth through LCAS? Can it automatically resize the VCG to suit the traffic pattern?

22. Can LCAS facilitate asymmetric bandwidth provisioning when working with non-LCAS nodes? If yes, what are the benefits of these features?

23. How are LCAS commands incorporated in H4 and K4 bytes? What are higher- and lower-order control packets?

24. What action is performed by the commands MST, resequence acknowledge (RS Ack), SQ, CRC-8, and GID?

25. How do the LCAS operations perform for lower-order V-CAT?

26. What were the reasons compelling the development of GFP?

27. What is GFP? Please explain with an illustration.

28. List the advantages of GFP.

29. How does a common encapsulation protocol help in efficient transport of data?

30. How is the need of stuff bytes in data packets avoided in GFP? What are the advantages gained?

31. GFP facilitates efficient data transport. How?

32. How is GFP optimized for specific services?

33. What are the network topologies supported in GFP?

34. Draw the general structure of a GFP frame. What is the main strength of the GFP? What is the error probability in the header?

35. Describe the functions of the payload header of GFP. What types of payloads are supported currently?

36. What is GFP(F)? Show the mapping of some popular protocols into GFP(F).

37. What is GFP(T)? Why it is required and what are its special features?

38. What protocols are currently supported in GFP?

39. Compare and tabulate the salient features of GFP(P) and GFP(T).

40. What is a RPR? Explain the concept with a diagram.

41. List the classes of service provided in RPR. What are the features of these classes?

42. What do you understand about fairness control in RPR?

43. What are the terms "steering" and "wrapping" in the context of RPR?
44. Discuss in brief the management features of RPR.
45. What are the advantaged of RPR over SDH?
46. Define MSPP, MSSP, and MSTN.
47. What are the WDM and OTN technologies? Explain briefly.
48. What do you think is going to be the technology of the future—SDH, pure data network, or OTN? Why?

Critical Thinking Questions

1. Plan and explain with schematic diagram a sample data traffic being carried over SDH.
2. What is your own perception of a development and deployment scenario of various technologies in the next few years?
3. Of the three improvements, V-CAT, LCAS, and GFP, which is the most important and why?
4. How would you load a data of 1.25 Gbps using V-CAT into VCs of various sizes to get maximum efficiency?
5. Why do we need a feature of dynamically changing the bandwidth? Is it really required in practical situations? By whom and why?
6. What are the reasons for catering for classes of service in data communication services?

Bibliography

1. W.H. Hayt, J.E. Kemmerly, and S.M. Durbin, *Engineering Circuit Analysis*, Sixth Edition, Tata McGraw-Hill, 2008.
2. *Optical Transmission in Communication Networks*, Web Pro-Forum Tutorials, International Engineering Consortium, 1998.
3. G. Saundra Pandian, *IP and IP-Over SDH*, Training course, Railtel Corporation of India, 2003.
4. P. Moulton and J. Moulton, *The Telecommunications Survival Guide, Understanding and Applying Telecommunication Technologies to Save Money and Develop New Business*, Pearson Education, 2001.
5. *Introduction to Synchronous Systems*, Training material, M/S Alcatel, 1994.

6. J. John and S.K. Aggarwal, *Proceedings of the International Conference and Exposition on Communications and Computing IIT Kanpur*, February 2005.

7. W. Stallings, *Data and Computer Communications*, Seventh Edition, Prentice-Hall, India, New Delhi, 2003.

8. D.E. Comer, *Computer Networks and Internets*, Second Edition, Pearson Education Asia, 2000.

9. J.H. Franz and V.K. Jain, *Optical Communications Components and Systems*, Narosa Publishing House, New Delhi, 2000.

10. A.S. Tanenbaum, *Computer Networks*, Fourth Edition, Pearson Education, 2003.

11. S.V. Kartalopoulos, *SONET/SDH and ATM, Communications Networks for the Next Millennium*, IEEE Press, 2003.

12. K.R. Rao, Z.S. Bojkovic, and D.A. Milovanovic, *Introduction to Multimedia Communications, Applications, Middleware, Networking*, John Wiley and Sons, 2006.

13. ITU-T Recommendation G.709, *Interfaces for the Optical Transport Network (OTN)*, International Telecommunication Union, Geneva, Switzerland.

14. ITU-T Recommendation G.7041, *Generic Framing Procedure (GFP)*, International Telecommunication Union, Geneva, Switzerland.

15. ITU-T Recommendation G.783, *Characteristics of Synchronous Digital Hierarchy (SDH) Equipment Functional Blocks*, International Telecommunication Union, Geneva, Switzerland.

16. ITU-T Recommendation G.7042, *Link Capacity Adjustment Scheme (LCAS) for Virtual Concatenated Signals*, International Telecommunication Union, Geneva, Switzerland.

17. ITU-T Recommendation G.7043, *Virtual Concatenation of Plesiochronous Digital Hierarchy (PDH) Signals*, International Telecommunication Union, Geneva, Switzerland.

15

Transmission Media for PDH/ SDH and OFC Technology

The transmission of plesiochronous digital hierarchy (PDH) and synchronous digital hierarchy (SDH) signals is possible on any of the prevalent media, i.e., copper wire pairs, coaxial cables, radio, or optical fiber, as the multiplexing technologies and hierarchies are totally independent of the media type. However, as the data rates increase, the higher bandwidth media, i.e., optical fiber, emerges as the natural choice. In fact, SDH has been designed with optical fiber cable (OFC) in mind, and thus, OFC is naturally best suited for SDH and high-rate PDH signals. As has been seen (Section 10.1) the PDH has already become obsolete with the advent of SDH now. In the initial days, PDH was and still is being transported on all types of media because the feeder tributaries to SDH are still the low-rate PDH signals by and large. As the OFC technology matured, all the long-distance circuits of high bit rates and high densities started shifting from their then preferred media, which was radio, to the optical fiber. Shifting to OFC boosted the possibility and hence the implementation of higher data rates and, consequently, many more numbers of user circuits. However, OFC has its limitations as well. Take two adjacent hills, a kilometer or so apart and separated by a valley, as a hypothetical example. Deployment of any type of cable including OFC is very difficult in such a situation. This is the reason for the radio media still being alive—for patching gaps that cannot be covered by the OFC. A similar situation is in metro city areas where laying cables through the buildings and roads is a stupendous task. Currently, the demand of radio media is driven more by such difficulties in metro city areas, rather than in barren territories.

Let us study the suitability of other types of media for PDH/SDH in more detail.

15.1 Types of Media for PDH/SDH Transmission

15.1.1 Copper Wire Pair

Copper wire twisted pairs in the form of multipair cables are the second generation of communication media, after the open wires made of galvanized iron or aluminum. They were deployed initially for connecting a subscriber to the local exchange, and for that application, they are still the only medium barring cellular mobile telephony, as mobile telephone signals are on wireless media.

However, when the need for intercity communication arose and copper wire pairs were deployed to carry multiplexed multichannel circuits on carrier waves, they could offer only a very small time solution. For carrying merely 12 voice channels on the copper wire pair, repeaters were required at every 6 km or so. With every repeater adding noise and distortion, it was not possible to maintain an acceptable quality for long distances, as it required large number of repeaters. The world then switched over to coaxial cables and radios, which we will discuss in the subsequent sections.

The PDH signals in the form of the initial E1, which was called a PCM, were carried on the copper wire pairs. The arrangement provided 30 voice channels multiplexed on two pairs of wires, but the distance covered for satisfactory reception without requiring a repeater was merely 2 km or so. Thus again, a long-distance link required a very large number of repeaters. However, because of the digital advantage, the noise and distortion did not build up at every repeater due to digital regeneration of signals (refer to Sections 2.4–2.8 to know how this occurs), the large number of repeaters proved to be too costly, and therefore, a search started for a better media that could provide the much-needed relief in repeater spacing.

The fundamental reason why the copper wire pair cannot carry the signal to longer distances is its frequency response characteristic. The transmission line, as it is called, can be modeled as being composed of small sections of elements of resistors, inductors, and capacitors (see Section 2.2.3.1). Due to the frequency-sensitive components, the attenuation (or losses) in the line are frequency dependent. In fact, the attenuation is proportional to the signal frequency. The higher the frequency, the higher is the attenuation. The higher the bit rate, the higher is the frequency band, and the losses in the copper pair transmission line at bit rates of 2 Mbps for a PCM were very high, limiting the transmission distance to 2 km or so, while the 3.4-kbps voice channel traveled through it quite comfortably. The distance covered by voice channels, being of low frequency, was limited mostly by the ohmic resistance of the wires.

Obviously, more than 2-Mbps data rates were not deployed on copper wire pairs. Hence, the question of carrying SDH signals on it does not arise. The copper wire pair is still, however, the only option available for connecting a landline telephone subscriber to the exchange. It is also almost universally deployed in its new avatar, i.e., unshielded twisted pair (UTP) cable in the local area networks (LANs) in data communication networks. In fact, it is so popular that so far six of its versions have hit the market, called UTP CAT-1 (category 1) to UTP CAT-6 (category 6). Although the LANs deploy coaxial cable and OFC for special high-data-rate applications, the UTP cable is by far the most widely used media that gives a data rate of up to 100 Mbps up to a distance of nearly 100 m.

15.1.2 Coaxial Cables

As the name suggests, the coaxial cable has two axes. It is a cylindrical cable consisting of two conductors. The first conductor is a wire in the center, and the second outer conductor is a solid or breaded sheath on the dielectric material surrounding the conductor. This construction keeps the central conductor at a constant distance from the cylindrical outer conductor for the whole length of the cable (Figure 15.1).

As seen in the previous section, the number of channels possible on the copper wire pairs was very limited aside from the system needing a very large number of repeaters. To overcome this problem, the coaxial cable was invented. It made use of one of the properties of conducting wires, which becomes prominent at high frequencies, called "skin effect." The skin effect leads to the concentration of current flow on the outer surface of the wire (or skin of the wire, so to speak) as the frequency increases.

Coaxial cables were initially deployed for the analog signals, and they were able to carry as many as up to 10,800 channels. Although the systems carrying a fewer number of channels had repeaters at comparatively larger distances, the more channel-packed systems had closely spaced repeaters. The number of channels and repeater spacing for some cases are shown in in Figure 15.2.

It clearly was a tradeoff between the number of channels and repeater spacing. Low density systems could be afforded in thinly populated areas, whereas most areas needed a high number of channels leading to close repeater spacing. The repeater spacing, however, remained in the range of 4 to 8 km only at the maximum. As can be seen from Figure 15.2, the repeater spacing decreases to 1.5 km from 4.5 km as the number of channels are increased from 2700 to 10,800. This is because coaxial cables, like the copper

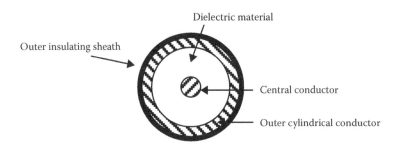

FIGURE 15.1
Coaxial cable.

Number of channels	Repeater spacing (km)
300	6–8
2700	4.5
10,800	1.5

FIGURE 15.2
Repeater spacing of typical coaxial cable systems.

wire pair, suffers transmission losses that are proportional to the frequency. As the number of channels becomes increases, the frequency range (bandwidth) also increases and so does attenuation.

The same coaxial cable was deployed for the PDH signals when digital technology developed. The systems included a mix of analog and digital signals, besides video and digital (PDH) channels. For example, a communication system on a 2.6/9.5-mm coaxial cable could carry 1800 analog channels and one E2 (6448 Mbps) PDH bit stream, with a repeater spacing of 4.5 km, whereas another system on the same coaxial cable could carry one E4 (140 Mbps) PDH bit stream, with similar repeater spacing of 4.5 km. The coaxial cables with higher diameters could obviously carry more numbers of channels or could cover larger distances as compared with those of lower diameters.

As can be seen, the enhanced bandwidth available on coaxial cables was used in packing more numbers of channels, rather than increasing the repeater spacing, and this might be economically the best suitable solution.

SDH, however, was not tried on coaxial cables, as the development of OFC technology preceded the development of SDH.

The coaxial cables are still in use but hardly ever for the long-distance transmission. They are mostly used as feeder cables for the radiofrequency wireless systems, in cable TV networks, and in LANs, to a limited extent.

15.1.3 Microwave Radios

The problem of a very large number of repeaters, and hence a large amount of noise and distortion, and frequent breakdowns due to occasional breakdown of one of the several repeaters, in the case of long-distance circuits, was resolved to some extent by the deployment of wireless communication systems. These wireless communication systems were popularly called "microwave radios," as they operated mostly in microwave frequency range (3–30 GHz).

The antennae of the microwave systems were installed at towers with heights of 100 m or so. The space wave communication (line-of-sight direct communication) facilitated a range of 50 to 100 km, depending upon the clearness of the territory in between (and location at hilltops and so forth) and the height of the towers. This increased repeater spacing provided the much-needed relief, and drastically improved the quality of long-distance communication. The analog microwave systems developed could carry up to as much as 6000 voice channels.

The microwave radios were used to carry the PDH signals as well. In fact, they were the only hope for good long-distance communication before the OFC technology was developed. The radios were popularly called digital microwave radios.

With the advent of OFC technology, the microwave radios started fading into the background, as the OFC took over the long-distance communication completely because of its many advantages. In fact, the OFC removed "all" the problems that were being faced by the communication engineers in the area of transmission. It facilitated very high densities of channels on a single pair of fiber (for example, 30,240 channels could be carried in an STM-16) and many fibers in a cable and repeater spacing, which was equivalent to or better than that of microwave radios.

Digital microwave radios have recently got a new lease on life in filling up the gaps that are left out by the OFC system. In metro city areas, it is sometimes very difficult to lay OFC in certain locations or it may have a huge cost or long execution time. In such locations, digital microwave radios come in handy. The radios available today can give from STM-1 (1890 channels) to STM-4 (7560 channels). The STM radios are also useful in such or any other difficult terrain to complete the protective rings of the OFC-based SDH network. Thus, as such, the microwave radios are suitable for PDH and SDH, but the capacities available are very limited as compared with the OFC-based systems, and hence they are likely to remain restricted to the role of filling the gaps as stated above or to provide the last mile connectivity where terrain or time are constrained for OFC provision.

15.1.4 Free Space Optics (Air Fiber)

Another option available today for the transport of SDH signals is "free space optics" (FSO), also called "air fiber" (so called because of its high carrying capacities, which are comparable to optical fiber). PDH hierarchical transport is not available on this system, as this has been developed after the PDH transport (not the tributary level) had already become obsolete, and the SDH was in full swing. In fact, the technology has seen practical deployment only in the last 4 to 5 years only.

FSO technology is a fairly recent development, as stated above. It is a wireless system working at infrared or laser frequencies. The driving force for the development of this technology is the metro area distribution of high bandwidth, to fill the gaps in OFC reach and provide a protective ring to the SDH (or even for the IP-based data traffic, as the case may be), working on optical fiber. It is virtually the same job that is being done by the SDH radios but with a difference. The main differences in FSO systems with respect to SDH radios are as follows:

(i) It has a very short range (up to 1 km or so) as compared with SDH radios (up to 50 km or so); thus, it is not suitable for high distances. The infrared and laser frequencies experience a high amount of absorption in foggy weather or during heavy rains; thus, FSO is not very suitable for areas where fog persists for long periods.

(ii) The equipment and antenna are very small, leading to very fast installation. The antenna can be mounted just outside the window (or even inside the window, beaming through glass). The whole installation procedure for establishing the link is a matter of a few hours. Redeployment elsewhere is equally easy, with no wastage whatsoever.

(iii) Systems are readily available for bandwidths up to STM-4 and GB Ethernet. Developments on further capacities are in progress, and higher capacities are likely to become available soon.

Due to the advantages of ease and speed of deployment, the systems are very attractive for patching the gaps of OFC reach in metro city areas, as explained earlier. They also prove to be very useful in extending the distribution layer, i.e., last mile, of the main SDH or data network. A FSO terminal installed on a huge building may be able to meet the broadband Internet and telephony requirements of hundreds of individual users or that of a few corporate ones.

15.1.5 Optical Fiber Cable

OFC is obviously the best suited media for the transmission of PDH or SDH signals. In fact, the high data rates of SDH such as STM-4 (622 Mbps), STM-16 (2.5 Gbps), STM-64 (10 Gbps), and STM-256 (40 Gbps) have been designed with OFC media in mind, as it is not possible to achieve the transmission of such high data rates on any other media.

Besides such high data rates, repeater spacing of up to 100 km is possible due to low attenuation, depending upon the transmitted laser power and receiver threshold. The OFC transmission, being on a nonmetallic and hence a nonconducting media, is free from many types of noises and interferences, which are a big nuisance in the metallic media. Also one cable may contain several fibers, permitting many parallel transmissions. An even more wonderful feature is that one pair of fiber can carry several wavelengths (presently up to 160 wavelengths are available). Each wavelength carries its own STM-N (N = 1, 4, 16, 64, or 256), thus multiplying the capacity manyfold; the

OFC technology is a true revolution for the capacity of information carry-ing media. On top of all these fascinating features, another and perhaps the most important factor is that the OFC has the lowest cost; in fact, for a given capacity, the cost of the OFC system (including media and transmitting and receiving equipment) is a very small fraction of the cost of other prevailing systems.

We will see some details about this fascinating technology in the following sections.

15.2 Optical Fiber Communication Technology

Optical fiber communication technology uses optical fiber for transmission of communication signals, instead of copper pair or coaxial cables or radio microwave. The optical fiber is made of glass (silica). Why is there a shift from copper to silica? Silica's low cost notwithstanding, there are a number of advantages offered by this new technology as well. In fact, there is no other technology that can surpass optical fiber in its role in current telecom-munication's massive needs, which include voice telephony, video transmis-sion, data transmission, video conferencing, Internet, etc.

Let us first see what this technology is all about.

15.2.1 Principles of OFC

The OFC for communication is made of glass. The construction of an optical fiber is shown in Figure 15.3.

As shown in the figure, the fiber is made of a core, a cladding, and a protec-tive jacket. The core and the cladding are both made of silica, but the core is doped with another material, such as germanium, to increasing its "refrac-tive index." The principle of propagation of light in this fiber is illustrated in Figure 15.4.

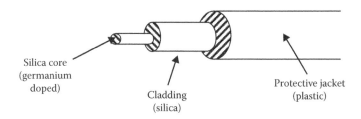

Silica core
(germanium
doped)

Cladding
(silica)

Protective jacket
(plastic)

FIGURE 15.3
Construction of an OFC.

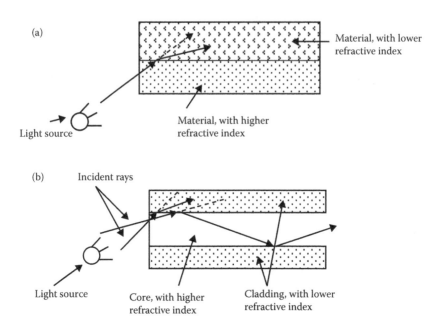

FIGURE 15.4
(a) Bending of a light ray as it enters the medium of a different refractive index. (b) "Total internal reflection" in the case of optical fiber.

A basic principle that the light follows is that whenever a ray of light enters a medium of higher refractive index, it bends slightly toward the medium of higher refractive index instead of traveling straight (Figure 15.4a).

The refractive index of the core is increased by doping the silica with another material, such as germanium, whereas the cladding is generally made of pure silica. Thus, the refractive index of core is kept more than that of the cladding. Hence, when the light rays are fed into the core, those rays that are directed toward the cladding bend toward the core while passing through the cladding. Now if the angle of incident light on the core is adjusted so that the rays that tend to enter the cladding are bent to such an angle that they actually fail to enter the cladding and get totally reflected inside the core, no rays will enter the cladding. Any angle less than this angle will produce the same result. This phenomenon is called "total internal reflection" of the light wave and results in the propagation of the whole of the light through the fiber. In the telecommunication applications, this angle is normally in the range of 5.7° to 11.5°.

While we can transmit light through the optical fiber, how does it help the communication of signals such as PDH, SDH, etc. To use this light transmission, we need to convert our electrical signals to the light (optical signal) and send them onto the fiber and at the receiving end convert them back into electrical signals to establish a communication link (Figure 15.5).

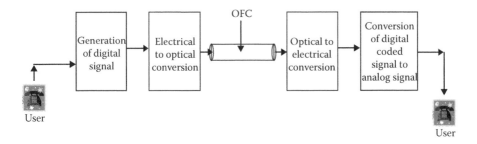

FIGURE 15.5
A communication link using optical fiber.

15.2.2 Optical Frequency/Wavelength Used

To understand the feasibility of the frequencies to be used for optical communication through optical fibers, let us again consider the fiber construction (Figure 15.6).

The propagation of the optical ray can take place properly as explained in Section 15.2.1, provided the wavelength (λ) of the optical signal is less than the radius of the core. Thus, for proper propagation

$$\lambda < d/2.$$

It is implied from the above equation that all the wavelengths lower than $\lambda = d/2$ will propagate, resulting in the optical fiber behaving like a high-pass filter and having infinite bandwidth as long as the wavelength is smaller than $d/2$. Hence, any frequency corresponding to or higher than a wavelength of $d/2$ can be used for the transmission, with no limitation on the bandwidth.

However, all things have a limit, and so is the case with optical fiber communication. There are practical limits forcing the use of particular wavelengths and the bandwidths are limited. Figure 15.7 indicates the attenuation versus wavelength characteristic of an optical fiber.

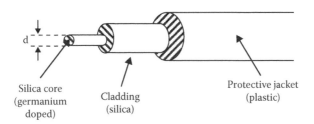

FIGURE 15.6
Construction of an OFC.

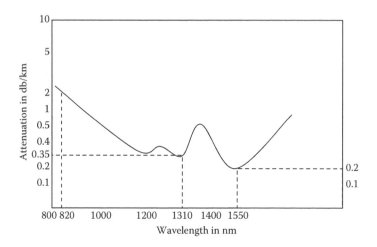

FIGURE 15.7
Attenuation characteristic of optical fiber.

As shown by the figure, there are two clearly marked dips in the attenuation characteristic of the fiber: one is at 1310 nm and another is at 1550 nm. It is definitely better to use these frequencies rather than any other to take advantage of lower attenuation and to have longer repeater spacing. The attenuation at 820 nm wavelength is 2 db/km, at 1310 nm wavelength, it is 0.35 db/km, and at 1550 nm, it is 0.2 db/km. The initial systems were developed at 820 nm; however, when devices were developed to operate at higher wavelengths, the low attenuation wavelengths 1310 and 1550 nm could be used. Bandwidths of the order of 20,000 GHz (20×10^{12} Hz) are possible at 1310- and 1550-nm wavelengths. The bandwidths are not infinite, of course, but such high orders of bandwidths are unimaginable in any other system. (When the optical fibers were first developed in the 1990s, they had attenuation of up to 2 db/km, using 820-nm wavelength. The technological improvements, however, brought this value down to 0.2 db/km in the 1980s with the use of 1550-nm wavelength.)

The wavelengths 1310 and 1550 nm fall in the "near infrared" region of the electromagnetic spectrum. Their corresponding frequencies are given by

$$f = c/\lambda$$

where c is the velocity of light (3×10^8 m/s, or approximately 2.998×10^8 m/s) and f is the frequency.

Therefore, for 1310 nm,

$$f = (3 \times 10^8)/(1310 \times 10^{-9})$$
$$= 2.29 \times 10^{14} \text{ Hz}$$
$$= 230 \text{ THz (approximate)}$$

For 1550 nm,

$$f = (3 \times 10^8)/(1550 \times 10^{-9})$$
$$= 1.93 \times 10^{14} \text{ Hz}$$
$$= 193 \text{ THz (approximate)}$$

The optical fiber communication uses 2.3×10^{14} and 1.93×10^{14} Hz frequencies, which fall in the "near infrared" region of the electromagnetic spectrum (close to red color of visible light), due to their low attenuation.

The above figures of attenuation are not always true, but they depend on the type of fiber. The stated values are for a "single-mode" fiber made of "glass." In the next section, we will see various types of fibers.

15.2.3 Types of Optical Fibers

The optical fibers may be classified on the basis of material of construction, mode of operation, and the type of refractive index deployed. On the basis of material of construction, the fibers may be divided in two categories:

(i) Plastic fiber
(ii) Glass fiber

Per the mode of operation, the fiber can be divided in three categories:

(i) Multimode fiber
(ii) Single-mode fiber
(iii) Dispersion-shifted fiber

Lastly, on the basis of the refractive index deployment, it can be divided in the following two categories:

(i) Step index fiber
(ii) Graded index fiber

Let us see about them in brief.

15.2.3.1 Plastic Fiber

In plastic fibers, the core is made of either plastic or glass, and the cladding is made of plastic. These fibers are more flexible physically, facilitating easy handling. The losses in plastic fibers are much more than those of the glass fibers; hence, they are mostly deployed for short-distance applications, e.g., within campus such as LAN or SDH connectivity. The lifespan of these fibers is also short.

15.2.3.2 Glass Fibers

In glass fibers, the core is made of silica doped with some other material such as germanium to increase its refractive index. The cladding is usually made of pure silica. The difference in the refractive index helps in achieving the total internal reflection and hence facilitates a proper propagation of the light wave as explained in Section 15.2.1. Glass fibers have much less attenuation as compared with plastic fibers and are almost invariably used for long-distance communication, but why are the fibers not breaking often, given that glass is very brittle? It does not break because a very high tensile strength is achieved when glass of highest purity is used and microscopic flaws are removed. The attenuation characteristic shown in Figure 15.7 pertains to glass fiber only.

15.2.3.3 Multimode Fiber

When the diameter of the core is sufficiently high, more than one "ray" of light is permitted to travel through it, and the fiber is called a "multimode fiber." The typical core diameter of multimode fibers is 50 to 125 μm. We have seen that the wavelengths popularly used in the present generation of optical fibers are 1310 and 1550 nm. Thus, diameters of 50 to 125 μm are approximately 38 and 95 times for the 1310-nm wavelength and 32 and 81 times for the 1550-nm wavelength, respectively. The phenomenon of multimode propagation is illustrated in Figure 15.8.

Due to the large diameter of the core, many light rays simultaneously get launched into the fiber. Besides a straight ray, many other rays also travel in the core after getting reflected/refracted from the cladding. The straight ray is called the "main mode," whereas the other reflected rays are called the "secondary modes." Due to difference in the path length, various modes reach the receiver at different times, leading to spreading of the signal waveform. This phenomenon is called "multimodal dispersion." Due to

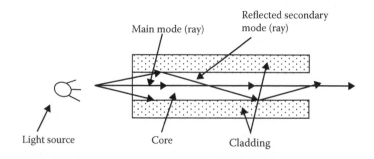

FIGURE 15.8
Multimode propagation.

this multimodal dispersion, the bandwidth of the signal and thus the data rates become limited.

To minimize the problem of multimode dispersion, either we can use single-mode fibers or we may use the "graded index fibers" (which we will discuss shortly in a subsequent section). The single-mode fiber needs to have a very small core diameter and hence is more expensive than the multimode fiber due to precision manufacturing requirements. However, with the volumes increasing due to high demand, the single-mode fiber has itself become very cheap, limiting the use of multimode fibers to low-data-rate applications only.

15.2.3.4 Single Mode Fiber

In contrast with the multimode fiber, the single-mode fiber allows only one, i.e., main, ray to propagate through the core. Single-mode propagation is achieved by reducing the diameter of the core substantially. A typical diameter of the single-mode fiber is 8 μm. Figure 15.9 shows the single-mode propagation.

The small diameter of the core allows launching of only the main (or principle) mode, doing away with the multimode dispersion and consequent pulse spreading. Better data rates (or bandwidths) are thus achieved. In almost all long-distance applications, only single-mode fibers are used, with multimode fibers restricted to local area (short distance) low-bandwidth applications. However, as stated in the previous section, with the cost of the single-mode fiber coming down very fast due to high volumes of production, the application of multimode fiber is further marginalized.

15.2.3.5 Dispersion-Shifted Fibers

We have seen that different frequencies travel through the transmission media at slightly different velocities causing phase or delay distortion, in the case of non-OFC communication systems (see Section 2.2.3). The same

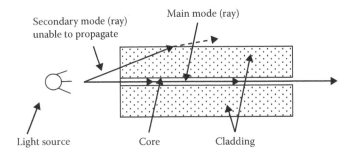

FIGURE 15.9
Single-mode propagation.

phenomenon takes place in optical fibers as well, but because of the involvement of optical frequencies, it is called "chromatic dispersion."

To minimize this dispersion, we should try to use the optical source that is as spectrally pure as possible, not allowing many wavelengths. We may also choose a line code with inherent ability to minimize the inter symbol interference, which can tolerate higher chromatic dispersion.

Although dispersion may not be a big problem in the usual systems carrying STM-16 or so on a single optical frequency, in the case of dense wavelength division multiplexing (DWDM), where a large number of wavelengths multiplexed together are transmitted as a single optical signal, the dispersion becomes a major bottleneck in achieving reasonable repeater spacing. The problem is more pronounced on 1550-nm wavelength, as it gives nearly 15 times more chromatic dispersion as compared with 1310 nm. Since most long-distance backbone links operate on 1550 nm due to higher distances of repeaters possible, because of low attenuation, the chromatic dispersion is very high.

Thus, to tackle the problem of chromatic dispersion, special fibers have been developed called "dispersion-shifted fibers." These fibers have substantially less chromatic dispersion than the normal single-mode fiber and are easily able to support the WDM applications. In these fibers, the point of minimum dispersion is shifted to a specific wavelength (say 1550 nm), which gives low attenuation by a manipulation of reduction in the core diameter, varying the relative refractive index, and change in material composition.

15.2.3.6 Step Index Fiber

As we have seen above, the refractive index of the "core" material is higher than that of the "cladding" material to achieve the phenomenon of total internal reflection. This change of refractive index is sudden or in a step, at the boundary, where the core and cladding meet; hence, it is called "step index," and the fiber is called "step index fiber." The need to name it so arises because in some fibers the index changes gradually in the core itself until it meets the cladding. Such fibers are called "graded index fibers," which we will discuss in the next section.

The single-mode fiber that we discussed in the previous section is a step index fiber, and the reasons will become clear as we discuss the graded index fiber in the next section.

15.2.3.7 Graded Index Fibers

In the case of multimode fiber, we have seen above that the multimode propagation results in multimodal dispersion, which results in signal waveform spreading and limits the throughput of the fiber. This problem can be overcome if all the rays, which may travel through different paths (direct or reflected), are somehow made to reach the receiver at the same time.

We know from the simple optics that the light travels faster in a medium of lower refractive index and slower in a medium of higher refractive index. Thus, if we design the refractive index of the core in such a way that it is more in the center and less on the edges, then the rays of light that travel through multiple reflections and take a longer time to reach the receiver can be made to travel faster as compared with the main or principle ray, which will travel slower due to its higher refractive index. Thus, if the distribution of refractive index is carefully chosen throughout the core diameter, it is possible to achieve a condition where the main and the reflected rays will reach the receiver at the same time. The more closely we achieve this condition, the lesser will the multimode dispersion be, and we will be able to achieve much higher throughput from the fiber.

Such fibers are called "graded index fibers" and are a variety of multimode fibers.

15.2.4 OFC System Components

The optical fiber communication technology does not change the signal generation process. For example, for a speech signal to be generated by a person, it has to be converted to a digital signal and necessary coding has to be done for transmission. This process, as in the earlier cases, generates the digital signal in the electrical form. As shown in the optical communication link diagram of Figure 15.5, the electrical signal needs to be converted to optical signal at the transmitter and back to electrical signal at the receiver. The process is similar to the modulation of a carrier or radiofrequency baseband digital signal (see Section 4.3), with a difference that the carrier signal is an optical signal in this case.

The optical devices are very different from the devices operating at radio frequencies; thus, we will have a brief look at the types of devices and their functioning.

The optical devices that we need are

(i) Optical source: carrier for transmission of digital signal
(ii) Photodetector: receive the optical signal and demodulate it into the electrical signal

15.2.4.1 Optical Source

Two types of optical sources are popularly used in optical fiber communication:

(i) LED
(ii) LASER diode

The LASER diode is technically far superior out of the two; however, as far as applications in OFC systems are concerned, there is no absolute superiority of one over the other, as each one has some advantages over the other.

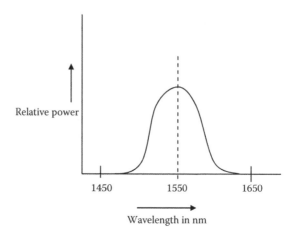

FIGURE 15.10
Frequency spectrum of an LED.

15.2.4.1.1 LED

The LED is the simplest optical source available for optical fiber communication. It is a low-cost device used with multimode fibers. It cannot be used with single-mode fibers because its frequency spectrum is much wider than warranted for single-mode operation. Figure 15.10 shows the frequency spectrum of a typical LED.

The launched power of LED is much less as compared to that of LASER diodes; hence, they are suitable for short distances only. (Launched power is a product of the output power and the coupling efficiency in the fiber. In case of LED, the coupling efficiency is very poor, of the order of 1% to 2%, as compared with that of LASER diodes, where it is up to 60%.)

Due to the above features, the low cost, and requirement of less complex interface circuitry, LEDs mostly find application in short-distance optical fiber communication systems requiring data rates of up to a few hundred megabytes per second. LEDs are generally more tolerant to environmental adversities and thus are more rugged.

15.2.4.1.2 LASER Diode

There are many types of LASERs, such as gas lasers, dye lasers, solid-state lasers, and semiconductor lasers. The optical fiber communication mostly used semiconductor lasers.

The basic requirement of an optical source for long-distance optical fiber communication is that it should be able to produce a very narrow frequency spectrum to make possible the single-mode transmission, as discussed earlier in this chapter. The distributed feedback (DFB) laser is able to produce

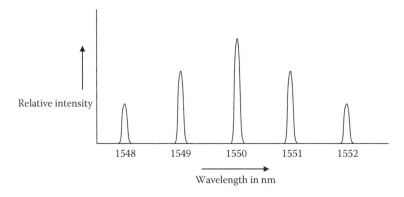

FIGURE 15.11
Frequency spectrum of a laser diode.

extremely narrow frequency spectrum. The spectral half widths of up to 0.0001 nm are possible in DFB lasers. (Spectral half width is defined as the width of the frequency spectrum at half of the maximum magnitude. In case of LED, this spectral half width is in the range of 50 to 100 nm.) The frequency spectrum of a typical laser diode is shown in Figure 15.11.

Laser diodes are much more expensive than the LEDs, but due to their ability to produce extremely narrow (coherent) frequency spectrum as shown above, they are invariably deployed for long-distance communication in single-mode operation. They need a much more controlled environment to maintain their parameters within desirable ranges. Their power output is much higher than the LEDs, and they have a much better coupling efficiency (up to 60%) due to the narrow spectrum. Thus, the "launched power" in case of laser diodes is much higher, making them suitable for achieving long distances without repeaters.

15.2.4.2 Photo Detector

The job of the photo detector at the receiver is to detect the incoming optical signal and convert it to an electrical signal. The main features that are sought in a photo detector are high sensitivity (to be able to receive a weak optical signal), fast response time (to be able to correctly reproduce the high data rate variations), can operate in the desired frequency range, ruggedness, and reasonable life and cost.

There are several types of photo detectors available in the market. Semiconductor devices such as the PIN diode and avalanche diode (ASD) are the ones generally used in optical fiber communication. Both these devices have their own pros and cons and are used depending upon their suitability, as described below.

The silicon-based PIN diodes are simple and low-cost devices. They are fast and reliable. However, they are available only in the 600- to 900-nm wavelength range, whereas the currently popular low attenuation wavelengths are in the range of 1300 and 1550 nm.

The avalanche photo diodes (APDs) based on InGaAs (indium gallium arsenide) are able to operate in the desired wavelengths of 1300- and 1550-nm range. These devices are slower than PIN diodes, being only half as fast. They have a big advantage on the PIN devices because they have an internal gain of several hundreds (20 to 35 db) due to the avalanche process, which gives rise to a great improvement in signal-to-noise ratio. The PIN diodes, on the other hand, have to use the required amplification externally, which increases the noise.

The photo detector needs to be followed by a "low-noise" preamplifier to amplify the very weakly received signal without adding much noise. The signal then needs to be amplified further to bring it to the levels acceptable by the equalizing and pulse-shaping circuits. After pulse shaping, the signal is sent to the decision making circuit, which decides whether the received signal at any moment is a 1 or a 0. Further digital-to-analog conversion delivers the signal to the subscriber. The optical communication link of Figure 15.5 is modified to include these factors and can be represented as shown in Figure 15.12.

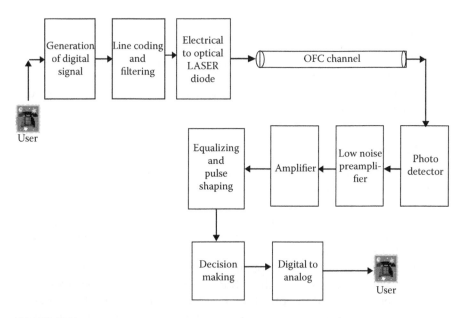

FIGURE 15.12
Optical communication link.

15.3 All Optical Networks

The electrical-to-optical conversion, and vice versa, is required in an optical communication link, as seen in the previous section. However, there are many situations where the dropping or insertion of the "Clint signal" is not required at a particular node, and the repeater just performs the function of regeneration of the whole of the incoming data by converting it from optical to electrical and back to optical and retransmits it to the next repeater. Such a situation is actually a common place rather than an exception, particularly in transoceanic undersea links or long-terrestrial links with cities far apart. In such situations, if the regenerative repeater can avoid the conversion to electrical and back to optical, a substantial cost can be saved. Although it appears to be impossible, it is possible.

Equipment known as optical amplifiers make this possible. The signal can be regenerated by simply amplifying the carrier optical signal, without going through the electrical stage, of course.

Thus, with the help of optical amplifiers, the signal can be taken to long distances without requiring conversion to the electrical domain, as long as no dropping or insertion is required. However, optical technology has more advantages. Even where we need to drop or insert, conversion of the whole of the optical signal to the electrical domain is not required. We can only drop the required wavelength, whereas the rest of the wavelengths can continue their travel purely in the optical domain. This is made possible by the optical technology equipment called "optical cross connects" (OXCs) and "optical add drop multiplexers" (OADMs). (We have discussed so far the optical communication links with a single wavelength. This type of "dropping or insertion" of a particular wavelength will be applicable in links that use many wavelengths transmitted on a single fiber. These systems are called wavelength division multiplexing, or WDM. We will discuss the WDM systems shortly in this chapter.)

Let us have a look at these optical devices in brief.

15.3.1 Optical Amplifiers

An optical amplifier directly amplifies the optical signal traveling through an optical fiber, eliminating the need of optical to electrical conversion and back, as brought out in the discussion in the preceding section. The regenerative repeaters (where no dropping or insertion of client signal is required) can be completely replaced by the optical amplifiers. The optical amplifiers require very little electronics and are very cost effective. Figure 15.13 illustrates the concept of deployment of optical amplifiers in an optical transmission link.

The optical amplifiers provide a gain in the range of nearly 20 db. If optical preamplifiers are used in the receivers, another 10-db gain can be obtained.

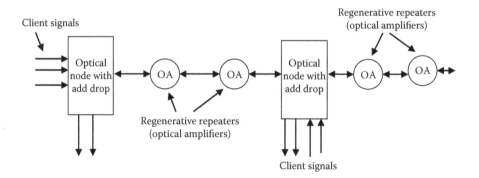

FIGURE 15.13
Optical link using optical amplifiers.

Keeping the same launched optical power and receiver sensitivity, this 30 db gain will increase our repeater spacing by nearly 120 km (with 0.25 db/km loss at 1550 nm band) if the distance is not limited by dispersion.

The optical amplifiers are simple devices made of semiconductor or fiber doped with rare earth materials. There are many types of optical amplifiers; however, the most popular one that provides good amplification at 1550 nm wavelength is the erbium-doped fiber amplifier. It is based on glass fibers doped with an appropriate amount of erbium.

A chain of optical amplifiers can enable establishment of very high distance links, without electrical repeaters. Undersea transoceanic links of thousands of kilometers are possible. One such example is the transatlantic transmission link, with an undersea segment length of 5913 km. This link has 133 optical amplifiers, each of them placed at a distance of 45 km, and it provides an STM-16 (10-Gbps) capacity among the UK, France, and the USA.

Other components of all optical networks are OXCs and OADMs deployed in WDM systems. Before discussing them, let us see what WDM is.

15.3.2 Wavelength Division Multiplexing

The capacity of an optical communication link can be increased manyfold, if instead on one, many wavelengths can be carried simultaneously on the same fiber. This is achieved through the process known as WDM. The process is similar to the analog frequency division multiplexing (FDM) applicable to the electrical domain (see Section 1.8 for FDM). The process is shown in Figure 15.14.

Each digital signal stream of 2.5 (STM-16) or 10 Gbps (STM-64), after its TDM processing, rides onto a particular wavelength. All these wavelengths, from 1 to n, are multiplexed together to make a combined optical signal by the DWM equipment. This combined optical signal is transmitted down the fiber. At the receiving end, a reverse process demultiplexes the combined

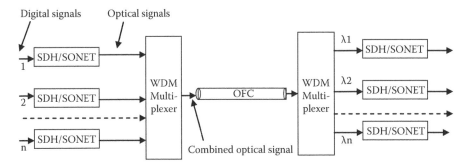

FIGURE 15.14
Wavelength division multiplexing.

optical signal into individual wavelengths, which are then further processed for obtaining back the original digital signal.

The WDM technology has emerged as the most important technology in telecommunication, after the invention of optical fiber. The capacity of a link can be multiplied manyfold (not just increased by some percentage) without any addition of the physical media whatsoever.

This has come as a tremendous boon to the network operators, whose traffic carrying capacities on the already in place optical fibers were exhausted. Without incurring any cost on fiber commissioning and without any loss of time, they are able to multiply their network capacities manyfold by simply deploying the WDM devices.

Although hundreds of wavelengths can be multiplexed together, the practical WDM systems are available in two types: coarse (CWDM) and dense (DWDM). The CWDMs generally consist of a few multiplexed wavelengths, say 4 to 10, whereas higher orders of multiplexing fall in the category of DWDM. DWDM systems of 40 λs, 80 λs, and up to 160 λs are currently available in the market. Imagine the phenomenal capacities being carried by these systems, for example, a 10-Gbps (STM-64) system with 160 λs will carry a whopping 1600 Gbps. With 40-Gbps systems (STM-256) coming up on the electrical side and the number of wavelengths increasing in the DWDM systems, multiterabit capacities are not far away.

The WDM devices are simple optical dispersive devices such as prism, grating, or dielectric thin film filters or they can be made of optical fiber waveguide.

The wavelength separation between any two consecutive wavelengths in the WDM is in the range of 0.8 nm for CWDM, whereas it is in the range of 0.2 nm for the DWDM applications. The spacing of 0.8 nm translates to quite a wide band of frequency. We can calculate it from the equation:

$$v = \lambda f$$

where v is the velocity of light (3×10^8 m/s), f is the frequency, and λ is the wavelength.

Therefore, for $\lambda_1 = 1550$ nm, $f_1 = c/\lambda = (3 \times 10^8)/1550 \times 10^{-9} = 193{,}548$ GHz.

For $\lambda_2 = 1551$ nm, $f_2 = c/\lambda = (3 \times 10^8)/1551 \times 10^{-9} = 193{,}423$ GHz.

Therefore, $f_1 - f_2 = 125$ GHz (for $\lambda_1 - \lambda_2 = 1$ nm).

To date, there is no competitor to DWDM in achieving ultrahigh transmission rates. The latest developments in transport technology such as optical transport networks use WDM technology integrated with the TDM multiplexing procedures to achieve very high capacities with flexibility.

15.3.3 Optical Cross Connect

In a WDM transmission system, an optical cross connect serves the purpose of routing the optical wavelengths to the desired ports. A four-port optical cross connect is shown in Figure 15.15.

The cross connect function allows any of the input ports on one side to be connected to any of the output ports on the other side. It is a kind of an automatic exchange switch where any subscriber can be connected to any other. Such cross connectivity is required to direct a selected wavelength in a particular direction or to a particular user or application connected on the cross connect port.

15.3.4 Optical Add-Drop Multiplexer

The role played by the ADMs in the electrical domain in SDH/SONET systems (see Section 12.1.3) is played by OADM in the optical domain transmission system, i.e., WDM system. The OADM allows any (one or more) wavelength to be inserted or dropped out of the number of wavelengths carried by WDM. The dropped wavelength is then processed for electrical conversion followed by the dropping of required client signal through SDH and other processing in the electrical domain. A reverse process is adopted for inserting a client signal tributary. The OADMs are also made of the devices used for making WDM systems.

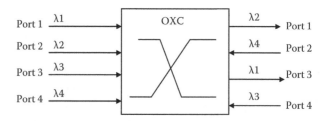

FIGURE 15.15
Optical cross connect.

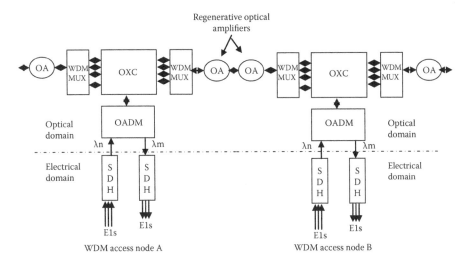

FIGURE 15.16
Optical WDM network.

Figure 15.16 illustrates the deployment of OXCs and OADMs in WDM networks.

As long as there is no need to drop or insert the user tributaries (client signal), the network can continue to operate purely in the optical domain, greatly increasing the capacity and distances covered and offering huge savings on cost as compared with single wavelength networks, as explained in the previous section on WDM.

15.4 OFC Link Budget

Similar to the case of wireline or radio systems, the OFC link has to be designed in such a way that the received power is not only above the threshold of the receiver but there is also enough margin to cover short-term, unforeseen losses due to degradation of any of the link components. Various components of an OFC link are shown in Figure 15.17.

The optical output power (or optical launched power) is fed into one of the fibers of the cable containing many such fibers, through a connector and a small piece of optical fiber called "patch cord." The cable is then taken outdoors. The cable is usually in drum lengths of 3 km each; thus, it is required to be jointed with another cable and so on until we reach the receiving node. The jointing of cable is done by jointing individual fibers to each other by a procedure known as "fusion splicing." (Although other methods such as mechanical splicing are also available, they are rarely used due to high

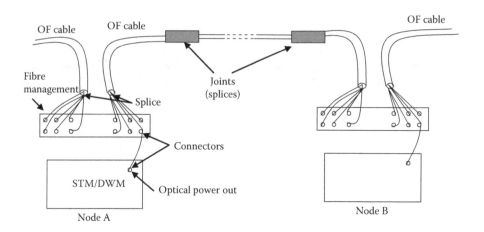

FIGURE 15.17
OFC link and its components.

losses.) Each fiber joint is called a "splice." Fusion splicing implies the joining of fibers using an electric arc that "fuses" both fibers together.

The transmitted signal gets attenuated in all the above components. The losses in each of them are

(i) *Connector losses:* The losses in each of the connectors are in the range of 0.2 to 1 db. The connector losses are only at the nodes; thus, the total contribution of connector losses in the overall link losses is not very significant. However, a poor connection may give rise to heavy losses and could well bring the whole link down. Connectors are required at the STM equipment as well as in the "fiber management boxes," which are like "fiber cross connects," where the fibers connecting the STMs can be changed whenever required for fault rectification or maintenance purposes.

(ii) *Fiber attenuation:* The losses or attenuation in the fiber itself are about 0.2 db/km for 1550-nm wavelength and 0.35 db/km for 1310-nm wavelength. This is the media loss, which is inescapable, and is the main factor in deciding the transmission length of the link.

(iii) *Splicing losses:* The splicing loss occurs at the connection of the OFC to the patch cord at the node, and at the places of OFC joints, which are normally 3 km apart. The splicing loss depends upon the quality of splicing. A carefully done splice may produce 0 loss, whereas a poorly made splice can give a loss of several decibels. A loss of 0 to 0.1 db per splice is normally permitted. For the purposes of calculations, splicing loss and connector loss are generally included in the fiber attenuation. For 1550 nm, the fiber attenuation, splicing losses, and connector losses are altogether taken as 0.25 db/km, and those for 1310 nm are taken as 0.4 db/km.

The margin in the power available above receiver threshold after meeting all these losses is called "loss margin." Let us see an example of illustration loss margin.

15.4.1 Loss Margin

The difference between the received optical power and the receiver threshold is called the loss margin (or the margin for unforeseen losses, or for losses due to sudden degradation of any component of the link). The more the loss margin, the more stable is the link. The lesser the loss margin, the more susceptible is the link to the environmental conditions and hence is more likely to fail. Hence, it is necessary to design the link with sufficient loss margin.

Let us take the following parameters for a link for calculations of loss margin:

Launched power = −0 dbm

Receive threshold = −30 dbm

Wavelength of operation = 1550 nm

Length of the link = 60 km

Hence,

Total permissible link losses: 0 − (−30) = 30 dbm.

Link losses in this case: 0.25 × 60 = 15 dbm.

Therefore, loss margin: 30 − 15 = 15 dbm.

The 15-dbm loss margin can be considered a very good margin in a link suffering a total attenuation of 15 dbm. In case the connector losses or splice losses suddenly increase due to poor quality of initial installation work or due to harsh environmental conditions such as dust or heat or even if the launched power gets reduced to some extent (reduction to half will be equal to 3-dbm loss), the loss margin will be able to sustain the link functionality comfortably.

15.4.2 Dispersion Limit

The calculations made in the previous section for loss margin presume that the dispersion is within tolerable limits, which is generally true for such short distance links with limited bandwidth. However, when the bandwidth of the optical signal increases (say in the case of WDM) or the length of the link is very high, the chromatic dispersion may not allow satisfactory operation. In such cases, the media to be used is "dispersion-shifted fiber," which is designed to have minimum dispersion.

The dispersion limit of the cable is decided by a factor called bandwidth-distance product (BDP). Thus, the higher the bandwidth, the lesser is the distance of the link.

A typical BDP for a single-mode fiber (nondispersion shifted) operating at 1550 nm is 75 Gbps/km.

Thus, if our bandwidth is 622 Mbps (STM-4), our dispersion limit for the distance will be

$$D = 75,000/622 = 121 \text{ km}$$

which is much higher than our link length of 60 km; thus, we do not have to worry on the dispersion front. However, if our payload is STM-16 (2.5 Gbps), then

$$D = 75,000/2500 = 30 \text{ km}.$$

This means our link length becomes limited by dispersion, and some kind of dispersion minimizing techniques will have to be applied.

Review Questions

1. Which type of media is required for PDH and SDH? Which media is most suitable and why?
2. What types of problems were faced with long-distance communication on copper wires for analog and digital signals?
3. Why is the range of high-speed communication restricted on copper wires?
4. In which applications is the copper wire pair used as the preferred media?
5. Why was the coaxial cable invented? Show the construction details with a diagram.
6. What is the relationship between the repeater distance and number of channels in a coaxial cable-based communication system?
7. What were the typical capacities of PDH channels in coaxial cables of 2.6/9.5 mm? How did the diameter matter?
8. What was the motivation for development of microwave radio systems for long-distance communication?
9. What were the typical tower heights, hop lengths, and capacities in microwave radio systems?

10. All the media related problems were solved by the optical fiber. Is this statement correct? Elaborate.

11. Can PDH and SDH signals be used on radios? What is the microwave radio deployed for in the era of OFC?

12. What is FSO or air fiber technology? What were the driving forces for its development?

13. Differentiate between FSO and microwave radio with respect to technology and application areas.

14. What are the current and potential applications of air fiber technology?

15. List the advantages of OFC technology.

16. Describe the construction details of an OFC.

17. Describe the principle of light travel through optical fiber through the phenomenon of total internal reflection with the help of suitable diagrams.

18. Show with the help of a block diagram how the PDH/SDH signals are transmitted on optical fiber.

19. What is the relationship between the core diameter and the frequency of transmission on an optical fiber?

20. Draw a graphical representation of attenuation characteristic of optical fiber. What is its significance? What are the favorable and applied frequencies and what type of bandwidths are possible?

21. What is the attenuation of OFC at different wavelengths practically deployed?

22. Calculate the frequencies for 1310- and 1550-nm wavelengths. In which part of the electromagnetic spectrum do they fall?

23. What are the types of optical fibers depending upon the type of material used, mode of operation, and type of refractive index deployed?

24. Compare the characteristics of: glass fibers with plastic fibers, multimode fibers with single-mode fibers, and step index fibers with graded index fibers.

25. Explain the principles of working of single-mode and multimode fibers with the help of suitable diagrams.

26. What is chromatic dispersion? What is dispersion-shifted fiber? Where it is used and why?

27. In addition to the normal components, what are the optical devices required in optical communication?

28. What are the main types of optical sources applied in optical fiber communication? What are the salient characteristics of LED source? What are its application areas?

29. Describe the salient characteristics of LASER diodes? What is their frequency spectrum? What are their application areas and why?

30. What is a photo detector? What are the popular types of photo detectors and what are their salient features and application areas?

31. Draw a block schematic of an optical communication link showing different electrical and optical stages and explain its functioning.

32. What are all optical networks and what are their advantages?

33. What are optical amplifiers? Illustrate their application in an optical communication link. What are the advantages of use of optical amplifiers? What kind of link lengths are possible with them and how?

34. What is WDM and why it is important? Explain the multiplexing with the help of a diagram.

35. Describe the variations available in WDM systems and their data carrying capacities.

36. What are the devices used for WDM? What is the wavelength spacing? Calculate the frequency separation corresponding to the wavelength spacing.

37. What is optical cross connect? Explain with the help of a diagram.

38. What are OADMs? Show the application of optical amplifiers, optical cross connects, and OADMs in a DWM network.

39. What is an OFC link budget? Describe the process of making a loss budget considering a practical OFC link. Explain various losses like connector losses, fiber attenuation and splicing losses, and how much they contribute to the total losses in the link.

40. What is meant by loss margin? Design an OFC link with a 15-dbm loss margin.

41. How does the length of a link vary with dispersion? Illustrate with an example calculation.

Critical Thinking Questions

1. Was SDH tried on coaxial cables? If not, why? What would have been the repercussions if it was tried?

2. Can you roughly work out the installation cost per voice channel for a distance of 100 km for coaxial cable analog system and SDH on optical fiber?

3. List more applications of optical fibers other than that in telecommunication. What properties of optical fibers are used in those systems?

4. How did microwave radios change from the time they were first installed up to now?

5. What could be the best combination of optical fiber, microwave radios, and air fiber for providing a comprehensive coverage to metro cities and countrywide networks?

6. What would have been the scope of optical fiber in communication in the absence of lasers?

7. What would make better sense for the technology developers, focusing on more electrical bandwidth on one wavelength or on more wavelengths on one fiber? Give reasons.

Bibliography

1. J.H. Franz and V.K. Jain, *Optical Communications Components and Systems*, Narosa Publishing House, New Delhi, 2000.
2. G. Kennedy and B. Davis, *Electronic Communication Systems*, Fourth Edition, Tata McGraw-Hill, New Delhi, 2005.
3. S. Haykin, *An Introduction to Analog and Digital Communications*, John Wiley and Sons, Asia, 1989.
4. D. Roddy and J. Coolen, *Electronic Communication*, Fourth Edition, Prentice-Hall, India, New Delhi, 2001.
5. J. John and S. Kumar Aggarwal, *Proceedings of the International Conference and Exposition on Communications and Computing IIT Kanpur*, February 2005.
6. A.F. Molisch, *Wireless Communications*, John Wiley and Sons, London, UK, 2005.
7. NIIT, *Introduction to Digital Communication Systems*, Prentice-Hall, India, New Delhi, 2004.
8. F.E. Terman, *Electronic and Radio Engineering*, Fourth Edition, McGraw-Hill Kogakusha, Tokyo, Japan, 1955.
9. J.D. Ryder, *Network Lines and Fields*, Prentice-Hall, 1975.
10. S. Ramo, J.R. Whinnery, and T. Van Duzer, *Fields and Waves in Communication Electronics*, John Wiley and Sons, 1994.
11. L.E. Frenzel, *Principles of Electronic Communication Systems*, Tata McGraw-Hill, New Delhi, 2008.

16

Introduction to Optical Transport Networks

With the advent of optical fiber technology, the telecommunication services not only improved immensely in quality but also became much more affordable. In fact, the cost of making a long-distance call today is a very small fraction of the cost 30 years back. In India, until a few decades ago, it was cheaper to travel to the intended long-distance destination by train and come back, as compared with making a telephone call to the place lasting for 10 minutes. As the affordability improved, the demand increased, which further boosted the volume of production of equipment and cables, making them cheaper again. Thus, the demand and cost have been driving each other for the benefit of the public. The demand was further boosted by the data traffic growth, which had been phenomenal in the last 10 years or so. Data-centric applications such as LANs, Internet banking, online ticket booking of railways and airlines, corporate WANs and VPNs, etc. multiplied the demand of data traffic, manyfold. The most important factor, however, in the growth of the data traffic, which projected a demand for convergence of voice and data traffic, was the transport of voice and video services as data such as VOIP, streaming video, triple play, video conferencing, etc. Data traffic had thus been getting increasingly important over the years. The scenario finally led to the demand of convergence of data and voice services and a transportation backbone that is more data-friendly rather than being voice-friendly.

On the ground, however, synchronous digital hierarchy (SDH), which was optimized for the voice traffic, was the only long-distance backbone carrier available. Forced by the demand of efficient transport of data services the next-generation (NG) SDH came out with a number of new features. These features are

(i) Virtual concatenation (V-CAT), which helped in much better utilization of capacity and added a large amount of flexibility for carrying data traffic.

(ii) Link capacity adjustment scheme (LACS), which provided facilities for dynamic adjustment of the bandwidth or features of bandwidth on demand and added resilience to the network.

(iii) Generic framing procedure (GFP), which facilitated efficient and smooth transport of data traffic of any type over SDH, be it Ethernet, multiprotocol label switching (MPLS), or VOIP. (All these features have been discussed in detail in Chapters 13 and 14.)

However, the enormous capacity of SDH has started falling short of the demand of today. Although 2.5-Gbps (STM-16) links dominate the majority of deployment, 10-Gbps (STM-64) links have a limited deployment so far, and 40-Gbps (STM-256) systems have just started hitting the market. Thus, for anything above 10 Gbps, the only option is to go for wavelength division multiplexing (WDM). Since WDM is able to carry a large number of 10-Gbps/2.5-Gbps streams on a single fiber, it is not only able to meet the requirement of capacity but exceeds it manyfold, with the possible number of wavelengths in the range of a hundred plus in the DWDM systems. However, WDM's management and protection capabilities are very limited in comparison to those of SDH.

Thus, to meet the current and future demand of the backbone networks, a system was needed that had the large capacity of DWDM systems and the strong management and protection features of SDH.

Accordingly, optical transport networks (OTNs) was defined by ITU-T through their recommendation G.709. The OTN has the best of both technologies: DWDM and SDH. The vast capacity of DWDM meets the strong management and resilient features of SDH in OTN. The multiplexing hierarchy defined in the OTN is called Optical Transport Hierarchy (OTH). The OTH is also known as "digital wrapper," as it can "wrap" around any type of communication protocol and transport it across the network. It is interesting to note that while in the case of SDH the popular name of the system is SDH, and not synchronous transport networks (STN), in this case, OTN is the popular name and not OTH.

One more important aspect of high-density communication has been handled effectively in the OTN. The OTN standards promote a very strong forward error correction (FEC). This results in reception of the signal with desired BER for much longer distances, reducing the total number of regenerative repeaters in the long-distance links, which helps in reducing the overall system cost.

Although the OTN was conceived in 1998, the recommendations have been published by ITU-T in 2001. A number of vendors have started offering OTNs during the last 3 to 4 years, and the deployment by the operators has just begun.

16.1 OTH Principles

As discussed above, the purpose of developing the OTN was to combine the best of both worlds, i.e., SDH and DWDM, into one system, to get highly advanced OAM features along with very high system capacities. In view of the exceeding demand of data traffic, it was necessary to facilitate direct mapping of data traffic instead of following the SDH route; thus, this aspect has been catered for in the design of OTN as well. Another important parameter that has been considered as a design principle is the provision of strong features of FEC to achieve higher repeater spacing.

With the underlying objectives as stated above, the architecture of the OTN has emerged to be the strongest contender for the new-generation transport networks.

The payload can be SDH signal, Ethernet, or IP, or any other data protocol that is mapped on GFP or the asynchronous transfer mode (ATM). The granularity (minimum capacity or resolution) of payload is 2.5 Gbps, which in the case of SDH is equivalent to STM-16. The hierarchy provides for generation of a payload unit of 10 Gbps by multiplexing of 4 numbers of 2.5-Gbps tributaries. Alternatively, a 10-Gbps (STM-64, 10-Gbps Ethernet, etc.) can be received directly. These 10-Gbps payload units are again multiplexed to form a 40-Gbps unit, or alternatively a 40-Gbps signal can be received directly. Each of these 40-Gbps units is converted into an optical channel (OCh) of a particular wavelength. Several such wavelengths are multiplexed together to form a CWDM or DWDM stream and transported across the network.

Provision exists for conversion of either 2.5- or 10-Gbps tributaries to be directly converted to an optical wavelength if the situation so desires. It is also possible to transmit a single wavelength (or "color" as it is frequently called) in case there are no more wavelengths to be multiplexed.

How exactly is the above multiplexing carried out? Let us see the details of the frame structure of OTN/OTH in the following sections.

16.2 Multiplexing Structure of OTN

The stages of multiplexing in an OTN are shown in Figure 16.1.

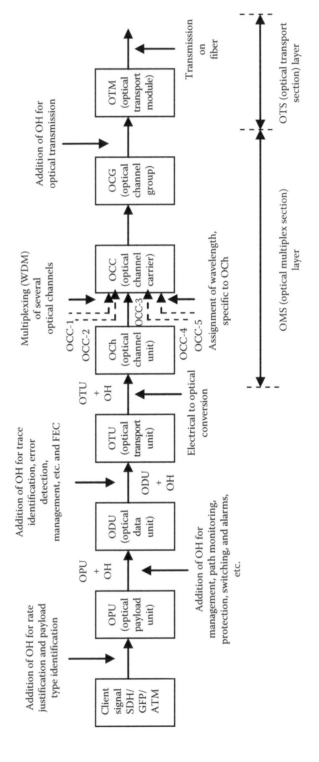

FIGURE 16.1
Multiplexing structure OTN.

16.2.1 Optical Payload Unit

This is the first information structure in the OTH. The client signal is accepted by this unit in the form of SDH, GFP, or ATM. The bit rate justification is done in the optical payload unit (OPU) by adding the necessary positive or negative justification bytes. Overhead bytes are added for payload type (PT) identification.

16.2.2 Optical Data Unit

The optical data unit (ODU) is formed by adding a number of overhead fields to the OPU for the purpose of management. Functions such as tandem connections monitoring, error detection through BIP-8, end-to-end path monitoring, and automatic protection switching (APS) are provided through various overhead bytes. The management through specific bytes and a mechanism of fault detection and reporting are also provided. The overhead functioning is equivalent to the path overhead of SDH.

16.2.3 Optical Transport Unit

The features of FEC and some more overheads for trace identifications, BIP-8 calculations for error detection, general management, etc. are provided to the ODU to form the optical transport unit (OTU). The OTU completes all the functionalities (OAM features) required and forms the last information structure on the electrical domain. The functionality of overheads of OTU is equivalent to that of the section overhead in the case of SDH.

16.2.4 Optical Channel

The electrical to optical conversion of the signal is performed by the OCh, after adding an overhead for management.

16.2.5 Optical Channel Carrier

The optical channel carrier (OCC) assigns a specific wavelength out of the WDM wavelengths to each OCh.

16.2.6 Optical Channel Group

Many OChs with their assigned carrier wavelengths are multiplexed together to form the optical channel group (OCG), the multiplexed optical WDM signal.

16.2.7 Optical Transport Module

OCG is added with an overhead called OTM overhead signal (OOS) to form the OTM, the transport module that is transmitted across the network.

16.2.8 Optical Multiplex Section

Optical multiplex section (OMS) is similar to the multiplex section of SDH, where adding or dropping of optical tributaries is carried out. The part of OTH hierarchy containing "OCC and OCG" as described above belongs to the multiplex section or multiplex section layer, when the optical link between two or more nodes is established.

16.2.9 Optical Transmission Section

The last component of OTH hierarchy, the OTM as described above, belongs to the optical transmission section (OTS) of an optical link established between two or more nodes. This is similar to the regenerator section layer of SDH.

16.3 Multiplexing Hierarchy of OTN

Multiplexing hierarchy of OTN (OTH) has been defined as a hierarchical set of digital transport structure, standardized for the transport of suitably adopted payloads over the optical transmission network. We have seen in Section 16.1 that the lower-order client signal tributaries are multiplexed into higher-order optical units. Let us see how it is defined by ITU-T. Figure 16.2 illustrates the stages involved.

The digital signals are accepted at the nominal bit rates of 2.5, 10, and 40 Gbps. The 2.5 Gbps signal is added with an overhead (as explained in the previous section for all overheads in brief) to form what is called an OPU level 1 (OPU-1). Adding more overheads converts this OPU-1 to the ODU or ODU-1. Four such ODU-1s are multiplexed in OD/TUG-2 (optical data tributary unit group) to form a higher-level payload unit called OPU-2. However, if there are no more ODU-1s to be multiplexed, the ODU-1 can directly go to the final module on the electrical domain called OTU-1, which forms by adding necessary overhead to ODU-1.

The OPU-2 formed by multiplexing four numbers of ODU-1s is treated as a payload by the next level of multiplexing and ODU-2 is formed by adding the necessary overheads in a similar manner as ODU-1 was formed from OPU-1. Alternatively, the OPU-2 is formed by directly mapping a payload of 10 Gbps nominal rate, be it data or SDH.

The third level of multiplexing involves multiplexing of this ODU-2 with three more such ODU-2s in the ODTUG-3 to form the higher level payload

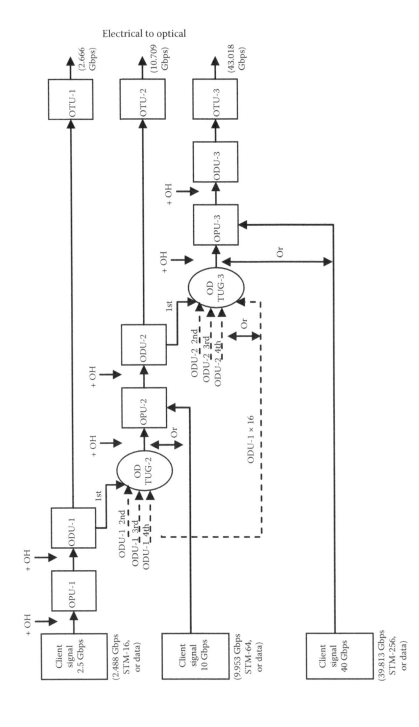

FIGURE 16.2
Multiplexing hierarchy of OTN: OTH and the nominal bit rates.

unit called OPU-3. Alternatively, the ODU-2 can be directly mapped on to OTU-2 by adding the necessary overhead if there are no more ODU-2s to be multiplexed. The OPU-3 can also be formed by directly mapping a 40-Gbps client signal, be it data or SDH, through additions of necessary overhead. It is also possible to multiplex 16 numbers of ODU1s in the ODTUG3 to form the OPU-3. The OPU-3 is mapped onto an ODU-3 and finally to OTU-3 in a similar manner, as is done for lower levels.

Thus, it is possible to have OTU-1, OTU-2, or OTU-3 available for conversion to optical domain; it is also possible to have a combination of any two or all three.

Subscript 1, 2, or 3 to OPU, ODU, and OUT, for example, OPU1, OPU2, or OPU3, is called "k." Thus, the units are described in general as OPU_k, ODU_k, or OTU_k, where k = 1, 2, or 3 represents the payloads of 2.5, 10, and 40 Gbps nominal rates, respectively.

The OTUs, in additions to the overhead mentioned above, also include a large number of bytes for FEC. We will see the details of FEC later in this chapter. The OTUs are then converted to the optical domain called OCh as shown in Figure 16.1, which depicts the process for a single level. Various stages of multiplexing up to the final module, i.e., OTM, are shown in Figure 16.3 for all three levels of OTUs.

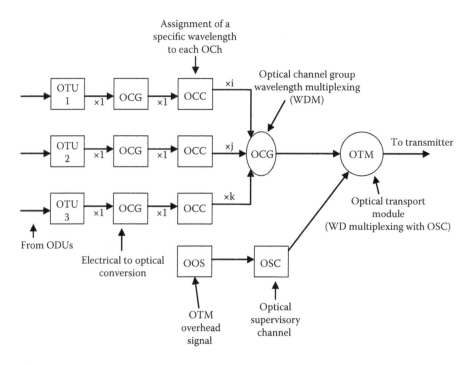

FIGURE 16.3
The multiplexing stages of OTH from OTU to OTM.

The tributaries of the OTU level formed from the client's signals could be any of the three levels, i.e., OTU1, OTU2, or OTU3. Each of these tributaries is allotted a particular wavelength as a carrier, after conversion from electrical to optical domain. The carrier OCC represents one of the wavelengths of the WDM band of operation. All the OCCs are multiplexed together using the WDM techniques to form the OCG. The number of OCCs (and thus the number of OTUs) can be anything subject to the total number of wavelengths that can be carried by the WDM system designed to operate.

The overhead for the control and management of the OChs is designated as the OOS. This consists of four parts functionally, which are

- OTSOH: optical transport section overhead
- OMSOH: optical multiplex section overhead
- OChOH: optical channel overhead
- COMMSOH: general management and communication overhead

Please note from Figure 16.3 that the OOS is carried on a separate channel called optical supervisory channel (OSC), which is not a part of the WDM channel group (OCG). This channel is separately multiplexed using WDM, directly at the OTM level. For this reason, this overhead is called "nonassociated overhead."

The details shown in Figure 16.3 are for the so-called full functionality OTM. However, the structure, mapping, and multiplexing details of OOS and its components as discussed above "have not been defined" by the ITU-T. This is keeping in view that the optical interfaces are dependent upon the optical technology, which is still emerging, and so they are likely to change as the technology changes; therefore, full functionality is being studied further.

The presently defined systems thus do not make use of the OOS (and so OSC) as shown in Figure 16.3. The systems are accordingly designated as "restricted functionality" systems. Only OTM-0 (with single wavelength) and OTM-16 (with 16 wavelengths) are defined, whereas other classes are being studied. In fact, they are defined as $OTM_{n,m}$ where n is the total number of wavelengths that the system can carry (at the lowest bit rate supported on the wavelength; for higher bit rates, fewer number of wavelengths will be supported) and m is the index indicating the supporting tributary rates, for example, m = 1 is for 2.5 Gbps, m = 2 for 10 Gbps, m = 3 for 40 Gbps, m = 12 for 2.5 and 10 Gbps, m = 23 is for 10 and 40 Gbps, and m = 123 if all three, i.e., 2.5, 10, and 40 Gbps. Thus, for example, an $OTM_{16,12}$ will support 16 wavelengths and tributary data rates of 2.5 and 10 Gbps. This is for full functionality systems. The restricted functionality systems are also represented in the same way, with the only difference that a letter (r) is added to indicate the "restricted functionality." For example, the above $OTM_{16,12}$ will become $OTM_{16,r,12}$. The functionality required for the management of the OChs in the restricted functionality systems presently defined is achieved through the use of overhead of OTU called OTU(v). Again, the frame structure and coding of OTU(v) is not yet defined, but basic functionality has been defined.

The details of restricted functionality systems have been avoided in Figure 16.3 to maintain clarity of understanding.

16.4 OTN Layers

Similar to SDH, the OTN has to achieve its functionality through various stages such as multiplexing, transmission, etc., which are called layers. Figure 16.4 depicts the layers of OTN.

OTS, which is almost equivalent to the regenerator section of SDH (Sections 10.9 and 10.10), is between two optical line amplifiers. The line amplifier could be a stand-alone amplifier or a part of the system at a node with multiplexing and optical and electrical cross-connect facilities.

The section between two nodes that have add/drop facilities (optical add drop multiplex, or OADM), which will necessarily have cross-connect facilities too, is called OMS, and the layer is accordingly called the OMS layer. The section between an OADM node and a repeater at the electrical level, i.e., 3R repeater (3R processing represents reamplification, reshaping, and retiming), is also called the OMS layer. The same section is also called the OCh layer

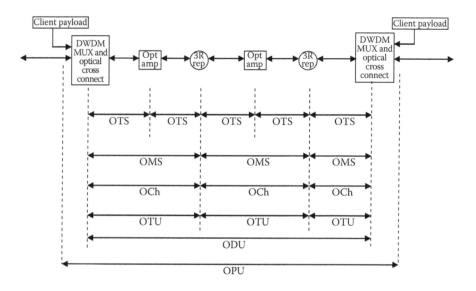

FIGURE 16.4
OTN layers.

because the OCh switching can take place due to the presence of optical cross-connect. This layer is similar to the multiplex section layer of SDH.

The OUT, and consequently, the ODU, can be dropped or inserted at any of the OMS nodes. However, the OPU/ODU carries a point-to-point client signal, and hence, it may cross several OMS sections before dropping. This is equivalent to the path layer of SDH.

16.5 OTN Domain Interfaces

OTN defines two types of interfaces in the networks.

16.5.1 Intradomain Interface

This type of interface is provided within the domain of a single network operator having the equipment of a single vendor. These interfaces need not provide for 3R processing at each end of the interface.

16.5.2 Interdomain Interface

These are the interfaces provided between the domains of different network operators or within the network of a single operator between the equipment of different vendors. These interfaces are required to provide for 3R processing at each end of the interface. Figure 16.5 illustrates the provision of these interfaces.

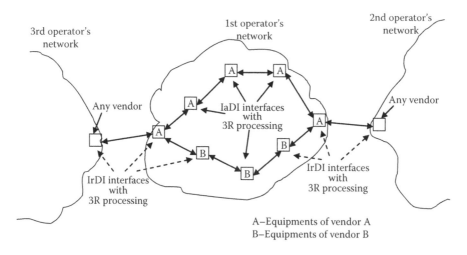

FIGURE 16.5
Interdomain and intradomain interfaces.

16.6 Advantages of OTN

There are several advantages that OTN offers over the SDH or any other contemporary system.

16.6.1 Very High Link Capacity

We have seen in the previous sections that OTN can carry several wavelengths. Each wavelength can carry data rates of up to 40 Gbps. Thus, the link capacity in OTN is enormous as compared with the existing SDH, which, currently, can carry only up to 10 Gbps or at the most 40 Gbps. Although DWDM systems provide high capacities, they lack the OAM and protection features of SDH. The OTN combines the advantages of both the systems.

16.6.2 Forward Error Corrections

OTN provides for very strong FEC features that enable the correction of up to as many as 128 consecutive errored bytes. This results in a very low bit error rate (BER). In other words, a gain of several decibels (5–6 db) is obtained by using FEC for a given BER. Thus, for a given BER, the repeater spacing can be increased, which results in cost reduction of the link.

16.6.3 Backward Compatibility with Existing Protocols

OTN allows any type of TDM or data traffic to be mapped into the payload area of its frames. Although the SDH and ATM can be mapped directly, all other data protocols such as Ethernet, IP packets, etc. can be mapped through GFP. This provides backward compatibility with all the existing systems, leading to no changes required before adoption of OTN. This mapping flexibility provides a completely transparent (without change of any bits/bytes, whether overhead or data) transport of the client signals. This becomes a very important feature when the OTN operator transports the signals of an SDH operator (or any other protocol for that matter), in which case the original clock of SDH nodes is also maintained.

16.6.4 Improved Efficiency for Data Traffic

Although the data communication traffic has been steadily growing over the years, the backbone for the long-distance transport of the data traffic continues to be SDH. Since SDH is optimized for TDM traffic, the IP or other protocol-based data traffic have to be mapped onto it using contiguous concatenation. Although substantial relief has been provided by V-CAT, L-CAS, and GFP features in the new-generation SDH systems, the process is quite complex, particularly for high bit rates, like 10-Gb Ethernet. Another alternative for efficient transport of data traffic was to use a direct wavelength of DWDM network for the transmission of Ethernet. The later option, however, loses highly on efficiency. If only 1-Gb Ethernet is to be transmitted on a 10*Gb DWDM channel, it will result in wastage of 90% of the bandwidth.

OTN provides a good solution to the above problem. With the data rates varying from 2.5 to 40 Gbps for the client signal, an efficient transport is provided for a high range of client data rates. To transport 1 Gbps Ethernet signal, for example, the efficiency will be 40% for direct mapping. However, by cleverly mapping it on the SDH using V-CAT, we can get efficiencies above 90% (see Section 14.3). The SDH payload is, of course, transported with the maximum efficiency on OTN, as the data rates are made to match with those of STM-16, STM-64, and STM-256.

16.6.5 Reduced Number of Regenerators

With OTN being an optical network, the number of electrical regenerators in the long haul links can be minimized by deploying optical amplifiers. The use of optical amplifiers reduces the cost of regeneration (the optical amplifiers are much cheaper than a normal repeater) and the complexity of the network designs.

16.6.6 Strong Management and Protection

The strength of the SDH has been its strong features of operations, maintenance, management, and protection switching. All these features are made available in OTN as well. Protection switching in the linear, ring, or mesh topology is available. Besides the error detection through parity check similar to SDH, a very strong error correction mechanism through FEC is available that is not available in SDH. Management capabilities are available through GCCO bytes, which are equivalent to the data communication bytes (DCC) of SDH, and the management of the optical layer is provided through the OOS. OTN also scores over SDH in providing much better features of tandem connection monitoring. There are six levels of tandem connection monitoring, as compared with only one level in SDH. Thus, the OTN truly offers a great amount of manageability besides tremendous flexibility.

16.6.7 Quick Localization of Faults

The tandem connections monitoring (TCM) is an important feature (see Section 16.7.3) for quick localization of faults. As compared with only one level of TCM in the SDH, there are six levels of TCM provided in OTM. These high numbers of TCM levels not only facilitate fast fault localization but also help in the enforcement of the service level agreements (SLAs) because the operator whose section is faulty is clearly identified.

16.7 Frame Structure of OTN

The OTN frame is 16,320 bytes long. The representation is usually done through a structure of 4 rows and 4080 columns, as shown in Figure 16.6.

The payload of each OTN frame is 15,232 bytes long. It neither corresponds to any of the SDH levels nor does it have any relationship with any of the data protocols. The SDH or data payload accommodates 15,232 bytes at a time in the OTN. Thus, the larger payloads have to cross the boundaries of the OTN payload area, leading to many OUT frames carrying the payload.

Similarly, there is no relationship of OTN frame duration with any other protocol. As against the 125-μs frames of SDH, the OTN frame durations are as follows for various bit rates (as the frame length is constant at 16,320 bytes) (Figure 16.7).

Let us see about the functions of various overheads bytes with reference to Figure 16.6.

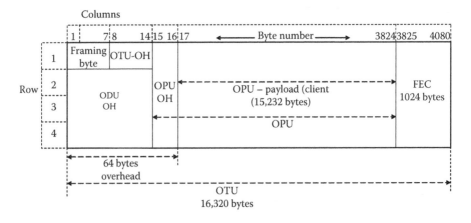

FIGURE 16.6
The OTN frame, indicating various overheads, payload, and FEC bytes.

Payload bit rate (Gbps)	Frame duration (μs)
2.5	48.471
10	12.191
40	3.035

FIGURE 16.7
Frame durations for various payloads in OTN.

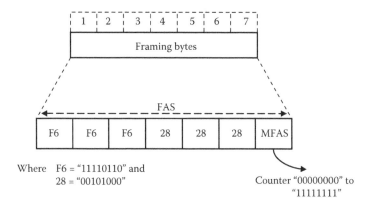

FIGURE 16.8
Assignment of the FAS word.

16.7.1 Framing Bytes

The frame alignment in OTN is performed by a 6-byte frame alignment signal (FAS) word. The receiver looks for the FAS sequence in the received signal and determines the beginning of an OTN frame. The code used for the FAS word is "F6F6F6282828" (hexadecimal notation) (Figure 16.8). The seventh framing byte is used for a multi-FAS (MFAS).

MFAS is used for counting the frame numbers from 0 to 255. This information of frame number is used for many of the overhead functions, which depend upon the information in many frames (of a multiframe), as we will see in subsequent discussions.

16.7.2 OTU Overhead

Next to the seven numbers of the framing bytes, there are seven bytes (numbers 8 to 14) that belong to the OTU overhead (refer to Figure 16.6).

The first three bytes (numbers 8 to 10) of the OTU overhead (OTU-OH) are used for section monitoring (SM). The next two bytes (numbers 11 and 12) are general communication channel (GCC0) bytes for general management. Functions of these GCC0 bytes are not presently defined. These bytes are similar to the data communication bytes of SDH (DCC bytes). It is contemplated that the GCC bytes will be used for generic MPLS type of control plane signaling and/or network management functions. Two bytes, 13 and 14, are reserved for future use (RES). Figure 16.9 gives the details of OTU-OH.

Of the three bytes of SM, the first byte is used for the "trail trace identifier (TTI)" function. The TTI function is used to monitor the signal from the source to the destination, similar to the function of J0 byte in SDH. It contains the information about the network elements, like source access point identifier (SAPI) and destination access point identifier (DAPI). In a multiframe

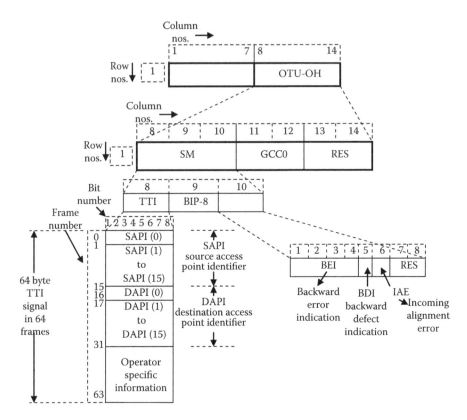

FIGURE 16.9
Details of OTU-OH. (From the International Telecommunication Union, Fig. 15.2 frame structure frame alignment and OTUk overhead, p. 41 of ITU-T rec G.709/Y.1331, December 2009. With permission.)

of 64 consecutive frames, this byte generates a 64-byte TTI signal. These 64 frames are repeated four times in a multiframe of 256 frames as mentioned in Section 16.7.1. The first 16 frames, i.e., 0 to 15, contain the SAPI, the next 16 frames, i.e., 16 to 31, contain the DAPI, and the rest of the frames, i.e., 32 to 63, contain information that is specific to the operator.

The second byte of SM (9th byte of OTU frame) contains the bit interleaved parity (BIP-8) for error check (for details of BIP check, see Section 6.7).

The third byte of SM (10th byte of OTU frame) contains a backward error indication (BEI) signal in the first four bits, which conveys to the upstream direction, the count of errored blocks detected by the BIP-8 parity check. The next bit of this byte is the backward defect indication (BDI), which is used to convey a signal fail status. The next bit (number 6) is an incoming alignment error (IAE) bit, which allows the ingress point of a node to inform the egress point of an error in the incoming signal. The next two bits of this byte (bits 7 and 8) are reserved for future use and are set to 00 values.

16.7.3 ODU Overhead

The overhead pertaining to the ODU (ODU-OH) is the largest chunk of overhead bytes in the OTH. A total of $3 \times 14 = 42$ bytes are allocated to this OH (refer to Figure 16.6).

The first three and last six bytes are reserved for future standardization. The details of overhead assignments are illustrated in Figure 16.10.

There are a total of six tandem connections monitoring (TCM) fields occupying three bytes each. The tandem connection monitoring feature breaks the end-to-end path of a client into several small (tandem) connections for monitoring the error performance. This allows a quick localization of fault (and hence a quick restoration), as the tandem connections having errors can be easily identified. The OTN is very rich in this feature, as it has six levels of TCM as compared with only one level in SDH. These six levels allow nested and cascaded domain monitoring, allowing parallel monitoring of a client signal that is configured over many operators' domains. The SDH in contrast permits only cascaded connection monitoring. It is possible (in OTN) to configure the tandem connections in such a way that the error performance of each network operator in a long chain of networks is monitored separately, for the end-to-end path of a client involving several networks. The identification of the defaulting operator can facilitate enforcement of the SLAs with the imposition of necessary penalties. Figure 16.11 illustrates the concept of nested and cascaded multilevel tandem connection monitoring in OTN. The functions of various bytes of TCMs are more or less similar to the functions of SM discussed in Section 16.7.2 and Figure 16.9. The only difference is in the last three bits of the third byte, where instead of one bit for IAE and the last two bits reserved for future use in the case of SM, the last three bits are used for monitoring the path status and are called status bits (STAT). These STAT bits indicate the presence of a maintenance signal. Figure 16.12 depicts the TCM bytes and their functions. The functions of all the TCM fields are the same.

Columns

		1	2	3	4	5	6	7	8	9	10	11	12	13	14
Rows	2	RES			TCM ACK	TCM-6			TCM-5			TCM-4			FTFL
	3	TCM-3			TCM-2			TCM-1			PM			EXP	
	4	GCC1		GCC2		APS/PCC			RES						

FIGURE 16.10
Assignments of the ODU-OH bytes. (From the International Telecommunication Union, Fig. 15.3 ODUk frame structure, ODUk and OPUk overhead, p. 42 of ITU-T rec G.709/Y.1331, December 2009. With permission.)

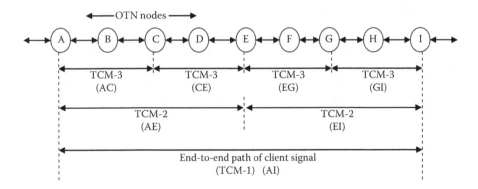

FIGURE 16.11
Tandem connection monitoring through cascaded and nested TCM groups.

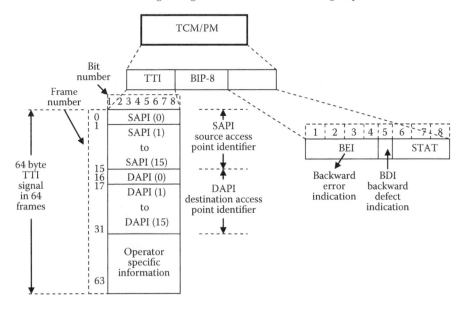

FIGURE 16.12
Structure and functions of TCM (1–6) and PM fields. (From the International Telecommunication Union, Fig 15.3 ODUk frame structure, ODUk and OPUk overhead, p. 42 of ITU-T rec G.709/Y.1331, December 2009. With kind permission.)

The field TCM ACK is reserved for future standardizations and intended to be used to activate or deactivate various TCM fields.

The path monitoring (PM) field facilitates end-to-end path monitoring. The structure of the PM field is the same as that of TCM fields, as shown in Figure 16.12.

The tandem connections (shown in Figure 16.11) AC–CE–EG–GI and AE–EI are cascaded connections with respect to each other, while AI–AE–AC, AI–AE–CE, AI–EI–EG, and AI–EI–GI are nested connections. Only three levels of TCMs have been shown here, but in the international connections involving a large number of intermediate nodes, the six levels of TCMs can be effectively utilized. In this example, we have seen the nested and cascaded connections that are overlapping. Both independent and overlapping connections are supported by OTN.

The two bytes named EXP (Figure 16.10) are not standardized and are left for experimental use by the vendors within their own network. The GCC1 and GCC2 fields containing two bytes each are for general communication, just like the GCC0 bytes of OTU-OH described earlier.

The functions of four bytes of APS/protection communication channel (PCC) are depicted in Figure 16.13.

The bytes provide for the APS and PCC. APS is supported on different monitoring levels. At present, only linear switching is supported while other aspects are being studied further.

There is a "fault type and fault location reporting communication channel" (FTFL) located in the 14th column of 2nd row (Figure 16.10). As the name suggests, this byte communicates the message about the fault type and location. These messages can be used for triggering the alarms or for activating the APS protection. The byte FTFL carries the messages in a multiframe of 256 frames. The first half of the multiframe, i.e., frame numbers 0 to 127, is called a forward field, and the second half, i.e., frame numbers 128 to 255, is called a backward field. The first byte of each of these fields is for "fault type indication." Byte numbers 1 to 9 identify the operator. Figure 16.14 depicts the various fields with their functions. In "fault type" fields, a signal fail is indicated by "00000001," signal degrade is indicated by a "00000010," and a value of "00000000" indicates no fault. Other codes (00000011 to 11111111) are reserved for future standardization.

The operator identifier fields provide information about the geographical origin of the signal and other identifications in frames 1 to 9 of the forward and backward fields.

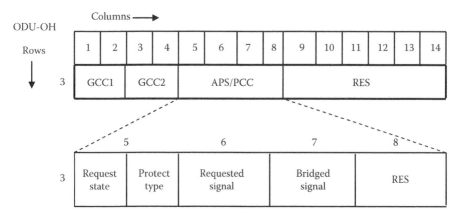

FIGURE 16.13
APS/PCC functions.

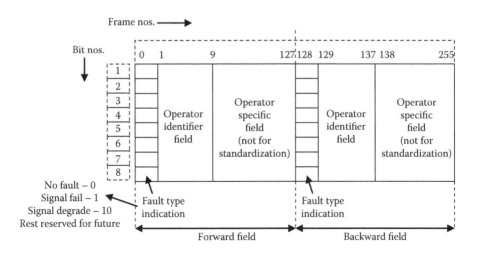

FIGURE 16.14
FTFL multiframe.

16.7.4 OPU Overhead

The OPU overhead (OPU-OH) is located in columns 15 and 16 of the OTH frame in all the four rows (Figure 16.6). Three bytes of the first column of the overhead are reserved for future international standardization.

The fourth byte in the first column is called a PSI byte (payload structure identifier). The PSI is a multiframe information contained in the 256 (0–255) frames of the OTU-MFAS (Pl, see Section 16.7.1). However, only the first frame ("0") is used at present for carrying the 1-byte-long payload-type information (PT), whereas the bytes of the other frames (1–255) are reserved for future international standardization.

The byte in the first frame of PSI is called PT and indicates the type of payload, for example, STM-N, GFP, ATM. Some values of this byte are also reserved for the future, and some are not used because their patterns are used in the maintenance signals of ODU.

The OPU is the OTNs interface with the client signals. The client signals may be asynchronous, having slightly different bit rates. The OPU provides for rate adjustment by the process of justification (for details of justification process, see Section 8.2.3). There is one negative justification opportunity byte available, i.e., "NJ0" (Figure 16.15), in the overhead that can carry the payload information in case of faster tributary rate. A positive justification byte "PJ0" is located in the OPU payload area (Figure 16.15), which may not carry the payload if the tributary rate (payload bit rate) is slower. These two bytes are enough to take care of the permitted rate variations. There are three bytes for justification control (JC bytes) that carry the information as to whether the justification opportunity byte

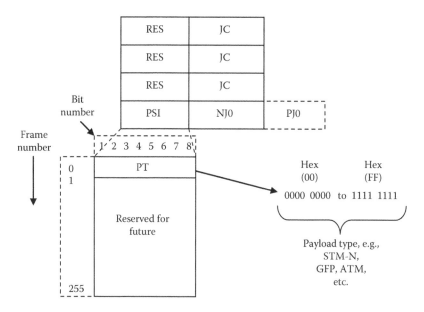

FIGURE 16.15
OPU-OH details.

is carrying data (payload) or not. A majority voting decision (two of three JC bytes confirming) is considered to take care of bit errors in the JC bytes during transmission.

Aside from asynchronous mapping mentioned earlier (unequal tributaries data rates), synchronous mapping (equalized tributary data rates) is also supported on the OPU.

16.8 Forward Error Correction

FEC is one of the strongest features of the OTN and has been considered to be one of the main justifications for OTN. FEC means "detecting and 'correcting' (not only detecting), the errors in the received signal." If we are able to correct the errors in the received signal, we get a phenomenal advantage. The expected signal quality of a received signal in a link is expressed in BER. The BER has a certain fixed value depending upon the quality requirement of the links, say 10^{-8} (1 bit error in 10^8 bits). The distance between regenerative repeaters are adjusted so that the BER is within this specified limit. If the repeater spacing is increased, the error rate will increase due to weakening and distortion of signal, addition of noise, etc. But if FEC can correct the errors, then for the same BER, the distance can be increased, which results in cost reduction, as fewer numbers of repeaters will be required. A penalty, however, has to be paid in terms of coding efficiency, as some extra bits have to be carried along with the signal bits. In the case of OTN, these extra bits are nearly 7% of the total payload. The gain achieved due to error correction is nearly 5 to 6 db. When expressed in terms of launched optical power, it results in nearly 20- to 24-km increased repeater spacing, considering the optical fiber attenuation to be 0.25 db/km. Figure 16.16 depicts the variation pattern or BER with or without FEC.

The FEC is achieved in the OTN by using a coding mechanism called "Reed–Solomon" code or RS code. It is a "block coding scheme" from the family of systematic linear cyclic block codes that uses n number of bits to transmit m number of signal data bits. It can correct up to $(n - m)/2$ bits in the received signal. In the OTN, we use RS (255–239) code, where n = 255 and m = 239. Hence, $(255 - 239)/2 = 8$ bytes are corrected in a bit sequence of 255 bytes. If this coding is used only for error detection, it can detect up to 16 errored bytes. The RS code is generated by using a device called the RS coder, which uses logic circuits to generate the code at the transmitter. A reverse process at the receiver decodes the signal.

The 239 data bytes are formed by picking up 1 byte from the overhead area and 238 bytes from the payload area. These 239 bytes generate 16 parity bytes through the use of RS coding to form blocks of 255 bytes each. The whole frame is thus divided into 64 such subframes, or each of the four OTU rows is divided in 16 subrows, which are byte interleaved. Byte interleaving is done by arranging the first row from the first byte of overhead and 238 bytes of payload and 16 parity bytes (parity bytes are

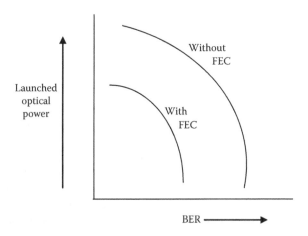

FIGURE 16.16
Variation pattern of BER with launched power with and without FEC.

calculated over these 239 bytes) and then picking up the second byte of overhead to form the second subrow and so on (Figure 16.17b). The 16 parity bytes for each subrow are transmitted in the byte numbers 240 to 255 of the same sub row. The number of bytes in each row of the OTU frame is such that they exactly match the number of bytes required/generated by 16 subrows.

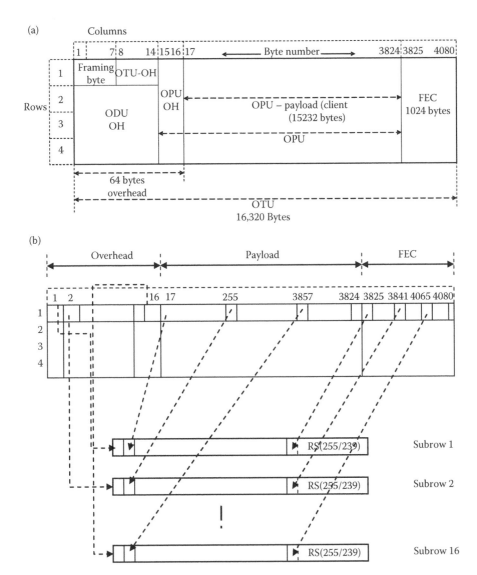

FIGURE 16.17

(a) The OTN frame, indicating various overheads, payload, and FEC bytes. (b) Formation of subrows out of OTU rows by interleaving and applying RS code for FEC.

16.9 Operations and Maintenance in OTN

The OAM features of the OTN are as strong as those of SDH and even better. In the digital domain, most of the alarms and indications are similar to those of SDH. The indications such as loss of signal (LOS), loss of frame (LOF), out-of-multiframe or loss of multiframe (OOM), alarm indication signal (AIS), trace identifier mismatch (TIM), etc., which are available in SDH, are also available in OTN at various multiplexing levels, such as OTU and ODU. The criteria for detection and clearance of these alarms and indications are of course different in bit pattern, as they are dependent upon the frame structure. These alarms and indications are, therefore, not discussed in detail here; interested readers may refer to the chapter on SDH/OAM for the principles of functioning and ITU-T recommendations G.798 for the details of criteria for detection and clearance.

However, since the OTN has a "managed optical layer" as well, there are many alarms and indications pertaining to the optical domain that do not figure in the discussion on SDH. The important ones are described in brief as follows.

16.9.1 Forward Defect Indication

This is the optical domain equivalent of the AIS signal and is transported in the OOS. Forward defect indication (FDI) is present at various layers such as in the OMS layer, OMS-FDI-P to indicate a defect in the payload, and OMS-FDI-O to indicate a defect in the overhead bytes. In the OCh layer, there are also OCh-FDI-P and OCh-FDI-O for similar functions.

16.9.2 Open Connections Indication

In the OCh layer, when a node observes that the upstream signal is not connected, an open connections indication (OCI) is sent in the downstream node, called OCh-OCI. A similar indication is present in the ODU called ODU_k-OCI.

16.9.3 Locked

When no signal is passed through in the upstream, a locked (LCK) indication is sent in the downstream by the concerned node. A locked signal is defined for ODU_k layer (ODU-LCK).

16.9.4 Payload Missing Indications

A payload missing indications (PMI) signal to a downstream node indicates that either there is no optical signal or the optical signal is without a payload at the source point of signal in any of the tributaries. PMI is generated at the OTS layer (OTS-PMI) and OMS layer (OMS-PMI).

16.10 Future of OTN

As of today, the world's telecommunication traffic backbone is built on SDH/SONET. More than 450,000 SDH/SONET "rings" are in operation worldwide, providing the telecommunication capacities to meet the needs of all types of telecommunication services, be it voice, data, or video.

However, the growth of data traffic has shadowed the growth in voice traffic in recent years. The SDH is optimized for voice; thus, at every access point, it needs to be converted to data protocols wherever the application is a data service. Moreover, the maximum capacity that the SDH/SONET could carry was only up to 10 Gbps (10 Gbps is STM-64's capacity; this is perceived as a limit because STM-256 is still not very popular), thus the increased demands are met by WDM, by carrying more than one wavelength, on the optical fiber, with each wavelength carrying a traffic capacity of up to 10 Gbps.

OTN solves both these problems through efficient transport of "data" traffic and facilitating very high capacities. We have seen earlier in this chapter, the advantages offered by OTN. It extends the extremely strong OAM features of SDH/SONET to the DWDM systems and combines both of them into a single management entity. In fact, the OTN equipment being offered today can have right from "DWDM" to "E1" capability in a single rack. A large number of wavelengths, each carrying presently up to 10 Gbps (and in the future up to 40 Gbps STM-256s), provide enormous capacity. While SDH/SONET can be mapped seamlessly, the data traffic is carried with equal ease in GFP and ATM formats on OTN. The backward compatibility with all the existing systems (SDH/SONET, GFP, ATM, etc.) easily embraces the existing services without calling for any replacement whatsoever. This backward compatibility becomes the biggest advantage to the project implementers and investors when it comes to the economics of service provisioning. The FEC feature permitting higher repeater spacing and multilevel TCM (tandem connections monitoring) are added benefits of OTN, which score over the features of SDH/SONET.

As such, the telecommunication operators are quite optimistic about the OTN, hoping it will provide all that is required for the "next-generation networks."

However, the network operators involved in providing more data services have a series of apprehensions about the OTN. They even feel that the SDH/SONET vendors have vested interests in pushing the OTN as the next-generation telecommunication network.

Their main apprehension is that the OTN is again a "point-to-point" network optimized for voice, just like SDH/SONET. What is needed today is a network that is optimized for data because even the voice and video can be, and are being, transported as data. The data networks are able to meet the necessary quality requirement of these real time services, for which the protocols are already proven. Thus, there is no point in converting the data packets to TDM, transporting them as TDM signals, and again converting them back to data packets, instead of having a data transport backbone itself.

The devices and products required for high-end synchronized TDM technologies are much costlier than those required for data transport; thus, the initial capital cost of the project will be low in the case of data transport networks. The operations and maintenance cost will also be less because much of the provisioning and reconfiguration required will be done automatically in the pure data environment, in contrast to each provisioning being carried out by experts, involving time and expense in the case of OTN. Instead of OTN, Ethernet OTN or a packet transport network is required for the next generation, according to these experts. A substantial amount of work is supposed to have been done by IEEE, ITU, and IETF in this field.

There is currently no alternative to OTN to meet the needs of the phenomenal growth of telecommunication traffic. It is now for the "time" to tell the future technology that will prevail.

Review Questions

1. What are the market forces driving a strong demand for convergence of telecommunication technologies?
2. What were the improvements made in the SDH to make it data-friendly in the so-called next-generation SDH?
3. What were the driving forces for the development of OTN standard by ITU-T? Which recommendation of ITU-T defines the OTN?
4. What had been the key strategies in the developments of OTN?
5. What are the general principles of OTH?
6. What are the terms OPU, ODU, OTU, OCh, OCC, OCG, OTM, OMS, and OTS? Explain them in brief. Draw the multiplexing structure of the OTN indicating the functions of all these at different stages of multiplexing.
7. Explain the complete multiplexing process of OTH in the electrical domain. Draw a block schematic showing all the stages and the activities at each stage.
8. Explain the multiplexing process of OTH in the optical domain with the help of a block schematic.
9. What is "nonassociated overhead"? Please explain its components and its functioning.
10. What is meant by full functionality and restricted functionality systems? How are they represented?
11. How is OTN divided in layers? Describe their roles in brief and also draw a schematic diagram clearly showing different layers.

12. What are the types of interfaces defined in OTN? Explain with illustration.

13. List the advantages of OTN.

14. How is the OTN able to increase repeater spacing by introducing FEC?

15. Is OTN backward compatible with the existing protocols? Please elaborate.

16. How is the efficiency of data traffic improved in OTN as compared with the alternatives?

17. How is the number of regenerators minimized in OTN?

18. What are the management and protection features of OTN? Which are the features that make it better than SDH?

19. OTN facilitates quick localization of faults and helps in SLA enforcement. Please explain how.

20. Explain the general frame structure of OTN. What are the frame durations for different payloads?

21. How is the frame alignment done in OTN?

22. Describe the functions of various bytes of OTU-OH with the help of a diagram showing all the details.

23. Describe the functions of various bytes of ODU-OH with the help of a diagram showing all the details.

24. What are the functions of TCM and PM byte? Explain with the help of a diagram showing details.

25. Explain the concept of nested and cascaded monitoring of tandem connections in OTN. How is this superior to SDH?

26. What are the functions of APS/PCC bytes?

27. What is the FTFL channel? How does it activate APS? What are the indications for signal fail and signal degrade?

28. Discuss the functions performed by various bytes of OPU-OH.

29. How is justification achieved in OTN and which are the bytes involved?

30. What do you understand about FEC? How does it help in improving repeater spacing? Explain with an example.

31. Show the relationship between launched optical power and BER with and without FEC graphically and explain.

32. Which type of code is used in OTN for implementing FEC? What penalty in the form of extra bandwidth is paid? How many bytes are corrected?

33. Discuss and illustrate with the help of a diagram the arrangement of rows and subrows in an OTN frame to implement the FEC.

34. How do the OAM features of OTN compare with those of SDH?

35. Describe in brief the terms FDI, OCI, LCK, and PMI.

36. What are the future prospects for OTN, particularly in view of a strong demand for a pure data backbone?

Critical Thinking Questions

1. It appears that the cost of SDH and other circuit-switched equipment is higher as compared with data communication equipment. Why?

2. Why is the granularity of OTN kept at 2.5 Gbps and not higher or lower?

3. Does OTN really provide the final solution to the problem of exploding growth of telecommunication traffic? What are your views?

4. Why do we need 3R repetition at some nodes and not at some other nodes?

5. How is FEC better than backward error correction? Is any of these present in SDH?

6. Why is the FAS in OTN 6 bytes long?

7. Do you know of any other code that could have served a better purpose for FEC than the Reed–Solomon code?

Bibliography

1. ITU-T Recommendation G.709, *Interfaces for the Optical Transport Network (OTN)*, International Telecommunication Union, Geneva, Switzerland.

2. ITU-T Recommendation G.798, *Characteristics of Optical Transport Network Hierarchy Equipment Functional Blocks*, International Telecommunication Union, Geneva, Switzerland.

3. J.H. Franz and V.K. Jain, *Optical Communications Components and Systems*, Narosa Publishing House, New Delhi, 2000.

4. K.R. Rao, Z.S. Bojkovic, and D.A. Milovanovic, *Introduction to Multimedia Communications, Applications, Middleware, Networking.* John Wiley and Sons, 2006.

5. P.V. Shreekanth, *Digital Transmission Hierarchies and Networks*, University Press, 2010.

6. A.S. Tanenbaum, *Computer Networks*, Fourth Edition, Pearson Education, 2003.

7. EXFO Inc, *The G.709 Optical Transport Networks, an Overview*, Application note, EXFO Inc, Québec City, QC.

8. D.E. Comer, *Computer Networks and Internets*, Second Edition, Pearson Education Asia, 2000.

9. W. Stallings, *Data and Computer Communications*, Seventh Edition, Prentice-Hall India, New Delhi, 2003.

Index

Page numbers followed by f indicate figures.